Introduction to
Geological Data Analysis

Introduction to Geological Data Analysis

A.R.H. Swan
SCHOOL OF GEOLOGICAL SCIENCES
KINGSTON UNIVERSITY

M. Sandilands
SCHOOL OF MATHEMATICS
KINGSTON UNIVERSITY

WITH STATISTICAL TABLES BY
P. McCabe

Blackwell
Science

© 1995 by
Blackwell Science Ltd
Editorial Offices:
Osney Mead, Oxford OX2 0EL
25 John Street, London WC1N 2BL
23 Ainslie Place, Edinburgh EH3 6AJ
238 Main Street, Cambridge,
  Massachusetts 02142, USA
54 University Street, Carlton,
  Victoria 3053, Australia

Other Editorial Offices:
Arnette Blackwell SA
1, rue de Lille
75007 Paris
France

Blackwell Wissenschafts-Verlag GmbH
Kurfürstendamm 57
10707 Berlin
Germany

Blackwell MZV
Feldgasse 13
A-1238 Wien
Austria

First published 1995

Set by Excel Typesetters Company, Hong Kong
Printed and bound in Great Britain
at the Alden Press Limited,
Oxford and Northampton

DISTRIBUTORS

Marston Book Services Ltd
PO Box 87
Oxford OX2 0DT
(*Orders*: Tel: 01865 791155
       Fax: 01865 791927
       Telex: 837515)

USA
  Blackwell Science, Inc.
  238 Main Street
  Cambridge, MA 02142
  (*Orders*: Tel: 800 251-1000
         617 876-7000)
         Fax: 617 492-5263

Canada
  Oxford University Press
  70 Wynford Drive
  Don Mills
  Ontario M3C 1J9
  (*Orders*: Tel: 416 441-2941)

Australia
  Blackwell Science Pty Ltd
  54 University Street
  Carlton, Victoria 3053
  (*Orders*: Tel: 03 347-5552)

A catalogue record for this title
is available from the British Library

ISBN 0-632-03224-3

Library of Congress
Cataloging-in-Publication Data

Swan, A.R.H.
  Introduction to geological data analysis/
  A.R.H. Swan, M. Sandilands
      p.   cm.
  Includes bibliographical references
  and index.
  ISBN 0-632-03224-3
  1. Geology – Statistical methods.
I. Sandilands, M. II. Title.
QE33.2.S82S93 1995
550'.72 – dc20

Contents

Introduction

Why analyse geological data?

Unlike most other sciences, geology does not have a strong tradition of numerical analysis. A great deal of the historical development of our understanding of geology was based on interpretation of field observations, with no perceived need for quantification or rigorous analysis. So the question, why analyse geological data? is still one that needs to be addressed. The question has a philosophical and a pragmatic answer.

1 Science progresses by the testing of hypotheses using an objective procedure. It is not good science if a geologist glances at a rock and states dogmatically: 'This is granite!' (even if he or she is a brilliant petrologist and is correct!). A scientific geologist has a set of criteria for identifying granites and applies these criteria in a way that is verifiable by other geologists (i.e. not just intuitive), and finally draws a conclusion. The use of qualitative criteria and assessment is inevitably subjective, so the problem should be stated and tested in terms of numbers.

2 In the past, geologists have primarily dealt with rocks on the basis of their appearance, and this has been a fruitful approach for field mapping and exploratory work. However, it is increasingly common for the primary geological information to be numerical. This can be ascribed to: (a) increasing use of digital recording instruments (e.g. wireline logs, satellite images, geophysics); (b) convenience of storing the enormous amount of data resulting from exploration in numerical form; and (c) appreciation of the importance of cryptic and subtle variations in rocks which can only be recorded by specialised instrumentation (e.g. trace element geochemistry, permeability). Years of geological research and exploration using traditional methods have discovered a lot of the relatively obvious theoretical principles and economic deposits; we have to use more sophisticated and sensitive techniques to uncover what remains!

Objectives

Geologists are now accustomed to having a computer on their desks and many have responded enthusiastically to the appeal of modern spreadsheet and other software. However, the development of such technology has outpaced geologists' ability to use it to full potential. This book aims to close the gap.

Like all other aspects of geology, data analysis is a subject with endless specialised branches and levels of sophistication. However, a high proportion of data analytical jobs can be completed with a modest knowledge and ability. This book provides explanation of the essential concepts together with description and exemplification of the most important techniques. We believe that we have included the core material essential for every modern geologist, plus some excursions into more advanced topics of particular importance and interest.

We assume the reader approaches the book with a reasonable knowledge of mathematics (concerned students should read Waltham, 1994), but we assume no previous experience of statistics. There are, perhaps, three levels of useful ability in the subject:

1 understanding of the full mathematical 'nuts and bolts' and derivations;
2 conceptual understanding of how methods work and how to apply them; and
3 pragmatic approach: little understanding but ability to follow procedures and make conclusions.

We aim for level **2**, but level **1** is attainable where the mathematics is not too advanced, and the book may be used to support operation at level **3**, though this can't be recommended!

The content of the book is designed to be suitable for a second-year undergraduate course in a geology or earth science degree programme, but includes material which would form a foundation for more specialised third-year or postgraduate courses. Inevitably, though, tutors will wish to customise the topics into courses of their own design. Regardless of the context in which it is used, we hope that the book explains the subject clearly and will improve the level of data analytical interest and ability amongst geologists.

Software

The reader will not have much use for a book such as this without access to a computer: less than half of the methods here can reasonably be attempted with only a pocket calculator, and computer use is the main *raison d'être* of the book. The computer software available is developing so rapidly that it would not be useful for us to recommend specific products for use in conjunction with this book. However, we can give a checklist of attributes and facilities that such software might have.

1 Basic data manipulation and storage facilities, including ability to apply arithmetic operations on columns of data.
2 Basic graphical capabilities: frequency histograms and scatter plots.
3 Standard statistical tests, especially *t*-tests, analysis of variance, correlation and regression (including multiple regression), plus some non-parametric tests.
4 Autocorrelation and Fourier transform.
5 A selection of spatial estimation algorithms.
6 Matrix operations (including eigenvectors) and/or principal components analysis, discriminant functions, cluster analysis.

These will not all be found in the same software product, so data transfer capabilities are important. Numbers **1** to **3** are essential to any data analyst; numbers **4** to **6** are required if the material in the book is to be followed, but some will be dispensable if the analyst is working in a restricted field.

How to use this book

It is crucial that the aspiring data analyst refrains from attempting complex analyses before understanding the basics, even though he or she may argue that the computer software is doing all the hard work! There are many ways that an analysis can go wrong literally before it has started. Chapter 1 should be compulsory reading before proceeding further, and Chapters 2 and 3 consist largely of basic foundation material. Each of the subsequent chapters draws heavily on concepts introduced in these first three parts. Chapters 4 to 8 can be investigated more independently: none is a clear prerequisite for any other, so the order of approach by the reader may be determined by need.

Often, however, students find the basics of probability and statistical inference (Sections 2.3–2.5) to be among the most difficult of all the topics here. If this is the case, we recommend that students observe the applications of these ideas through the remainder of the topics, and return to the theory later.

The text is organised into a main narrative plus peripheral boxed material. The 'boxes' are optional to the reader and contain worked examples or subsidiary advanced topics and explanations: they are separated so that the main text can be followed without diversion. The crucial equations and methods are highlighted within the text for clarity.

We anticipate that the worked examples will be among the most useful aspects of the book: they provide a model for the reader's own analysis and show how real geologically useful results can be obtained. In order to avoid repetitive statements of the number of decimal places used, the reader should note that numerical results have been expressed using the number of decimal places that are appropriate in the context of the problem.

We have not attempted to make this book a scholarly review of relevant research papers, but we refer to a selection of key sources from the geological literature so that students can investigate case studies and more advanced techniques.

FURTHER READING

Waltham D. (1994) *Simple Mathematics: a Tool for Geologists*. Chapman and Hall, London.

Students who are not confident that their mathematical background is sufficient should read this.

Acknowledgements

We are very grateful to a number of publishers and individuals for granting us permission to reproduce data sets, specifically: Geraint Owen for Appendices 3.5 and 3.9; Paul Raglan for Appendix 3.6; Roger Hewitt for Appendix 3.8; Richard Reyment for Appendix 3.12; the American Association of Petroleum Geologists for Appendix 3.2; Elsevier Science for Appendix 3.7 and part of Figure B6.7.1c; Plenum Publishing Corporation for Appendix 3.10; J. Wiley and Sons, Inc., for Appendix 3.11 and Saskatchewan Energy and Mines for Appendix 3.14.

Patrick McCabe calculated and allowed us to use many of the statistical tables.

We are indebted to a number of anonymous reviewers who grappled with a rather imperfect manuscript and provided corrections and helpful suggestions. Also, Michael House contributed useful comments on part of the first draft.

This book was commenced with the encouragement of Navin Sullivan and Mike Brown and came to fruition with guidance from Simon Rallison. In the intervening time, we have benefited from the tolerance, advice and encouragement from many among our families and our colleagues at Kingston University.

Inevitably, this book draws on the accumulated work of generations of geologists and statisticians. It is not feasible to cite this material exhaustively, but we are very aware of the debt that we owe and can only hope that this book will be a worthy addition.

1   Data Collection and Preparation

1.1   DATA QUALITY

Data analysts must have confidence in the quality of data before commencing an analysis: the phrase 'garbage in – garbage out' is a cliché, but it is inescapably true. This necessitates some degree of knowledge of or involvement in the process of measurement and recording. Unfortunately, no amount of analysis can prove that data are or are not garbage: for example, randomness in data analytical results may either be due to genuinely natural randomness in the geological setting or to failure of the measurement procedure, whether it is a state-of-the-art spectrometer, a camera on a satellite or a student with a ruler! Conversely, strong positive results may result merely from bias in the measurement procedure. We cannot give instruction here on the proper use of an infinity of different measurement devices, but we can provide guidance on other aspects of the planning and execution of a data retrieval exercise in order to arrive at a valid result.

It is of prime importance to ensure that the data are, in principle, capable of producing the required type of result. Normally, the geologist will have formulated a hypothesis to test – consider the following examples: (a) is there any significant difference between the compositions of granites from two intrusions? (b) is there regular cyclicity in a thin-bedded limestone sequence? and (c) can we estimate sandstone permeability from down-hole wireline logs? The constraints on the data required to resolve these and other types of questions will become apparent in subsequent sections of this book: for example, the problem cited in (a) above cannot be solved by taking and analysing one rock specimen from each intrusion – these may not be representative and we would not be aware of any variation within each intrusion; in problem (b) the analysis may be invalidated by incompleteness of the data sequence; in (c) it would be essential to start with a controlled set of definitive permeability measurements.

A data analytical project must always commence with careful thought and planning. The analyst must understand the nature of the geological data *and* the statistical technique *and* the constraints under which the two may be linked. This book seeks to engender a particular understanding of the latter.

The use of any measurement device must be accompanied by awareness of precision and accuracy: see Box 1.1.

> **Box 1.1 Precision and accuracy**
>
> *Precision*
>
> A measurement is precise if repeated measurements of the same geological entity are similar. Precision can be quantified in terms of the statistics of the repeated measurements, but it must be ensured that the replicates are identically prepared and from the same source. This is often done in geochemistry by thorough crushing and mixing before division into replicates. If the procedure is properly done, the error involved is random.
>
> *Accuracy*
>
> A measurement is accurate if it is close to the true value. In geology, the true value is usually unknown, although there are standard, well-characterised geochemical samples which are used for calibrating analytical equipment.
>
> All combinations of high/low precision and high/low accuracy are possible. Examples are given below; these may be imagined to be repeated compass measurements of the strike of the same bedding plane by the same student using the same equipment. We have to imagine that the student forgets the previous measurements between each attempt!
>
> | | | Precision | |
> |---|---|---|---|
> | | | High | Low |
> | Accuracy (suppose true value is 50°) | High | 49, 50, 50, 52, 50, 49, 51, 50 | 55, 47, 50, 52, 44, 53, 57, 47 |
> | | Low | 54, 55, 55, 57, 55, 54, 56, 55 | 60, 52, 55, 57, 49, 58, 62, 52 |
>
> A student performing in the 'high precision' category would be competent in the operational procedure of measurement; 'low precision' would indicate sloppy and inconsistent use of the compass.
>
> The 'high accuracy' category would indicate that the compass was correctly adjusted; 'low accuracy' might (typically) result from incorrect declination adjustment, giving a systematic bias.

1.2 HANDLING DATA

1.2.1 Objects and variables

The bulk of a typical data set will consist of numerical measurements. Each number will be a measurement of a specified property of one particular entity. The property may be something like percentage of quartz, gold in parts per million, grain size, length of pygidium, etc. In statistical terminology these are the variables. The entity is some geological object or sample (but note there are two meanings of the word 'sample': see Section 1.5), such as a hand specimen of granite, a soil sample, a thin section of a sandstone or a trilobite specimen. A single data set will typically contain information on many variables and many objects. Conventionally, we set up a column for every variable and a row for every object. It is usually easy to ensure that each measurement on a row has genuinely come from the same object, but great care must be taken to ensure that all the data in one column relate to the same variable. The entire contents of a column will normally be analysed together, and, for this to be valid, we must be sure that we are comparing like with like; the numbers must be measured under the same strict definition of the variable. This must include the following.

1 The units must be the same in every entry.

2 The same data retrieval procedure should be used throughout: for example, grain size percentage based on weights of sieve residues will not be comparable to grain size percentage based on thin section point counts. Note, though, that it is sometimes convenient to place results of different methods in the same column if the purpose is to compare methods, and if the different methods are identified by codes in other columns.

3 Unique to palaeontology, there is a problem of defining what are genuinely comparable measurable characters – the problem of homology. Comparable characters must be derived from the same structure in a common ancestor.

1.2.2

Compiling the data set

Missing data

The data analyst hopes to have numerical values for every variable for every object. If this is not possible, for example due to equipment failure, then the easiest option is to eliminate the affected rows and/or columns from further consideration. If this is unacceptable, the missing data should be represented by an unambiguous entry: '0' clearly cannot be used, and an absence of any figure or symbol may result in software reading the remaining entries out of step; '*' is often correctly interpreted by software as missing data – otherwise an absurd value such as −99.99 can be used, in the expectation that any future analyst will recognise that it isn't a real value!

Reference numbers, labels and codes

As well as measured variable values, a data set may need to include information to identify the individual rows and columns. This often causes data format problems (see below), but these can be minimised by using numerical codes arranged as extra columns. Any aspect of a measured object which may be interesting or necessary to treat separately, such as source location or type of sample preparation, can and should be conveyed by an extra column with appropriate coded values. The location of origin of a specimen, expressed as a grid reference, is a special property that can be used just as an identification or as a variable in spatial analysis: in the latter case, it should be split into easting and northing and stored as two variables.

Data entry and data formats

Data can be entered and stored on a computer using a text editor, word processor, spreadsheet or other software. Any software used to store a data set should be able to read it in the format in which it was stored. However, if it is desired to move data around between software packages, data format becomes an important consideration. The most widely transferable format is to have entirely numerical data stored as normal (American Standard Code for Information Interchange, ASCII) characters in neat rows and columns (perhaps separated by commas), but many other standards exist. If column labels and other alphabetical information need to be stored, more specific formats apply and a data file will become much more difficult to transfer. It is often advisable to maintain two versions of a data set, one with just the numerical values in a general format for transfer to diverse software, and another including the full annotations specifically for use with one favoured software package.

See Box 1.2 for an example of a typical data set.

1.3   TYPES OF GEOLOGICAL DATA

There are a variety of categories of data which may be encountered in geology. These are of varying quality, and data type frequently constrains the data analytical techniques which can be applied. The main data types are as follows (refer to Fig. 1.1).

Ratio scale data

The best-quality and most versatile data type: ordinary measurements such as length or weight are in this category. There are no special problems with this type of data.

Interval scale data

Interval scale data differ from ratio scale data in that the zero point is not a fundamental termination of the scale. The classic example is temperature measured in °C or °F: the zero point on these scales, though well defined, is not so fundamental as zero length or zero weight. Unlike °C and °F,

---

Box 1.2 Example of a data set

A	B	C	D	E	F	G
1	630	319	26.1	35.4	1	1
2	630	319	27.7	36.8	2	1
3	621	315	30.3	40.9	1	1
4	621	315	30.9	43.4	2	1
5	625	332	37.1	58.4	1	1
6	625	332	39.0	61.1	2	1
7	642	322	23.3	38.6	1	2
8	642	322	24.2	36.0	2	2
9	633	307	30.4	44.0	1	2
10	633	307	*	*	2	2
11	611	324	35.3	53.6	1	2
12	611	324	*	*	2	2
13	619	339	37.5	52.4	1	2
14	619	339	38.9	53.9	2	2
15	618	310	21.8	38.6	1	2
16	618	310	23.2	38.1	2	2

* would indicate that the measurement was unavailable, perhaps due to equipment failure.

Each column gives information for one variable or a label or code. Each row gives the information for one object (sample or specimen).

These numbers might represent the following:

A   sample reference number;
B   grid reference: easting;
C   grid reference: northing;
D   measured value, e.g. percentage feldspar;
E   measured value, e.g. percentage quartz;
F   measurement or treatment code, e.g. 1: point count; 2: geochemical analysis; and
G   geological category code, e.g. 1: intrusion or formation X; 2: intrusion or formation Y.

The italicised items (sample reference numbers and, particularly, column labels) could be dispensed with during data analysis, and transfer of the data between software packages is likely to be easier without them.

A great variety of the data analyses described in this book could be applied to such a data set in order to yield useful and interesting results (especially if the data set were larger!).

temperature in K is on the ratio scale, as 0 K represents an absolute lack of heat. The crucial difference for data analysts is that, as the name suggests, ratios can be used with ratio data but not with interval data. A ratio 2:1 may represent doubling if it is length, weight or temperature in K, but the same ratio in °C or °F is quite meaningless.

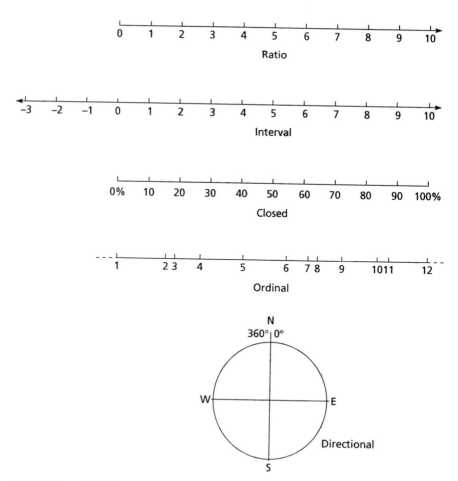

Fig. 1.1 Concepts of data types.

Closed data

These are data in the form of percentage, parts per million (ppm), or any other way of expressing a proportion of a fixed total: this is ubiquitous in geochemistry. Such data require cautious treatment, especially in bivariate and multivariate methods, because variables are fundamentally interdependent; this is discussed in Section 3.2.2.

Directional data

Directional data are data expressed in angles, for example bearings from north. Like interval scale data, ratios are not useful, but directional data require special methods (Chapter 5) because the numerical values cycle around through 360°, so that we have the arithmetically awkward situation that $0° = 360°$.

Ordinal scale data

This category is of considerably lower quality than the above: whereas intervals along the scale are constant in interval scale data, with ordinal data the scale is not regular: the only purpose of the scale is to place observations in relative order. Consequently, it is not valid to apply addition or subtraction, as well as division, to interval scale data. Examples in geology include Moh's hardness scale, Richter earthquake scale where assessed from field damage, and time where assessed by stratigraphic position. Ordinal data require the use of non-parametric methods (Chapter 4), which analyse data on the basis of rank order only.

Discrete data

All of the above data types are normally on *continuous* data scales: all points on the scale may, in principle, be used. On *discrete* scales, data can only have certain specific values, usually the integer values. The most common example is data in the form of counts of objects, for example numbers of fossils in a $1\,m^2$ quadrat. Discrete data have special frequency distributions (Section 2.1.1).

Nominal or categorical data

It is normal and preferable to work with numerical data, but information is sometimes presented in the form of names, perhaps a list of minerals or fossils: these are known as nominal or categorical data. We do not wish that such information should be immune from rigorous analysis, so it should be converted into numerical form. Most often, this is done in binary form: taking each category (fossil or mineral species) as a separate variable, we can record 1 for present and 0 for absent. Analyses may then be done using special multivariate methods (Section 8.6). Binary data, though, are usually a poor substitute for the quantity of information actually available in a geological situation: rather than just recording presence/absence, the analyst should aim to obtain counts.

1.4 TYPES OF ANALYSIS

Statistics and data analysis

Ideally, a data analytical job is initiated by a question such as: is the grade of ore high enough to be economical? is the sandstone sufficiently permeable to be a good reservoir? is the rock a tholeiite? what is the shape of the sand body? or does the trilobite lineage evolve?

Statistics is a term widely used for any type of 'number crunching', but statistics *sensu stricto* is a special discipline in which answers to questions such as those above are found by calculating estimates of the true values. The reliability of estimates is often expressed in probabilistic terms; for example: we are 95% sure that the ore is economical to mine. This formal, rigorous approach is not always necessary; often, the result is obvious after cursory examination of the data. There is an enormous variety of data analytical techniques available for examining data so that the required judgement can be made; these include graphical methods, special exploratory data analysis (EDA) techniques, surface estimation, eigenvector methods and many others. Techniques covered in this book include a mixture of general data analytical and rigorous statistical approaches. The analyst must always be aware that the general data analytical methods do not give definitive answers, but leave some interpretation to the user, and that rigorous statistical methods make special demands on the data and impose a formality on the exercise.

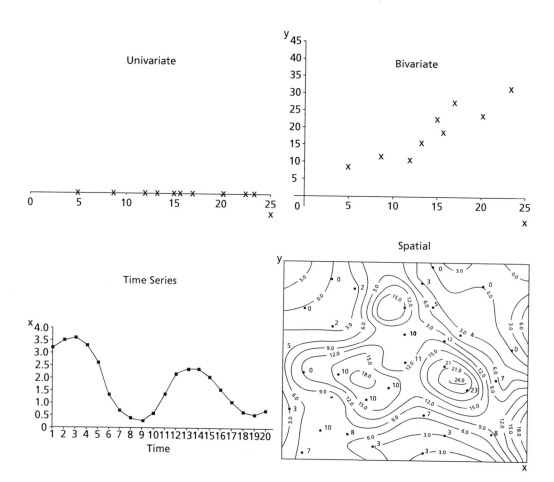

Fig. 1.2 Graphs illustrating the data distributions described by various types of analysis, based on number and type of variable involved. For the spatial distribution a Z value is recorded at each x, y coordinate.

Statistical and other data analytical techniques can be classified according to the number and type of variables involved. This is the division used as the basis for the structure of this book. (Refer to Fig. 1.2.)

Univariate methods

Each variable analysed in isolation. The data can be portrayed as a series of points along an appropriately scaled line. Univariate methods allow the distribution of points along the line to be described and statistical statements about the distribution made.

Bivariate methods

Two variables analysed together. The two measurements made on one object give coordinates of a point in a two-dimensional (2D) space, and a data set can be portrayed as a 2D scatter. Bivariate methods describe and analyse the shape of the scatter for the purpose of investigating the relationship between the data points and/or the relationship between the variables.

Time series methods

Sequences of data in time (or space) can be in various forms. Some can be treated as simple bivariate data, with one variable happening to be time. Often, the situation is conceptualised as a continuously varying curve.

Spatial analysis

Three (or four) variables analysed together, two (or three) of which are spatial coordinates: grid references or latitude/longitude, with or without altitude or depth. The other variable is a geological measurement of interest, and is regarded as varying continuously over the area. The data may be imagined as points in three dimensions, and analysis often has the objective of constructing a smooth surface to describe spatial variation.

Multivariate analysis

General methods applicable to any number of variables analysed simultaneously, and usually applied to more (often many more) than three variables. If these are m variables, the data may be imagined as points in an m-dimensional space. The prime objective is to reduce the dimensionality so that the shape of the data scatter can be viewed. Relationships between variables can also be investigated.

1.5   POPULATIONS AND SAMPLES

1.5.1   Definitions

When we make geological measurements and calculate statistics we are not primarily interested in the properties of the bits of rock (or whatever) that we have measured: we are interested in the properties of the large body of rock that we hope our measurements are representative of. For example, the permeability of a few small cylinders of sandstone in a machine in the laboratory is not in itself important; what is important is the permeability characteristic of the reservoir from which the cores were collected. The sandstone reservoir in this case is the population: this is the entity that we want to find out about. The actual measurements we record constitute the sample.

The population

The population is the total set of measurements which could, hypothetically, be taken from the entity being studied. The limits of the population (which, in geology, are usually spatial limits) are defined by the geologist at the beginning of the project. So, for example, the population may be defined as: the set of all possible permeability measurements from one sandstone formation in one oilfield in the North Sea; composition of all granites world-wide; size of all specimens of a species of trilobite; or trace element com-position of all stream sediments within 10 km of a copper mine.

The statistical sample

Confusion is unavoidable between the statistical and geological meanings of the word 'sample'. In geology, a 'sample' usually means one lump of rock or sediment (though the term 'specimen' is used if it is a fossil or a mineral). In statistics, the sample is the collection of objects or measurements which are a subset of the population of interest and are taken to represent that population. The sample measurements constitute the data that are actually available for analysis.

Bias

For any data analytical investigation to be valid, it is essential that the sample is an unbiased subset of the population. There are an infinite number of ways in which bias can creep into the sampling process, for example: rejecting crumbly cores from a permeability survey; collecting weathered lumps of granite; only collecting the big, conspicuous fossils; or collecting stream sediments only where it's easy to dig. This category of problem is, in principle, avoidable by careful application of discipline, though this is often

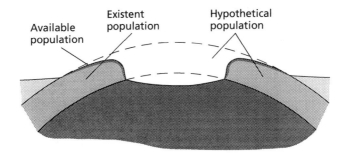

Fig. 1.3 Types of geological population. The population required may be of grain sizes of a sandstone formation, seen here in cross-section, but this can be defined in three ways. The existent is the best to use, but circumstances often dictate that only a restricted available population is samplable. Note also that it would be easy to oversample the higher part of the formation if restricted to surface outcrop.

not easy. There is always concern in science that discipline may consciously or subconsciously lapse in order that the result confirms the researcher's preconceived ideas!

In addition, there are always pitfalls in the measurement process due to instrumental or procedural error. Apart from making sure that the equipment is properly calibrated and in working order and that the operator is competent, error can be controlled by careful experimental design – see Section 2.6 on analysis of variance.

Defining the geological population

It is useful to consider three categories of population (see Fig. 1.3).

1 The hypothetical population. This incorporates all of the geological entity that ever existed, for example grain size measurements of a sandstone formation which outcrops at the surface and has clearly been partly destroyed by erosion. Investigations into the properties of the hypothetical population are statistically dubious and only of theoretical interest.

2 The existent population. This is the entire remaining portion of a geological entity, and is the normal, sensible choice of a target population.

3 The available population. This is the subset of the existent population which is capable of being sampled, considering the logistical constraints that pertain at the time of the study. In amateur projects, this will probably constitute just a thin surface layer exposed at outcrop; in exploration for a valuable resource, drilling may make virtually any part of a rock unit suitable for sampling. Where the available population is tightly constrained, it should be remembered that statistical conclusions should not be extended to the existent population. In particular, where the available population is at the exposed surface, the biasing effects of weathering and other shallow processes must be considered. This may severely limit the usefulness of the result!

1.5.2   **Sampling strategies**

Sample size

The only generalisation that can be made about the sample size required for a job is that it depends on the degree of subtlety of the problem being addressed, and this is often not known in advance! For example, suppose that it is important to know whether or not a sandstone formation has a porosity greater than 5%: if five randomly taken cores give values of 15%, 18%, 16%, 20% and 18%, we would already be confident of the answer, but, if the values were 8%, 4%, 5%, 9% and 5%, we would need a larger sample size before coming to any conclusion. It is good practice to investigate the properties of a small pilot sample before deciding on the definitive sample size required: see Section 2.5.3.

Spatial sampling schemes

In geology, the main decision about sampling populations is where to sample. Sample locations can be distributed according to a number of schemes: see Fig. 1.4. (For definitions and methods of analysing point distributions, see Section 7.1.)
1   Random: these may be planned by converting random numbers (from tables or computers) to grid references; unplanned sample point distributions may also prove to be effectively random.
2   Uniform: planned by randomisation within grid squares.
3   Regular or gridded: planned on rectangular or triangular grid.
4   Clustered: not usually planned, but often forced by patchy distribution of rock exposure.
5   Traverse: often forced by access and exposure constraints (roadsides, cliffs, rivers) or by logistics (predefined route, no time for detours!).
The advantages and disadvantages of these schemes are considered next; these depend on whether variation within a population is of interest, or only overall population parameters are required.

Characterising overall population parameters

Any geological population will have some internal variation: this may be random or systematically arranged (e.g. by stratification). However, we may wish to calculate statistics relating to just the overall characteristics of the population. If a copper orebody is to be mined by open-cast bulk extraction, the small-scale variations may be irrelevant and all we may be asked for is an overall estimate of the grade. In this case, we only need to be sure that the sampling scheme does not systematically over- or under-sample the different subdivisions of the population. For example, if we wish to characterise soil lead content in an area containing vertical galena-rich veins, a grid

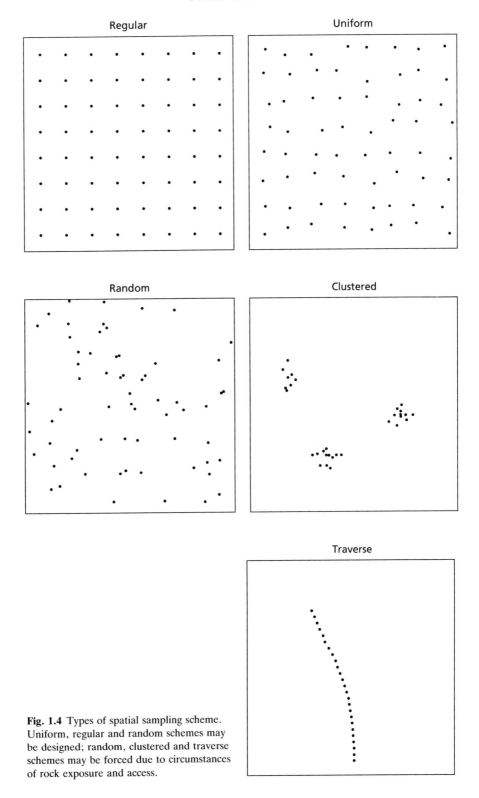

Fig. 1.4 Types of spatial sampling scheme. Uniform, regular and random schemes may be designed; random, clustered and traverse schemes may be forced due to circumstances of rock exposure and access.

sampling scheme may give a biased overestimate if a grid line happens to coincide with a vein; conversely, a biased underestimate will result if the grid points systematically 'miss' the vein. In general, a random or uniform distribution is best for eliminating bias. However, if the heterogeneity in the population is systematic and is visible in advance, for example stratification on a cliff face, we can plan a traverse scheme which samples each portion of the heterogeneity equally.

Characterising variation within populations

If heterogeneity in the population is visible in advance, the whole should be broken into subpopulations and sampled and analysed separately. In the special case of gradients or trends of variation, this will be definable by traverses parallel to the gradient. If we wish to define variation which is not visible in advance, a different approach is needed: all parts of the area need to be investigated equally. Random distributions may be too uneven, and a uniform or gridded scheme is to be preferred.

2 Statistics with One Variable

In this chapter we describe techniques of univariate analysis, in which only one measured variable is analysed at a time. Many of the exploratory and descriptive methods introduced here are recommended as routine ways of investigating properties of new data, even if the final analysis required is bivariate, spatial or multivariate. This section also introduces fundamental concepts that are generally applicable in data analysis.

The methods described are applicable only to interval and ratio scale data, unless specified otherwise.

2.1 GRAPHICAL AND NUMERICAL DESCRIPTION

2.1.1 The frequency distribution and the histogram

Univariate data can be presented as a list of numbers and imagined as a one-dimensional array of points along a scaled line (see Fig. 1.2). In order to present the data in a more digestible and interpretable way, it is useful to calculate summary descriptive statistics and to use a more 'user-friendly' graphical representation.

The properties of the data can be conveyed in terms of the changing density of points along the scale. This can be assessed by dividing the scale into a number of equal intervals, known as classes, and counting the number of points in each class. The count in each class is the frequency and the series of counts describe the *frequency distribution* of the variable. A visual impression of the changing density is given by a *histogram* – a diagram in which *areas* of rectangles represent frequencies in classes – see Fig. 2.1(a). Note that, if we have equal class intervals, the height of rectangles is also proportional to frequency. For discrete (e.g. integer only) variables the appropriate type of graph is a *line chart*, in which the lengths of lines are proportional to frequencies, as in Fig. 2.2.

In sedimentological grain size analysis and occasionally in geochemistry, the *cumulative frequency distribution* is plotted. This has the same horizontal axis, divided into classes, for the measured variable, but the frequency axis shows the percentage of the total sample having values less than or equal to that point on the scale – see Fig. 2.1(d). Cumulative curves have the advantage of using a standard (percentage) frequency scale and have some specialised uses (see Sections 2.5.6.2–2.5.6.4), but they do not show the properties of the frequency distribution so clearly.

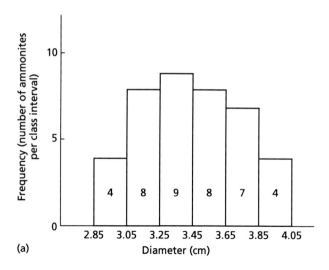

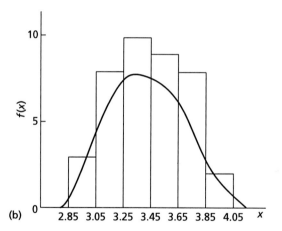

Fig. 2.1 Graphical representations of frequency distributions for continuous variables. (a) A histogram to illustrate the frequency distribution of the diameters of 40 ammonites. The area of each rectangle is proportional to the number of specimens in a class. (b) The histogram may be smoothed out approximately by a curve which can be described by a mathematical function.

2.1.2   **Properties and parameters of the frequency distribution**

The essential properties of univariate data are expressed in the frequency distribution and are displayed in the histogram. These are:

1 location: the 'average' position of the whole distribution along the scale;

2 dispersion: the extent to which the distribution is spread out along the scale, how much values vary from some central value; and

3 shape: the symmetry and pattern of the frequency distribution. (Refer to Fig. 2.3.)

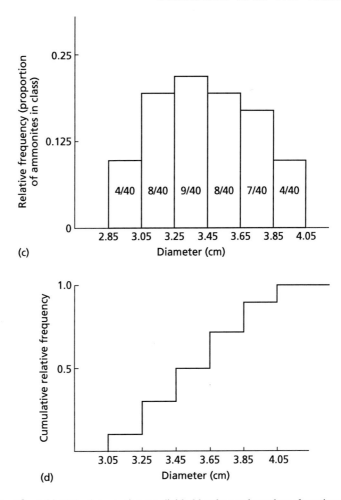

Fig 2.1 *Continued* (c) If the frequencies are divided by the total number of specimens, we obtain a relative frequency scale. (d) The cumulative frequency distribution shows how many values are less than or equal to any given value *x* of a variable.

The outline of a histogram can be approximated by a smooth curve (Fig. 2.1b), which may be described by a mathematical function called a *frequency density*, often denoted by $f(x)$ where *x* stands for values of the variable. Such mathematical functions may be used as idealised models for describing histograms and for conceptualising hypothetical population distributions. It is common to express ideas and summarise descriptions by using sketches of smooth frequency density curves.

The position, dispersion and shape of curves depend upon one or more constants called *parameters* of the population or frequency distribution. In Fig. 2.4 the parameter is λ and it governs the steepness of the fall in the curve. These mathematical parameters are related to statistical measures,

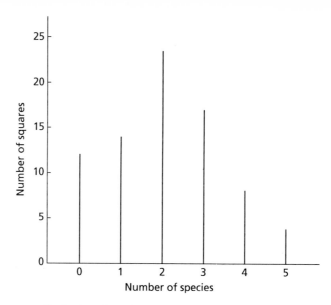

Fig. 2.2 Frequency distribution of numbers of species of fossil found in squares during a quadrat study of abundances. Frequencies of values of discrete variables are represented by heights of lines in a line or bar chart.

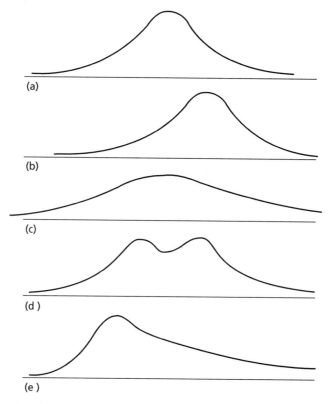

Fig. 2.3 Examples of frequency distributions: (a) symmetrical, unimodal; (b) differs from (a) in location only; (c) differs from (a) in dispersion only; (d) bimodal, otherwise similar to (a); (e) positive skew, otherwise similar to (a).

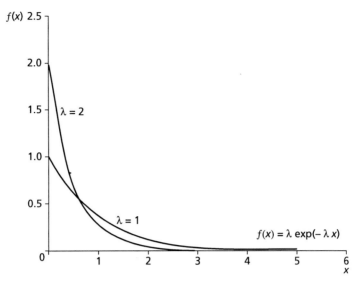

Fig. 2.4 Two exponential distributions with different parameters. Here, the parameter governs the rate of decline of density.

such as averages and measures of variability. A population can be summarised, then, by a mathematical function and the values of its parameters.

When dealing with population descriptions for which real frequencies are unknown, it is more meaningful to deal with proportions (*relative frequencies*) of values in classes; this idea will be developed in Section 2.4. All the foregoing remarks apply but frequency is replaced by relative frequency. A relative frequency distribution is shown in Fig. 2.1(c): the relative frequencies are obtained by dividing frequencies by the size of the population, 40. The total area under the appropriate histogram is 1 and the same must be true of the area under the curve described by any relative frequency density (Fig. 2.1b).

These diagrams and functions are useful for envisaging the kinds of value which are likely to be encountered when specimens are taken from a population with a specified frequency distribution. For example, the population represented by Fig. 2.3(e) has a small proportion of relatively high values so the values encountered in sampling will tend to be low, with only a few exceptions. The population is said to be *skew to the right*. When we sample from the symmetrical population in Fig. 2.3(a), however, we are likely to obtain roughly equal proportions of high and low values and most will be midway between the extremes. It is important to distinguish here between the attributes of the statistical sample and of the source population: they are obviously related but they are not identical; if this is unclear, refer back to Section 1.5.1. We will explain the parameters used for describing each, and how this allows us to make more precise statements about the outcome of sampling from a population. Of course, in real life we do not know the values of these population parameters and must estimate them from our data; but if we can see what kind of samples are to be expected from specified populations we will understand more easily how to infer something about the population from which a sample is taken.

Parameters of the population frequency distribution

In this section we describe how the frequency distributions of populations are quantified, but it must be remembered that this is done only for conceptual purposes: in geology, we never measure the entire population. The values actually calculated are the sample statistics: see Section 2.1.4 below.

The most important features of a population are the average value and the extent to which values are dispersed about the average. Several measures are available for each of these features and the choice of which to use must depend on their relevance to the physical problem under consideration as well as on statistical demands. We also need to consider the shape of the distribution, but this is more often conveyed by qualitative description.

(Notation: By convention, values of the variable in a population are denoted by upper-case Roman letters, e.g. X, Y, and the values for the N individuals in the population by letters with subscripts, e.g. X_1, X_2, ..., X_N. Population parameters are symbolised by Greek letters, e.g. μ (mu).)

2.1.3.1   Measures of location or position (average values)

If we imagine the values to be arranged on a straight scaled line, an average tells us the general whereabouts of the values on the line; for this reason averages are often called *measures of location* or position. Location is quantified as a single value of the measured variable which is in some way representative of the data. Representative values are clearly important in geology, for example in characterising grain size of a sandstone. Location can be defined in a number of different ways; each has its own rationale and geological usefulness.

Population arithmetic mean, μ

This is obtained by adding the values for all individuals and then dividing by the size of the population (the number of individuals in the population):

$$\mu = (X_1 + X_2 + X_3 + \ldots + X_N)/N$$

or

$$\mu = \frac{1}{N} \sum_{i=1}^{N} X_i$$

(This form of equation may be new to you and require explanation. There are N values of X in the population and these are symbolised by X_1, X_2, X_3 and so on, up to X_N. In the equation, X_i is used as a general symbol for all of these. The symbol Σ means 'sum', and the right-hand part of the equation is read as 'sum the values of X from X_1 to X_N'. This is then divided by N, or multiplied by $1/N$).

The arithmetic mean is typical of the population in the sense that it is the value which every individual would have if they were all the same: the total value is shared equally between all. It has the advantage of being simple and it is easy to interpret when the population is symmetric.

Population geometric mean

Sometimes the population is skewed to the right but the frequency density of *logarithms* of the values is symmetrical. This is very characteristic of populations representing concentrations of trace elements such as gold, cadmium and uranium. In such cases the arithmetic mean is too high to be typical and it is usual to compute the arithmetic mean of the logarithms of values and take the antilogarithm of the result. The resulting measure is called a geometric mean: it can be shown that the same result is obtained from multiplying all the values together and taking the Nth root of the product:

$$\text{Geometric mean} = (X_1, X_2 \ldots X_N)^{1/N}$$

Median

Another way of dealing with skew populations is to use the the middle value as the average: individual values are arranged in increasing order and the value in the middle of the list is the median. If the number of individuals is even, the arithmetic mean of the two middle values is used. Half the population have higher and half have lower values than the median (see Fig. 2.5a).

Mode

The mode is the most frequently occurring value in a population; in terms of the frequency or relative frequency density it is the value of the variable for which the height of the curve is greatest (see Fig. 2.5b).

Comparison of the mean, median and mode

In perfectly symmetric populations, the mean, median and mode are identical, but when a population is skew the arithmetic mean lies farthest into the tail and the median lies between the mode and mean, as illustrated in Fig. 2.5(c). The mean and median will be used most often in this book.

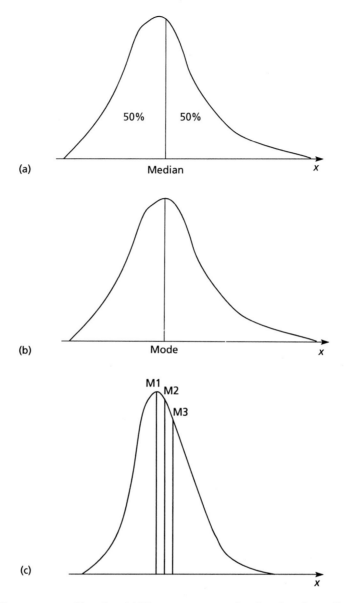

Fig. 2.5 Three measures of location. (a) There are as many values less than the median as there are greater. (b) The mode is the value which occurs most frequently. (c) In a skew distribution the median lies between the mode and mean. M1, mode; M2, median; M3, mean.

2.1.3.2 **Measures of dispersion**

Another important feature of a population is the variation between the individuals which it comprises. A high degree of variability obliges us to make more observations in order to obtain a reliable estimate of an average than is the case when population values are close. Sometimes, it may be justifiable (or just convenient) to regard dispersion as random 'error', distorting values away from the 'true' central value. The central value may reflect the attributes of the causative geological process; the dispersion may be the result of all the messiness inherent in nature and in sampling and measurement. More often, though, dispersion is just as important as position in geological interpretation, for example in defining the sorting of a sandstone. Once again, there are a number of different measures.

Range

This is the difference between the highest and lowest values in a population. It ignores all other values and as the extreme values are not usually representative in a large population its direct application is limited.

Interquartile deviation

The difficulty associated with the range is often overcome by ignoring the top 25% and bottom 25% of values in a population and calculating the range of the remainder. The values at which cut-off is made are called the first and third quartiles, denoted by Q_1 and Q_3: one-quarter of the population has values which are less than or equal to the first quartile and three-quarters have values less than or equal to the third. The difference between them is called the interquartile deviation, denoted by IQD (Fig. 2.6):

$$IQD = Q_3 - Q_1$$

Variance and standard deviation

The variance of a population is a measure of the scatter of values about the mean. Suppose that the values associated with the N members of the

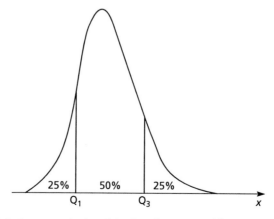

Fig. 2.6 The IQD is the range of values left when the upper and lower quarters of the distribution have been removed. It avoids the influence of extreme, unrepresentative values. IQD, Q_3-Q_1.

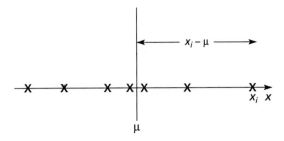

Fig. 2.7 Deviations from the population mean. The mean of the squares of these deviations is used as a measure of the variability of values in the population.

population are $X_1, X_2, \ldots, X_N$ and that their mean is μ. A small population of values scattered around its mean μ is illustrated in Fig. 2.7.

The variance is then defined to be

$$\sigma^2 \quad \text{or} \quad \mathrm{Var}(X) = \{(X_1 - \mu)^2 + \ldots + (X_N - \mu)^2\}/N$$

That is, the variance of the values in a population is the mean of the squared deviations of the values from the mean μ.

More concisely:

$$\sigma^2 = \frac{1}{N}\sum_{i=1}^{N}(X_i - \mu)^2$$

The units of variance will be the squares of the units of measurement so that, for example, if the variable is measured in megayears (My) the variance has the units of My squared. It is easier to interpret the square root of the variance because it has the same units as the variable. This quantity is called the *standard deviation* of the population and is of course denoted by σ.

2.1.3.3 Description of shape

As we have seen (Fig. 2.3), distributions with the same location and dispersion may still be different in shape. Inevitably, shape is a more subtle property and typically has more subtle geological explanations. Variations in modality, skewness and kurtosis are the most useful.

Number of modes

The term unimodal is used to describe distribution with one peak and consistent declines in frequency on each side. If there are two 'humps', the distribution is called bimodal (Fig. 2.3d); if more, polymodal.

Skewness

Frequency distributions may be symmetrical or skew. Distributions which are skew to the right (have a long tail of high values) are also described as having a positive skew; the mean is greater than the median. The reverse is true when a distribution is skew to the left or negatively skewed. This is exploited by the Pearson measure of skewness, defined by

$$\frac{\text{Mean} - \text{Mode}}{\text{Standard deviation}}$$

which is positive or negative according to whether the population is skew to the right or left.

Another useful measure uses the sum of cubes of deviations of population values from the mean, which can also be shown to be positive when the skew is positive and negative when the skew is negative. It is Fisher's measure of skewness, defined by

$$\frac{\dfrac{1}{N}\sum_{i=1}^{N}(X_i - \mu)^3}{\sigma^3}$$

Kurtosis

This is the 'peakedness' of the distribution: high kurtosis implies an attenuated modal peak frequency; low kurtosis describes a plateau-like distribution (Fig. 2.8). Kurtosis is less often needed as a useful descriptor.

2.1.4   Statistics of the frequency distribution of sample data

In practice, we can only use the attributes of samples to make estimates of the characteristics of the population as a whole. If the sample is large we can learn something of the shape of the distribution by drawing a histogram or line chart. Other diagrams, such as the dot plots described in Section 2.2.1, can often indicate whether the population is symmetric or skew, even when

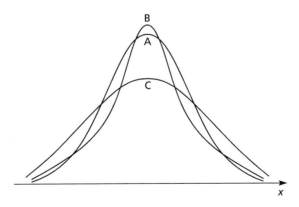

Fig. 2.8 Three distributions with different degrees of peakedness (kurtosis). Excessive peakedness or flatness may invalidate certain statistical procedures. A, zero kurtosis; B, positive kurtosis; C, negative kurtosis.

the sample is small. In any case we can calculate *statistics* from the data which will estimate the parameters of the population. It is important to realise that different samples from the same population will produce a different value for any statistic: we never know what the 'true' values of parameters are. Some justification for the choice of statistics to be used as estimators of parameters is given in Section 2.5; at present we may note that they are very similar to the definitions of the parameters themselves.

2.1.4.1

The sample histogram

In order to discover what pattern (if any) there is in the data, we begin by producing a frequency distribution and representing it graphically with a histogram (or by a line chart with discrete data). It is strongly recommended that histograms are plotted for all variables at the initial stage of all investigations. As we shall see, geological interpretations can sometimes be made directly, and the histogram will always indicate which procedure can be applied next (see Section 2.4.1).

For a given set of data, there are an infinite number of possible histograms: the class width (and hence the number of classes) and the class boundary positions can be changed. Although all combinations are, in principle, valid, there are strongly recommended rules of thumb which should be followed — see Box 2.1. Once the classes to be used have been chosen, frequencies are obtained by working through the data, observation by observation, and recording occurrences with tally marks. Statistical packages have commands to produce histograms: the classes will be selected for you but you should be prepared to override the default settings.

Any histogram with varying class widths should be viewed with suspicion: wider classes will look disproportionately strong. This is a popular way of distorting the true properties of data! Alternatives to histograms involving 'joining the dots' with straight or curved lines are also not recommended: these give the misleading graphical impression that there are inferences to be made about intermediate points.

Box 2.1 also contains information on cumulative frequencies. Such graphs will be used later to try to determine the distribution from which data come (Section 2.5.6.2).

We now need to describe the statistics that are calculated from real data in order to describe the data and to estimate population parameters.

(Notation: Names of variables are denoted by upper-case italic letters and sample values by the corresponding lower-case letters. The sample size is denoted by n. Sample statistics also have lower-case italic symbols. Individual observations are distinguished by subscripts. For example if the sample consists of observations of a variable X on n items the data will be denoted by $x_1, x_2, \ldots, x_n$.)

Box 2.1 Histogram construction

The following are data on the diameters in cm of 40 specimens of ammonite:

3.2	3.7	2.9	3.9	3.4	3.1	3.1	3.9	3.5	3.3
3.6	3.8	3.7	3.0	3.5	3.2	3.5	3.7	3.9	3.6
3.4	2.9	3.2	3.4	2.9	3.6	3.7	3.3	3.4	4.0
3.8	3.7	3.3	2.9	3.1	3.2	3.6	3.5	3.3	3.4

This gives the following frequency distribution:

Length (cm)	Frequency
2.85–3.05	4
3.05–3.25	8
3.25–3.45	9
3.45–3.65	8
3.65–3.85	7
3.85–4.05	4
Total	40

which in turn allows the construction of the histogram in Fig. 2.1.

The number of classes must be large enough to bring out the shape of the distribution but not so large that empty classes are scattered through it. A common method of choosing the number is to use one which is approximately equal to the square root of the number of observations. In this case we have used six classes. The boundaries of the classes are selected with the following points in mind.

1 All observed values must be covered.

2 It must be impossible for an observation to coincide with a boundary. Here, values are given to one decimal place, so we have specified boundaries to two places.

3 The distances between boundaries (the class widths) should, as far as possible, be equal. If we can achieve this then the heights of rectangles in the histogram, as well as their areas, will be proportional to frequencies of classes.

The steps required are: divide the range of the data by the proposed number of classes to find an approximate class width; choose a convenient width; and then begin the lowest class at a suitable point.

In this case we have

Range $= 4.0 - 2.9$
       $= 1.1$
Range/6 $= 0.18$ (approximately)

A convenient width near this value would be 0.2 and in order to accommodate the lowest observation we begin the first class at 2.85.

continued on p. 28

Box 2.1 *Continued*

If we divide the frequency in each class by the total (40), we obtain relative frequencies, which always have a total of 1.

Length (cm)	Relative frequency
2.85–3.05	0.100
3.05–3.25	0.200
3.25–3.45	0.225
3.45–3.65	0.200
3.65–3.85	0.175
3.85–4.05	0.100
Total	1.000

If we successively sum the relative frequencies, starting from the first class and including one extra class at a time, we obtain cumulative relative frequencies. These give the proportion of observations with values less than each class boundary.

Length (cm)	Cumulative relative frequency
3.05	0.100
3.25	0.300
3.45	0.525
3.65	0.725
3.85	0.900
4.05	1.000

2.1.4.2 Measures of location

Sample arithmetic mean

This is the most popular measure, and corresponds to the common concept of 'average'. It is calculated as the sum of all the data values for the variable divided by the number of data values (the sample size):

Equation for sample arithmetic mean

$$\bar{x} = \frac{1}{n} \sum_{i=1}^{n} x_i$$

Worked example: see Box 2.2.

Box 2.2 Worked example: sample statistics

The following data give diameters (mm) of clasts from a conglomerate:

23   24   27   29   29   30   33   33   34   38   45   60   60
88   126   221   256

Median

There are 17 observations so the median is $(n + 1)/2 = $ 9th in the rank order. The median is 34.

Mean

The sum of all 17 diameters is 1156, so the mean is $1156/17 = 68.0$. This is rather high as a representative value; this has resulted from the influence of the few very high values.

Geometric mean

The calculation

$23 \times 24 \times 27 \times 29 \times 29 \times 30 \times 33 \times 33 \times 34 \times 38 \times 45 \times 60 \times 60 \times 88 \times 126 \times 221 \times 256$

gives a value of

1.62928×10^{27}

We need the 17th root of this (so raise to power 1/17); the result gives a geometric mean of 39.87.

Some error will be involved in dealing with such large numbers on a calculator; nevertheless, the result seems to be a reasonable descriptor of the data.

The sample mean gives a good idea of the value of the population mean μ. Note that it uses all the information in a sample and may be affected by extreme values. For example, the sample mean of the data

3   1   5   4   7

is $20/5 = 4$, but that of the data

3   1   5   4   27

is $40/5 = 8$.

The sample mean is the value that is typically used to characterise the crucial properties of geological entities. For example, in mineral resource evaluation, the sample mean ore grade would be used; in estimating volume of fluids in sandstone reservoirs, we might use sample mean porosity.

Sample geometric mean

This is used for positively skewed data, for which the arithmetic mean may be unrepresentatively high. It is becoming increasingly popular as a descriptive statistic in porosity/permeability studies and in geochemistry.

> *Equation for sample geometric mean*
>
> Geometric mean $= (x_1 \, x_2 \ldots x_n)^{1/n}$
>
> Worked example: see Box 2.2.

Sample median

A measure which is less affected by uncharacteristically high or low observations is the sample median, which is the middle value when the data are arranged in order (or the average of the two middle values if n is even). This

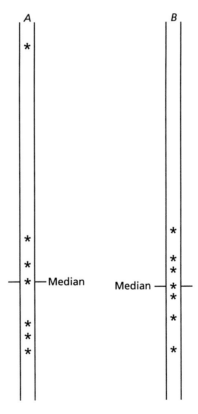

Fig. 2.9 Suppose that the asterisks represent positions in boreholes A and B at which a pollen species was found. The median positions are similar: the median in A is not affected by the single high occurrence. This may be a useful property, as that single occurrence may be anomalous or the result of fast sedimentation rates in the top part of A. We may not wish such occurrences to have a strong influence on the measure of location.

general definition renders it applicable to ordinal data as well as interval and ratio scale data: see Fig. 2.9.

Worked example: see Box 2.2.

Sample mode

The mode is a point in the class with the highest frequency on the histogram. Although this may not be centrally placed, it cannot be denied that the most popular value is a reasonable representative measure, at least for some purposes. In small data sets there may not be a mode, except by accident, or there may be several, usually spurious, modes.

2.1.4.3   ## Measures of dispersion

The degree of variation of data values can be diagnostic in geology, for example grain size sorting as an indicator of environment of deposition or geochemical variation indicating phases of magmatic activity.

It is also important to obtain a measure of the variability of values in a population in order to assess the number of observations we need to draw useful conclusions and to assess the reliability of our estimates of parameters.

Range

This is the difference between the highest and lowest values in a data set. It is reliable when it is calculated from small samples but very likely to be affected by unrepresentative extreme values in a large sample. At present it is not much used in geology.

Interquartile deviation and Folk parameters

Sedimentologists use modifications of the IQD, using a variety of 'cut-off' points, to quantify grain size frequency distributions. These result in Folk parameters: see Section 2.5.6.4 and Box 2.28.

Measures of deviation from the sample mean

Other measures are based on the scatter of the data about the sample mean and the computations use differences between the sample mean and the individual sample values. The mean of these differences or deviations from the sample mean is always zero because the positive deviations are balanced by the negative. One way of overcoming this difficulty is to find the mean of the magnitudes of the deviations, $|x_i - \bar{x}|$; that is, we treat all differences from the mean as if they were positive. The result is the *mean deviation about the mean* or simply the mean deviation. Using again the artificial data

3   1   5   4   7

whose mean is 4, the mean deviation would be

$(1 + 3 + 1 + 0 + 3)/5 = 1.6$

The most commonly used measures, however, use the squares of the devi-ations about the sample mean, and the first step in the calculations is the computation of a quantity called the *corrected sum of squares* (CSS),

$(x_1 - \bar{x})^2 + (x_2 - \bar{x})^2 + \ldots + (x_n - \bar{x})^2$

which may be written more compactly as

$\Sigma(x_i - \bar{x})^2.$

The meaning of the notation may be illustrated with the same artificial data set. The CSS is then

$\Sigma(x_i - \bar{x})^2 = (1 - 4)^2 + (5 - 4)^2 + \ldots + (3 - 4)^2 = 20$

This is very cumbersome if no computer is available and most textbooks give another, algebraically equivalent formula. If the more convenient form is used in a computer program, however, a serious loss in precision may result because computers are limited in the number of significant figures that they can retain; some data sets have been known to produce negative sums of squares. Consequently, we will not present the bad formula here! It was helpful in the past because it provided a comparatively simple way of

---

Box 2.3 Using calculators to obtain s^2

It is not necessary for you to perform the calculations implied by the formulae for s^2 or s because they are done by electronic calculators for small data sets and by standard computer packages when necessary. We should point out that certain calculators (notably those in the Casio range), which are excellent in every other way, use a notation which does not conform to statistical convention. They have two buttons for displaying standard devi-ations, one of which is labelled σ_n and the other σ_{n-1}. The first gives the result of using n as the divisor in the calculation of the sample standard deviation, the second uses $(n - 1)$. It is the second which we shall require but notice that the label should properly be s: we use σ for the population parameter and s as its sample estimate. The sample variance is obtained by squaring the standard deviation. (Press the x^2 button after displaying the value of s.)

Details of the calculation vary from machine to machine and they are described in the instruction books. As a check, enter the data 1, 5, 7, 4, 3 and show that the sample mean is 4 and that the variance is exactly 5. (The sample standard deviation will be given as 2.236067977 on a 10-digit display.)

calculating the CSS but the ready availability of electronic calculators removed the need for it (see Box 2.3).

One of the measures based on the CSS is the sample variance, s^2.

Equation for sample variance

$$s^2 = \frac{1}{n-1} \sum_{i=1}^{n} (x_i - \bar{x})^2$$

It is often called the sample estimate of the variance because it provides an estimate of the population variance σ^2. By analogy with the definition of population variance in Section 2.1.3, we might expect the divisor to be n, the size of the sample, but it can be shown that the resulting statistic tends to underestimate σ^2 (it is said to be biased) whereas s^2 as defined here is unbiased. Notice that the units of s^2 are the units of measurement *squared*, so that if the units are My, the sample variance is measured in My^2. The square root of this quantity will have the same units as the units of measurement and is easier to interpret; it is called the sample standard deviation, s. (It is a biased estimator of σ but has other properties which compensate.)

Equation for sample standard deviation

$$s = \sqrt{\frac{1}{n-1} \sum_{i=1}^{n} (x_i - \bar{x})^2}$$

2.1.4.4   ## Histogram shape: description and interpretation

The terms used for describing histogram shape are the same as for population frequency distributions (Section 2.1.3.3). Sample histograms will be rough approximations to the shape of the source population frequency distribution: there will be irregularities due to the vagaries of random sampling. It is common practice to overlook such effects when describing general shapes of histograms: we draw attention to those aspects of the shape that are sufficiently pronounced to lead us to believe that they reflect properties of the source population. Only these will be worth attempts at interpretation. Interpetations fall into a number of categories (refer to Box 2.4).

Normal distributions

As we will see in Section 2.4, the normal distribution is a special unimodal, symmetrical distribution. It can be derived from probability theory on the basis of random deviation around a preferred mean value: this is a situation that we might expect to encounter commonly in the natural world. A geological process, such as transportation of sediment in a river, may have

Box 2.4 Some geological interpretations of histograms

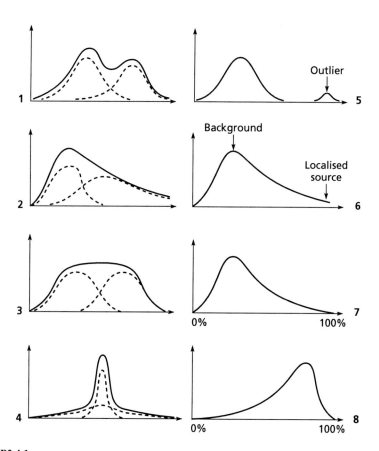

Fig. B2.4.1

1   Bimodality: due to mixture of two subpopulations with different means.
2   Positive skew: due to mixture of two subpopulations with different means and standard deviations.
3   Low kurtosis: due to mixture of two subpopulations with different means.
4   High kurtosis: due to mixture of two subpopulations with different standard deviations.

All of these interpretations as mixtures are sometimes invoked by sedimentologists. Pre-existing distinctive grain size populations may be mixed by storm or other high energy events.

5   Outlier: may be a very small subpopulation with a different mean (as bimodality above), or it may be an erroneous or otherwise anomalous measurement.
6   Positive skew: in geochemistry, the 'tail' can be interpreted as representing the rare local occurrences of high concentrations, perhaps near the source of an element (e.g. mineralised area, volcano). The peak represents

Box 2.4 *Continued*

the concentration that results when the element has been thoroughly dispersed in the system.

7 Positive skew: with compositional (closed) data, this often occurs in minor components (e.g. trace elements), resulting from the dilution effect of large components. It is an artefact of data closure.

8 Negative skew: with compositional data, this is the converse effect of the above, and occurs with major components (e.g. SiO_2). It too is an artefact.

characteristic controlling values, such as current velocity, but we do not expect all the water to have exactly the same velocity: there will be turbulence. So we have a 'preferred value' which is the average velocity, but with random dispersion around that value. Our initial expectation might be that the dispersion would be symmetrical, so a normal distribution might result. If we find that our data are approximately normally distributed, no special interpretation is necessary.

Mixtures

One way of interpreting departures from the normal distribution is by decomposing the distribution into more than one normal distribution with different means and/or standard deviations. This is a reasonable procedure if there is clear bimodality, but mixtures can be contrived that result in any type of modality, skewness or kurtosis. Some of these may stretch credulity! However, we will expect mixtures of discrete populations if we are sampling across geological boundaries. Sedimentologists are able to invoke storms and turbidities as mechanisms for mixing material from different energy regimes (and hence different grain sizes), and conglomerates are often bimodal in grain size due to the clast/matrix division: deposition in clast interstices occurs in different conditions from those in the conglomerate as a whole.

Outliers

Outliers are data points having anomalous values, usually anomalously high. They should be carefully investigated: they often represent errors in processing or some other spurious effect (perhaps a rusty nail in the sediment sample!). Where they are genuine, the typical interpretation is that they are isolated representatives from a minor population having extreme values. This may be of great interest: it could represent the source location of a trace element, for example, and point to an exploitable resource.

Data closure

Analysis of closed data (in percentage, ppm, etc.) reveals many artefacts that result from the way the data are prepared rather than from the natural system. Perhaps the most serious of these are discussed in Section 3.2.2, but there are also symptoms in univariate analysis. Simple simulations can demonstrate that, for minor components such as trace elements, the effects of variation of major components such as SiO_2 cause an artificial positive skew. The major components, conversely, tend to have a negative skew, centred high on the percentage scale and tailing off into lower percentages. The recommendation in Section 3.2.2, the use of the log ratio transformation, is also a suitable preparation for univariate analysis.

Exponents

We may argue that it is expected that the normal distribution can describe the type of random variation inherent in nature. Even if this is true, it may still be the case that the data we collect have a complex relationship with some underlying normal distribution. For example, if current velocities in a river are normally distributed and the transportation and deposition of sediment grains are governed by grain weight and local current velocity, then we expect the grain weight frequency distribution in resulting sediments to be normal. However, if we measure grain diameter, the data will be related to the cube root of grain weight (volume being proportional to radius cubed), and the resulting frequency distribution will be skew. This is a simplistic context, but geological materials are often related to processes in ways describable by power functions, and any exponent applied to a normally distributed variable will cause a skew.

2.2 GRAPHICAL SUMMARIES: EXPLORATORY DATA ANALYSIS

It is vital that new data should be examined in simple ways before they are subjected to more sophisticated statistical analysis. Preliminary examination of data may, for example, lead to the following results.
1 The data may include wrong or misleading values.
2 The nature of the variation can be readily made apparent. For example, we may find that the distribution is skew or that the data come from two or more sources which differ in some important respect.
3 We may see that sophisticated analysis is not required: geologically interesting effects may be blatantly obvious, in which case tests of statistical significance may be redundant.
4 Geological hypotheses and models may be suggested (which must be tested with fresh, independent data).
   The calculation of statistics is an important part of both describing the

Box 2.5 Examples of the use of dot plots

1 The concentrations of cadmium (ppm) found in 10 specimens of rock were found to be

15   11   24   25   24
28   30   49   38   24

and they produced the dot plot in Fig. B2.5.1. We can see at a glance that values tend to be clustered at the lower end and that the distribution is skew to the right.

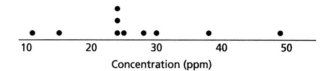

Fig. B2.5.1

2 Figures B2.5.2 and B2.5.3 are dot plots obtained from the lengths of 20 trilobites. Figure B2.5.2 shows the presence of two values of 99.9, which had been used to represent missing values. Once these are replaced by standard missing-value symbols, the distribution of real values becomes clear.

3 Dot plots are also useful for comparing two or more samples. Figure B2.5.4 represents widths of brachiopods taken from three sources. They are plotted on identical scales to make comparison easier. Note that the scales must include the lowest value observed in all three samples and the highest

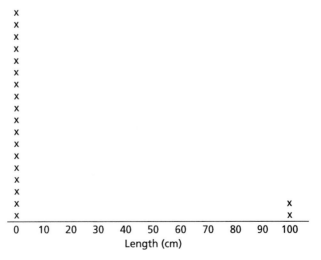

Fig. B2.5.2

continued on p. 38

Box 2.5 *Continued*

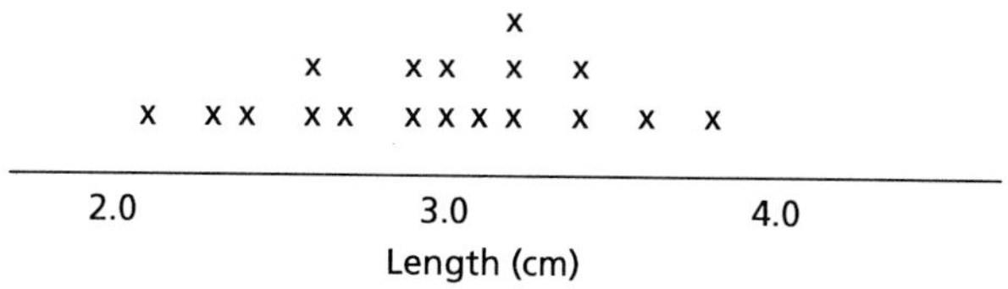

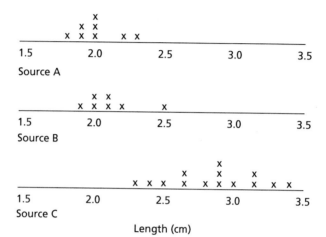

Fig. B2.5.3

Fig. B2.5.4

in all three. It is easy to see that samples *A* and *B* are similar in both their means and the degree of scatter, whilst sample *C* differs from them in both respects.

sample data and making inferences about the population, but their use may be misleading when the data include erroneous values or are mixtures from more than one population. Such features are often detected easily by graphical means, so we now consider four kinds of diagram which have been found particularly useful for monitoring the quality of data: the dot plot, stem-and-leaf diagram, box-and-whisker plot and scatter diagram.

2.2.1 The dot plot

The data from a single sample are displayed as dots on a scale, with repeated values represented by dots placed on top of each other. Computer-generated versions have limited resolution along the scale, so the 'repeated values' need not be identical. This is similar in concept to the histogram, except that we are not so concerned with grouping data into classes. The examples in Box 2.5 show the usefulness of the plot.

2.2.2 Stem-and-leaf diagram

This is essentially a histogram in which the individual values are presented. Each observation is divided into the most significant part, called the stem, and the rest, called a leaf. For the following data, representing ppm of an element in rock:

| 60 | 87 | 84 | 53 | 71 | 62 | 92 |
| 57 | 75 | 78 | 62 | 66 | 52 | 63 |

it is convenient to divide into stems consisting of 'tens' and leaves composed of 'units' and the stem-and-leaf diagram is:

```
5    2 3 7
6    0 2 2 3 6
7    1 5 8
8    4 7
9    2
```

Provided that each leaf is given the same space, the lengths of the lines will be proportional to the frequency of values with a given stem. So, in this form, we can observe both the general frequency distribution and the details of the data.

2.2.3 Box-and-whisker plots (or box plots)

These are useful for indicating suspicious values and for comparing sets of data. Using a horizontal scaled axis, we mark the positions of important descriptors: the main body of the data is marked by a box, the tails of the distribution are indicated by lines known as whiskers, and the extreme, anomalous values are represented by point symbols.

The ends of the box correspond to statistics called 'hinges', which are similar to quartiles. The lower hinge is the value which lies midway between the observation immediately below the median and the lowest value; the upper hinge is located similarly between the median and the highest value. The median is indicated by a line inside the box.

Lines (whiskers) from the ends of the box extend to the extreme values in the data set and positions of suspicious or otherwise interesting values may be marked and labelled, often by 'O' for 'outlier'.

Figure 2.10 shows a box-and-whisker plot illustrating the data used in the example of a stem-and-leaf diagram.

A set of such diagrams on the same scale provides an excellent comparison of two or more samples and as in the case of dot plots may be all that is required to establish similarity or difference.

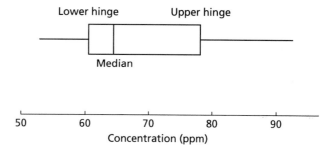

Fig. 2.10 Box-and-whisker plots give a clear indication of skewness: here the skew is positive.

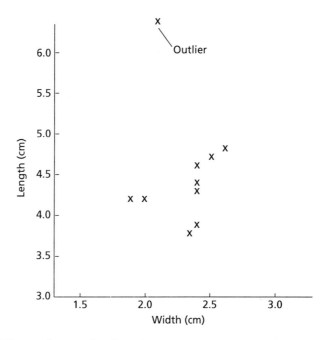

Fig. 2.11 Scatter plots reveal outliers, clusters and the presence or absence of relationships between pairs of variables. The data here are lengths and widths of 10 fossils:

Length (cm): 4.3 3.8 4.6 4.7 4.2 4.2 3.9 6.3 4.4 4.8
Width (cm): 2.4 2.3 2.4 2.5 2.0 1.9 2.4 2.1 2.4 2.6

2.2.4   **Scatter diagrams**

Geological entities are usually characterised by several variables. For instance, a rock may be described by percentages of oxides and a fossil by a number of dimensions. Relationships between these variables are studied with the aid of techniques such as correlation and regression analysis, described in Chapter 3, but a start should always be made by plotting pairs of variables against each other. Figure 2.11 is a plot of length against width for a sample of 10 fossils. Unusual observations show up well in these diagrams and the plot can indicate what kind of relationship, if any, exists between the variables. If it is clear that there is no relationship we need not try any more sophisticated analysis.

2.3 **PROBABILITY**

Probability is the basis for the formulation of many statistical methods, and probabilities often need to be quoted in statistical results. We now need to introduce the basic concepts of probability to ease the understanding of what follows.

2.3.1 **Interpretations**

There is no generally agreed definition of probability but there are several interpretations, each of which is useful for solving a particular type of problem. In order to introduce them, consider the following examples of hypothetical questions which might arise in geological contexts.

1 A set of 50 thin sections includes three which show crystals of apatite. If a slide is picked up at random what is the probability that it includes apatite?

2 What is the probability that an ammonite chosen at random from a particular horizon is larger than 6 cm?

3 A fault is known to intersect a 100 m geophysical traverse AB, but the exact point of intersection is unknown. What is the probability that the point lies within 20 m of end B?

4 A company must complete a drilling contract within 1 month or pay a penalty. What is the probability that a penalty will have to be paid?

The classical interpretation

In **1** there is a fixed number, 50, of possible outcomes. Any of the 50 sections may be picked up and there is no reason why one should be chosen rather than another. This is a situation in which the classical interpretation is helpful:

> *The classical interpretation*
>
> If a trial has n exhaustive, mutually exclusive and equally likely outcomes of which n_A constitute event A then the probability that event A occurs is
> $Pr(A) = n_A/n$

('Exhaustive' means that every possible outcome has been accounted for; 'mutually exclusive' means that only one of the outcomes can occur at any time.)

As there are 50 possible outcomes when a single section is taken at random and three of these constitute the event that apatite occurs, the probability that the selected thin section will contain apatite is 3/50 or 0.06.

Using the term 'equally likely' begs the question of what we mean by probability; it merely states that every outcome has the same probability, whatever that means. Nevertheless, this interpretation is helpful in problems about sampling from collections of items when the number of possibilities is known and there is no reason why one outcome should occur more than any other.

The relative frequency interpretation

Example **2** is a case where the number of items is not known and the outcome, size of ammonite, is a value from a continuous scale. Here we use an interpretation in terms of the proportion of times in which an event occurs in repeated random trials. We can, in principle, find the relative frequency with which specimens of this particular species of ammonite are found with a size exceeding 6 cm. It is a matter of experience that, provided sampling is random, the proportion of times a particular event occurs settles down around a fixed value; this value is taken to be the probability of the event (see Fig. 2.12). This interpretation does not permit the question 'What is the probability that the next ammonite chosen at random has a size in excess of 6 cm?' We can only ask about the proportion of cases in a large random selection.

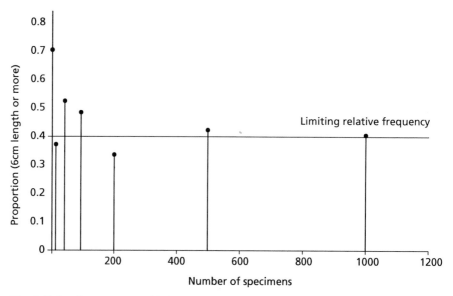

Fig. 2.12 In a long sequence of independent random trials the relative frequencies of outcomes settle down to values which are regarded as their probabilities.

The relative frequency interpretation

The probability of an event E is a limiting value of its relative frequency in a long sequence of independent, random trials.

The geometrical interpretation

Example **3** above requires a third interpretation, this one in terms of space. The probability of a randomly chosen point being in the interval XB is taken to be the length of the interval expressed as a fraction of the total length of the transect; here it will be 20 m/100 m or 0.2. In two dimensions, probability is the ratio of two areas and in three dimensions it is the ratio of two volumes. For example, if a region of 100 km^2 contains a batholith which has an area of outcrop of 15 km^2 and a point in the region is chosen at random, the probability of the point being in the feature is 15/100, i.e. 0.15.

The geometrical interpretation

The probability of a randomly selected point lying in a region of size v when the total available space is of size V is v/V.

Subjective probability

The last example, **4**, is one in which there is no symmetry and no trials have been made. Here, probability is interpreted as a degree of belief and is called subjective probability. In any situation we shall have our own ideas of the probability of a given event; it will depend upon our knowledge and the way we use it. Nevertheless, it is still possible to perform meaningful calculations, incorporating knowledge of a given subject. In fact, some statisticians believe that this is the way in which probability should always be interpreted and it is the basis of an important school of statistics: Bayesian statistics. Unfortunately, it will not be possible to treat the subject in this book.

2.3.2   **Sample spaces**

Whatever interpretation of probability we use for a given problem, it is important to identify the set of all possible outcomes of a trial – the *sample space* – because its nature governs the way in which we calculate probabilities of events. Table 2.1 gives examples of sample spaces associated with various trials. In the first two trials the number of outcomes is finite; in the third there is no clear upper limit and we take it to be infinite but say that the number is countable. In all three cases we can calculate probabilities of individual outcomes. The values in the sample space for the fourth trial are

Table 2.1 Examples of sample spaces associated with various trials

Trial	Type of outcome	Sample space
Choose one section from the 50 in example **1**, 2.3.1	Which section?	The list of all 50 sections
Choose five sections from the 50	Number showing apatite	0, 1, 2, 3
Observe tremors in a region for 25 years	Number of tremors	0, 1, 2, . . .
Determine the proportion of Al_2O_3 in a rock	Percentage of Al_2O_3	All values between zero and 100

on a continuous scale and again the number of possible values is infinite; we shall be able to calculate probabilities of obtaining values in some interval but will take the probability of obtaining any one value to be zero.

2.3.3 Elementary properties of probability

The interpretations of probability discussed above have numerical features in common and we shall use these to motivate some simple rules for calculating probabilities of more complex events.

2.3.3.1 Scale

The ratios used for describing probability in the classical, relative frequency and geometric probabilities all lie in the interval from 0 to 1, so we shall impose this condition on probabilities: for any event E,

$$0 \leqslant Pr(E) \leqslant 1$$

If an event is impossible, its probability is zero; if it is certain to occur, then its probability is 1.

Next, it is certain that one of the outcomes described by a sample space will occur: from what we have just said, the total probability of all the points in a sample space is 1.

2.3.3.2 Addition law for mutually exclusive events

Consider the following situations.

1 Suppose that a collection of 50 hand specimens of a rock consists of three which contain no particles of gold, seven which contain a little gold and 40 which are of high grade. If a specimen is chosen at random what is the probability that it contains little or no gold? The event of interest can occur in $3 + 7$ ways and its probability is 10/50 or 0.2. We note that the events are mutually exclusive.

Table 2.2 Proportions of beds of three rock types in a long sequence

Rock type	Number of beds	Proportion of beds
Sandstone	45	0.225
Mudstone	80	0.400
Limestone	75	0.375
Total	200	1.000

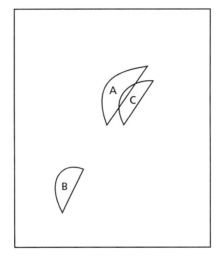

Fig. 2.13 Probability of a randomly chosen point lying in one or both of two regions. When the regions overlap (*A* and *C*) we must make allowance for points which lie in both regions and which may therefore be counted twice. If the regions are not connected physically (*A* and *B*) their areas may simply be added when calculating the probability of a point lying in one of them.

2 We examine a long sequence of beds consisting of sandstone, mudstone and limestone; the proportion of beds which are sandstone or mudstone is the total of the proportions of beds which are of these sediments (Table 2.2).

3 In Fig. 2.13, which represents the outlines of three hydrocarbon-bearing reservoirs in an oilfield, the probability that a vertical borehole located at random will hit *A* or *B* is the total of the areas of *A* and *B* divided by the total area of the region and this is the same as the sum of the separate fractions.

$$\frac{\text{Area of region } A + \text{Area of region } B}{\text{Total area}} = \frac{\text{Area of } A}{\text{Total area}} + \frac{\text{Area of } B}{\text{Total area}}$$

These examples motivate the following rule.

Addition rule for mutually exclusive events

If $E_1, E_2, \ldots, E_k$ are mutually exclusive, the probability that one of them occurs is obtained by adding their individual probabilities.

We write

$$\text{Pr}(E_1 \text{ or } E_2 \text{ or } \ldots \text{ or } E_k) = \text{Pr}(E_1) + \text{Pr}(E_2) + \ldots + \text{Pr}(E_k)$$

In mathematics, 'or' is often denoted by a plus sign when the outcomes or events are mutually exclusive:

$$\text{Pr}(E_1 + E_2 + \ldots + E_k) = \text{Pr}(E_1) + \text{Pr}(E_2) + \ldots + \text{Pr}(E_k)$$

General addition law

Suppose that we require the probability of a randomly chosen point lying in area A or C or both, where A and C overlap to some extent, as they do in Fig. 2.13. If we add the area of A to that of C, the overlapping area is included twice and to allow for this we must calculate the probability as

$$\text{Pr}(\text{Point in } A \text{ or } C \text{ or both}) = \frac{\text{Area of } A + \text{Area of } C - \text{Overlap}}{\text{Total area of region } R}$$

$$= \text{Pr}(\text{Area } A) + \text{Pr}(\text{Area } C) - \text{Pr}(\text{Overlap})$$

Similarly, suppose that we choose one fossil from a collection described by Table 2.3 in which specimens are classified by species and source. What is the probability that the specimen is of species A and comes from source Y?

This could be calculated from (Number of specimens of species A + Number from Y − Number of specimens satisfying both conditions)/100, i.e.

$$39/100 + 51/100 - 24/100 = 0.66$$

Table 2.3 Specimens classified by species and source

Source	Trilobite			
	A	B	C	Total
X	15	23	11	49
Y	24	14	13	51
Total	39	37	24	100

2.3.3.3   **Complementary events**

The interpretations of probability are all consistent with the idea that the probability of an event E not occurring is given by $1 - Pr(E)$. The event 'E does not occur' is called the complement of E and is denoted by $\bar{E}$ or by E_c. This gives the addition rule for mutually exclusive events.

$$Pr(\bar{E}) = 1 - Pr(E)$$

The use of this rule often simplifies calculations when the event of interest is complex and its complement is simple: we calculate the probability of the complement and subtract it from 1.

For example, we may be interested in the probability of *at least* one eruption of a volcano in a given time – this appears to require us to calculate and add the probabilities of 1, 2, 3, . . . etc. eruptions. It is, in fact, easier to calculate the probability of no eruptions occurring and then subtract the result from 1.

Thus in the example on specimens containing much, little or no gold the probability that the selected specimen will be of high grade is $1 - 10/50$ or 0.8. In the example of the sequence of rock types, the proportion of beds which are limestone will be:

$1 -$ (the proportion of beds which are sandstone or mudstone)

For the areas shown in Fig. 2.13, the proportion of area occupied by the region outside zones A, C is:

$1 -$ (the proportion of area in zones A, C)

2.3.4   **Conditional probability**

The probability of a particular event occurring will in general depend upon circumstances. For example, the probability of finding oil in a region is likely to depend upon the depth of the source rock; also, we think that the presence of oil is more likely when we find appropriate trap structures.

Consider this example. One hundred and fifty hand specimens of rock are classified according to the region from which they were taken and to whether their copper content was low, medium or high. The numbers of specimens in each category were as shown in Table 2.4.

Table 2.4 Hand specimens classified by region and grade

Copper	Region		
	A	B	C
Low (L)	30	19	8
Medium (M)	10	20	17
High (H)	15	18	13

If a specimen is chosen at random:

1 what is the probability that it contains a high level of copper? and

2 what is the probability that it contains a high level of copper if we are told that it came from region A?

In the first case, where we are given no information, the required probability is 46/150, i.e. 0.31. In the second, however, we are told that it came from region A so we need only consider specimens from that region. The probability of a specimen in this group containing a high level of copper is therefore 15/55 or 0.27. We write

$$Pr(H|A) = 0.27$$

and describe this as: 'the conditional probability of the specimen containing a high level of copper, given that it is from region A'. Note that the event of interest is to the left of the upright line and helpful information appears on the right. The symbol $Pr(A|H)$ would mean 'The probability that the specimen comes from region A, given that it contains a high level of copper' and its value would be 15/46 or 0.33.

We have expressed a conditional probability as a ratio of two numbers of ways in which events may occur, but we shall usually need to obtain it as a ratio of probabilities. In fact, we could write

$$Pr(H|A) = \frac{15/150}{55/150}$$

and this is indeed $Pr(H \text{ and } A)/Pr(A)$.

> *Conditional probability*
>
> The conditional probability of an event A, given that B has occurred is
>
> $$Pr(A|B) = \frac{Pr(A \text{ and } B)}{Pr(B)}$$

As another example, let us consider the probability that a randomly chosen point in Fig. 2.13 lies in region A, if it is known that it lies in C. We need only think of the points which lie in region C and in (A and C) and we obtain

$$Pr(A|C) = \frac{\text{Area of } (A \text{ and } C)}{\text{Area of } C}$$

2.3.4.1   **Multiplication rule for probabilities of joint events**

The definition of conditional probability involves the probability of events A and B both occurring and it enables us to write down an expression for this probability:

$$Pr(A \text{ and } B) = Pr(A|B) \cdot Pr(B)$$

We could also write

$$Pr(A \text{ and } B) = Pr(B|A) \cdot Pr(A)$$

The form we choose will depend upon the available information – that is, which probabilities we know: $Pr(A|B)$ and $Pr(B)$ or $Pr(B|A)$ and $Pr(A)$.

2.3.4.2   Total probability

If the probability of event A occurring depends on whether or not B occurs and we do not know if B has occurred, how can we calculate the overall probability of A? The answer lies in calculating an average value of the conditional probabilities of A, using the probabilities of B and $\bar{B}$ as weights:

$$Pr(A) = Pr(A|B) \cdot Pr(B) + Pr(A|\bar{B}) \cdot Pr(\bar{B})$$

It is instructive to see how this result can be obtained from earlier ones. First note that A occurs in one of two ways:

(A occurs and B occurs)   or   (A occurs and B does not occur)

In symbols, writing AB to mean 'A and B'

$$A = AB + A\bar{B}$$

As the events on the right-hand side are mutually exclusive we can add their probabilities to obtain the probability of A:

$$Pr(A) = Pr(AB) + Pr(A\bar{B})$$

and, applying the multiplication rule to the probabilities on the right-hand side,

$$Pr(A) = Pr(A|B)Pr(B) + Pr(A|\bar{B})Pr(\bar{B})$$

There are frequently more than two possibilities for the conditions under which event A may occur; again, we calculate an average probability for A. The process is like the calculation of overall proportions.

For example, suppose that a region is divided into blocks B_1, B_2 and B_3 which account for 20%, 10% and 70% respectively of the total area. The proportions of these blocks which may be exploited for hydrocarbons are 40%, 30% and 5% respectively. What proportion of the region is exploitable?

We add the proportions of exploitable area from the three blocks:

40% of 20% + 30% of 10% + 5% of 70% = 14.5%

or, in terms of decimal fractions,

$0.4 \times 0.2 + 0.3 \times 0.1 + 0.05 \times 0.7 = 0.145$

This is equivalent to answering the question, 'if a point is chosen at random in the region, what is the probability that it will be in an exploitable area?' If

A may occur under any of the mutually exclusive conditions $B_1, B_2, \ldots, B_k$ then the total probability of A is

$$Pr(A) = Pr(A \mid B_1)Pr(B_1) + Pr(A \mid B_2)Pr(B_2) + \ldots + Pr(A|B_k)Pr(B_k)$$

2.3.4.3   **Bayes' theorem**

We have seen how to calculate the probability of an event A which occurs with one of the mutually exclusive events $B_1, \ldots, B_k$. The reverse problem of calculating the probabilities of the mutually exclusive events is of great practical importance. For example, A may represent a description of the silica content of a specimen of rock (the rock contains more than 17% of silica, say) and we may know that the rock is one of the types $B_1, \ldots, B_k$: the question of interest is, 'What are the probabilities of the rock being a specimen of $B_1, \ldots, B_k$?' In practice (when we would use more complete information on the rocks) we might accept that the rock is of the type which has the highest probability. Such problems are examined in Section 8.4 on 'Discrimination'. The theorem for this situation was the work of an English clergyman, Thomas Bayes, who was interested in the problem of causes. In science we have no way of proving causes from effects in the sense that mathematics can prove conclusions from premises, but we can calculate probabilities of causes. The theorem can be written down simply from our definition of conditional probability, for the conditional probability of an event B_i, given that A is observed, is

$$Pr(B_i \mid A) = \frac{Pr(B_i \mid A)}{Pr(A)}$$

where $Pr(A)$ is the total probability described in the previous section. Also, the multiplication rule tells us that $Pr(B_i A) = Pr(A \mid B_i)Pr(B_i)$ so we can write

$$Pr(B_i \mid A) = \frac{Pr(A \mid B_i)Pr(B_i)}{Pr(A)}$$

$Pr(B_i)$ is called the prior probability of B_i because it is the value assigned before we are given any further information; $Pr(B_i \mid A)$, the value calculated in the light of the knowledge that A has occurred, is called the posterior probability of B_i; $Pr(A \mid B_i)$ is known as the likelihood of A.

Bayes' theorem

If an event A always occurs with one of the mutually exclusive events $B_1, \ldots, B_k$ then the conditional probabilities of these events, given that A has occurred, are given by

$$Pr(B_i \mid A) = \frac{Pr(A \mid B_i)Pr(B_i)}{Pr(A)}$$

Box 2.6 Worked example: Bayes' theorem

Measurements are made on a fossil which is known to be a specimen of one of three species. The following table gives the probabilities of obtaining such a set of characteristics in each species:

Species	Pr(characteristic A │ species B_i)
B_1	0.8
B_2	0.3
B_3	0.6

The probabilities of obtaining a specimen of each species are:

Species	Probability of occurrence
B_1	0.1
B_2	0.4
B_3	0.5

We would like to give an opinion on the species to which the specimen belongs. It is convenient to set out the calculation in tabular form:

Species	Prior probability	Likelihood $Pr(A \mid B_i)$	Product $Pr(A \mid B_i)Pr(B_i)$	Posterior $Pr(B_i \mid A)$
B_1	0.1	0.8	0.08	0.16
B_2	0.4	0.3	0.12	0.24
B_3	0.5	0.6	0.30	0.60
Total	1.0		0.50	1.00

Column 1 is a list of possible species and column 2 contains their prior probabilities $Pr(B_1)$ to $Pr(B_3)$. Likelihoods of the observed characteristics are given in column 3; if we multiply these values by the prior probabilities and sum the products the result will be the total probability of observing the characteristics A. The final column contains the posterior probabilities $Pr(B_i \mid A)$ for $i = 1, 2, 3$; note that these probabilities will sum to unity.

Species B_3 has the highest posterior probability, so unless conflicting evidence was offered we would assign the specimen accordingly.

In practice, if we were solely interested in assigning the specimen the final column could be omitted because each entry has the same divisor. We could base our judgement on the values of the products $Pr(A \mid B_i)Pr(B_i)$.

> where $\Pr(A) = \Pr(A \mid B_1)\Pr(B_1) + \ldots \Pr(A \mid B_k)\Pr(B_k)$
> Worked example: see Box 2.6.

2.3.5 Independence

When the probability of event A does not depend upon the occurrence or non-occurrence of event B then A, B are said to be *independent*. Table 2.5 conveys an idealised case where clasts have been classified as coming from conglomerate I or II and as being longer or shorter than 5 cm. The proportions of clasts less than 5 cm long are the same in both groups. Whether we take one at random from the conglomerate I or the conglomerate II group or choose one at random from the combined set, the probability of obtaining a clast shorter than 5 cm is 3/8 (15/40 = 9/24 = 3/8).

When events A, B are independent we can write $\Pr(A)$ instead of $\Pr(A \mid B)$ because the information on the right of the upright line is irrelevant. Then the multiplication rule for finding the probability that both A and B occur becomes

$$\Pr(A \text{ and } B) = \Pr(A)\Pr(B)$$

Conversely, if we find that this form of the multiplication rule is true for A, B then the events are independent.

When more than two events are involved the situation is a little more complicated. It is true that if A, B, C are independent then the probability that all three occur is

$$\Pr(A \text{ and } B \text{ and } C) = \Pr(A)\Pr(B)\Pr(C)$$

Worked example: see Box 2.7.

The fact that this rule holds, however, does not establish the independence of the three events. It is necessary to show also that the events are independent in pairs, that is

$$\Pr(A \text{ and } B) = \Pr(A)\Pr(B)$$

and

Table 2.5 Sixty four clasts classified by length and conglomerate

	Conglomerate	
Length	I	II
<5	15	9
>5	25	15
Total	40	24

Box 2.7 Worked example: independence

Suppose the probability that a specimen of iron ore contains more than 20% of magnetite is 0.15. If three specimens are obtained independently what is the probability that all three specimens contain more than 20% of magnetite? What is the probability that the first two contain more than 20% and the third contains less? It is convenient to introduce some notation to reduce the amount of writing needed. Let us write A_1 to denote the event that the first specimen contains more than 20%, A_2 to mean that the second specimen contained more than 20% of magnetite and so on. Then, in the second part of the question, we shall use $\bar{A}_3$ to mean that the third specimen did not contain more than 20% magnetite.

In the first part the event of interest is:

(All three contain more than 20% magnetite) = (A_1 and A_2 and A_3)

Taking probabilities of both sides and using the information that sampling was independent, we have:

$$
\begin{aligned}
\Pr(\text{All three contain more than 20\%}) &= \Pr(A_1 \text{ and } A_2 \text{ and } A_3) \\
&= \Pr(A_1)\Pr(A_2)\Pr(A_3) \\
&= (0.15)^3 \\
&= 0.00375
\end{aligned}
$$

Next, consider the event:

(the first two contain more than 20%, the third contains less)

which we represent by (A_1 and A_2 and $\bar{A}_3$). Now as sampling is independent,

$$
\begin{aligned}
\Pr(A_1 \text{ and } A_2 \text{ and } \bar{A}_3) &= \Pr(A_1)\Pr(A_2)\Pr(\bar{A}_3) \\
&= 0.15 \times 0.15 \times (1 - 0.15) \\
&= 0.019
\end{aligned}
$$

$$\Pr(A \text{ and } C) = \Pr(A)\Pr(C)$$

and

$$\Pr(B \text{ and } C) = \Pr(B)\Pr(C)$$

This has consequences for the analysis of categorical data when more than two types of category are involved and in the analysis of multivariate data, but for the present we will concentrate on the consequences of independence rather than tests for independence.

The idea of independence is extremely important. Inferential procedures will depend for their validity upon samples being taken in such a way that the selection of one specimen does not affect the choice of any other — that is to say, that specimens are selected independently and that we do not allow the result of one measurement to influence another. In this way we avoid bias, and each specimen will provide fresh information.

2.3.6 Random variables

Results of experiments and surveys can always be expressed in terms of numbers, even when they are categorical. Examples include analysing specimens of rock for oxides, measuring dips and strikes and measurements made on fossils. Even when the outcome is not inherently numerical, we can assign numbers to them: for example, if we are noting whether or not the fossils in a sample have some special feature or not we can give each specimen a score of 1 if the feature is present and zero if it is absent.

We cannot predict with certainty what the result of a measurement or other recording will be, so the trial we make is said to be random. A *random variable* associates a numerical outcome with a random trial. Some examples are given in Table 2.6.

Although we cannot say in advance what value the random variable will take, we may be able to assess the probabilities of observing particular values or values in given intervals: it may be possible to calculate the probability of detecting 20 α-particles in a 30-second interval, for example, or the probability of obtaining a specimen of rock whose gold content lies in the interval 5%–10%. The manner in which the total probability of 1 is shared between – distributed over – possible values of a random variable is called a *probability distribution*.

For example, suppose that five points on a section are examined at random with a microscope and the number of times X that a crystal of olivine is observed under the cross-hairs is recorded. Here the random variable is discrete and can take values 0, 1, 2, 3, 4, 5. It can be shown by methods to be described in Section 2.3.7.2 that if the proportion of olivine

Table 2.6 Examples of random variables

Trial	Outcome	Random variable
Place a detector near a radioactive source	Number of α-particles	Number of particles 0, 1, 2, . . .
Observe feldspar crystals in a transect (10 points)	Number of feldspar crystals	Number of crystals 0, 1, 2, . . . , 10
Analyse a rock for gold	Grade of rock	Percentage gold $0 < \%$ gold < 100
Test a rock for the presence of a mineral	Mineral absent or present	$X = 0$ for absent $X = 1$ for present
Classify a sediment as mud, shale or sand	Type of sediment	$X = 1$ for mud $X = 2$ for shale $X = 3$ for sand

Table 2.7 Discrete probability distribution

Number of crystals observed X	Probability $Pr(X = x)$
0	0.3277
1	0.4096
2	0.2048
3	0.0512
4	0.0064
5	0.0003
Total probability	1.0000

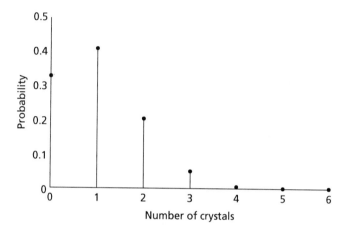

Fig. 2.14 Line chart to represent the probability distribution of finding 0, 1, . . . , 5 olivine crystals in a transect when the proportion of olivine is 20% and the mineral is scattered randomly through the rock.

in the rock is 0.2 then the probabilities of observing these numbers of crystals are those shown in Table 2.7. This table is an example of a discrete probability distribution. Note that all possible values of X are stated and that the probabilities sum to unity. The distribution is illustrated by a line chart (Fig. 2.14).

Suppose that a point is chosen at random along a 1 km transect AB and its distance from A is measured. The random variable X is the distance of the point from A. In this case X is a continuous random variable; the probability that it is exactly at a given distance x from A is taken to be zero, but the probability that it will lie in some interval will be proportional to the length of the interval. In fact,

$$Pr(x_1 < X < x_2) = (x_2 - x_1)/1$$

The distribution is illustrated in Fig. 2.15. Like the relative frequency histogram for data on continuous variables, the total area under the curve

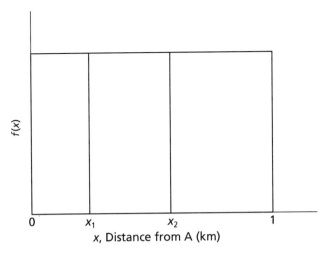

Fig. 2.15 The uniform distribution of distances of points along a transect, measured on a continuous scale.

between $X = 0$ and $X = 1\,\text{km}$ represents unity and probabilities of the kind given above are represented by the area under the curve between appropriate limits.

Some of the most important distributions are discussed in Sections 2.3.7–2.3.8. We can also summarise the properties of random variables in terms of mean values, variances and other measures, as we summarised frequency distributions in Sections 2.1.3 and 2.1.4. This is covered in the next section, on expectation.

2.3.6.1 Expectation of a random variable

The term 'expectation' arose in the context of gambling. A player's expectation in a single game was calculated from the possible winnings and their probabilities of occurring: each gain was weighted by its probability so that more importance was attached to outcomes which were likely to happen than to those which probably would not occur. The products of gains and their probabilities were then summed to produce a single index, the expectation or mean gain in a single game.

Suppose the game had five possible outcomes and that the gains and their probabilities were as in Table 2.8.

The gambler's expectation would be

$$(-10) \times 0.3 + (-5) \times 0.35 + 0 \times 0.20 + 10 \times 0.10 + 20 \times 0.05 = -2.75$$

Table 2.8 Probabilities of gains in a game

Gain	−10	−5	0	10	20
Probability	0.3	0.35	0.2	0.1	0.05

On average he or she would lose 2.75 units per game, if foolish enough to play.

As a very much simplified example, a company may reckon that, if it exploits a certain oilfield, the probabilities of making gains of $-£10m$, $£20m$ and $£30m$ sterling are 0.2, 0.7 and 0.1 respectively. Then the company's expectation from the exploitation would be

$$-£10m \times 0.2 + £20m \times 0.7 + £30m \times 0.1 = £15m$$

The expectation of a discrete random variable is calculated in exactly this way, for it is defined to be

$$E(X) = \Sigma x Pr(X = x)$$

where Σ means, 'sum over all possible values of X'. For the distribution in Table 2.7, the expectation or mean value of X is

$$E(X) = 0 \times 0.3277 + 1 \times 0.4096 + 2 \times 0.2048 + 3 \times 0.0512 + 4 \times 0.0064$$
$$+ 5 \times 0.0003 = 0.9999$$

Compare this calculation with that of the sample mean of a set of data such as 7, 7, 7, 8, 8, 9, 9, 9, 9, 9; we would write

$$\bar{x} = \frac{3 \times 7 + 2 \times 8 + 5 \times 9}{10}$$

and this is equivalent to (though we would not use it in practice)

$$\bar{x} = \frac{3 \times 7}{10} + \frac{2 \times 8}{10} + \frac{5 \times 9}{10}$$

where each observed value is weighted by its relative frequency.

2.3.6.2   **The variance of a random variable**

Variance is a measure of the extent to which values of a random variable tend to differ from the mean and it is defined to be

$$Var(X) = E\{(X - \mu)^2\}$$

the mean value of the squared difference between possible values of the random variable and the mean μ. In statistical formulae it is usually denoted by σ^2, sometimes with a subscript to indicate the name of the variable.

When the random variable is discrete, the variance is calculated from

$$Var(X) = \Sigma(x - \mu)^2 Pr(X = x)$$

so that again the differences which occur only rarely are given a low weighting. We illustrate the nature of this calculation in Table 2.9.

It is usual to show that the calculation can be performed in a more convenient way but we will not do so: the important point is to understand

Table 2.9 Calculation of variance of a random variable (x)

x	$x - \mu$	$(x - \mu)^2$	$Pr(X = x)$	$(x - \mu)^2 Pr(X = x)$
−10	−7.25	52.625	0.3	15.7688
−5	−2.25	5.0625	0.35	1.7719
0	2.75	7.5625	0.2	1.5125
10	12.75	162.5625	0.1	16.5663
20	22.75	517.5625	0.05	25.8781
Variance				61.4976

the nature of variance, not to be skilled in deriving well-established results mathematically.

2.3.6.3

Means and variances of simple functions of random variables

We often need to obtain means and variances of random variables which are derived from others; the most important cases are:

1 the addition of a constant term to a random variable X or the multiplication of a random variable by a constant; and

2 the addition of random variables or of multiples of them.

The second of these groups of cases is a dominating feature of multivariate analysis and will be discussed and illustrated in Section 8.2. In this section we shall state the results without proof.

The effects of constant terms

If all the possible values of a random variable X have a constant a added or are multiplied by a constant term c, the mean value will be modified in the same way:

$$E(X + a) = E(X) + a \quad (a \text{ is added to the mean})$$

and

$$E(cX) = cE(X) \quad (\text{the mean is multiplied by } c)$$

The addition of a constant does not change the variability and so we have

$$Var(X + a) = Var(X)$$

but multiplication by a constant will have an effect since differences of values from the mean will be multiplied by the constant. In fact, the squares of these differences will be multiplied by c^2 so

$$Var(cX) = c^2 Var(X)$$

Addition and subtraction of random variables

If a number p of random variables are added, the mean of the total is the total of the means. The mean proportion of total metal in blocks in a mine will be the total of the proportions of the individual metals. Similarly the mean difference between two random variables is the equal to the difference in their means. The variance of a sum or difference of random variables is less straightforward and we must consider two sets of conditions.

First, suppose that we form the sum of two random variables which are statistically linearly related in the following sense:

$$Y - \mu_Y = k(X - \mu_X) \quad \text{(where } k \text{ is a constant)}$$

that is, the difference between Y and its mean tends to be a multiple of the difference between X and the mean of X. For example, we might consider the sum of lengths of heads (cephala) and bodies (say, thorax plus pygidia) of trilobites so that $X + Y$ represents total length. Fossils whose heads are longer than the mean for the species will tend to have bodies which are longer than the mean. Variables X, Y are said to be positively correlated and the variation in $X + Y$ will be greater than the sum of variation in X and Y: the variance of $X + Y$ will be greater than the sum of (a) the variance of X with fixed Y; and (b) the variance of Y with X fixed.

$$\text{Var}(X + Y) \geqslant \text{Var}(X) + \text{Var}(Y)$$

The excess can be shown to be

$$2 \times \text{E}\{(X - \mu_X)(Y - \mu_Y)\}$$

which we will write

$$2\text{Cov}(X, Y)$$

where the expectation term is called the covariance of X, Y because it provides a measure of the extent to which X, Y vary together in a linear relationship. It will be treated in more detail in Section 3.2. Meanwhile we note that if X, Y are positively correlated

$$\text{Var}(X + Y) = \text{Var}(X) + \text{Var}(Y) + 2\text{Cov}(X, Y)$$

In some contexts, the constant k is negative so that increased levels of X are associated with decreased levels of Y and the variables are then said to be negatively correlated. This may be so among certain constituents of rocks, for example. As high values in one variable are compensated for by low values in the other, the variation in the total will tend to be dampened. In fact, for negatively correlated variables X, Y

$$\text{Var}(X + Y) = \text{Var}(X) + \text{Var}(Y) - 2\text{Cov}(X, Y)$$

The second of the two sets of conditions is that there is no statistical linear relationship between the variables. In this case, variables are said to be uncorrelated and then the variance of their sum is simply the sum of their individual variances:

$$\mathrm{Var}(X + Y) = \mathrm{Var}(X) + \mathrm{Var}(Y)$$

These arguments can be applied to sums of any number of random variables and to linear compounds of the type

$$a_1 X_1 + a_2 X_2 + \ldots + a_p X_p$$

We may now summarise the results which are most important for our purposes.

When X is a random variable and a, c are constants,

$$E(X + a) = E(X) + a$$
$$E(cX) = cE(X)$$
$$\mathrm{Var}(X + a) = \mathrm{Var}(X)$$
$$\mathrm{Var}(cX) = c^2 \mathrm{Var}(x)$$

When $X_1, \ldots, X_p$ are random variables,

$$E(X_1 \pm \ldots \pm X_p) = E(X_1) \pm \ldots \pm E(X_p)$$

If the variables are uncorrelated,

$$\mathrm{Var}(X_1 \pm \ldots \pm X_p) = \mathrm{Var}(X_1) + \ldots + \mathrm{Var}(X_p)$$

If X_1, X_2 are correlated,

$$\mathrm{Var}(X_1 \pm X_2) = \mathrm{Var}(X_1) + \mathrm{Var}(X_2) \pm 2\mathrm{Cov}(X_1, X_2)$$

2.3.7 Discrete probability distributions

Many geological studies involve counting events or recording numerical characteristics which can only take particular values in an interval. As examples of counting we may consider numbers of major tremors in a region during a given length of time or the numbers of faults observed in a transect. In these types of cases we may obtain formulae called probability functions for calculating the probability of any value of the variable and these formulae are determined by the mechanism of the process we are observing. For example, we shall discuss the use of a particular distribution (the Poisson) for describing events which occur independently and randomly, and see how it can be applied to spatial and temporal phenomena.

2.3.7.1 Discrete uniform distribution

A random variable which can only take a finite number of values and has the same probability of taking each one is said to have a discrete uniform probability distribution. If the variable can take 10 values, for example, then

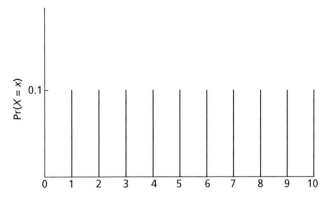

Fig. 2.16 Uniform distribution of n discrete values when each has the same probability, $1/n$, of being selected.

the probability of any particular value occurring is 1/10 (see Fig. 2.16). In general, if there are n possible values of X the probability function is

$$\Pr(X = x) = 1/n$$

Although it is not encountered in nature, the distribution is very important for its use in taking random samples of material and in computer experiments which try to imitate natural processes. For example: suppose that 20 specimens of rock are to be examined at points along a transect of length 0.5 km at whole numbers of metres from the start. We could write a computer program to generate the numbers 0, 1, 2, . . . , 500 in such a way that in the long run they occur virtually equally frequently. When the program is run each number produced would direct us to a point on the transect. (If a number occurred more than once we would ignore the second and later occasions and generate extra values.) In this way we would be directed to specimens and not allowed scope for conscious or accidental bias in our selection procedure.

A second application is described in Section 6.1 but we can give a simple example here: suppose that we wish to simulate a sequence of deposits of sandstone, mudstone and limestone in some locality and that we are interested in a case where probabilities of deposition of these sediments are thought to be 0.3, 0.25 and 0.45 respectively. Numbers are generated by machine, some of them representing the occurrence of sandstone, some mudstone and the rest limestone. Again we may write a program to generate, for example, the 1000 numbers 0.000, 0.001, 0.002, . . . , 0.999 in an apparently random fashion. We reserve 30% of them to represent sandstone (numbers 0.000–0.299), 25% to represent mudstone (0.300–0.649) and the remaining 45% limestone (0.650–0.999). The sequence of generated numbers 0.214, 0.328, 0.711, 0.025 would, for example, represent the deposition of sandstone, mudstone, limestone and sandstone.

The mean and variance of the discrete uniform distribution depend upon the possible values of X and, as they are of lesser interest in the remainder of the work, we shall not give examples of these parameters.

2.3.7.2 The binomial distribution

This is one of the distributions which are used when each of our trials has just two possible outcomes. A hand specimen may or may not contain a fossil, a body of ore may or may not be exploitable, an instrument may or may not be functioning satisfactorily, a crystal may or may not be olivine and so on. Such trials are called Bernoulli trials (after the Bernoulli brothers, who were Swiss mathematicians and carried out early work on probability). In the abstract, the possible outcomes are usually called 'success' and 'failure'. Success does not necessarily imply that something good has happened; if we are counting defective items in a batch then observations of such items are successes.

We will also need the binomial distribution in the work on non-parametric inference (Chapter 4). Sometimes our measurements are such that it is more appropriate to ask if one value is greater than another or not, rather than try to measure the difference between the values. In this case there are two outcomes: the first value is greater than the second, or it is not. This will be dealt with in Chapter 4. For the present we shall confine ourselves to details of the distribution.

The conditions for using the binomial distribution are as follows:
1   a finite number n of independent Bernoulli trials; and
2   the probabilities of the two outcomes are constant in every trial.

The result of interest, the random variable, is the number of successes in n trials – the number of hand specimens which contain fossils, the number of ore bodies which are exploitable, the number of instruments which are functioning satisfactorily, the number of crystals in a transect which are olivine.

When n is 1 the distribution is called the Bernoulli distribution. If the probability of success is p then the probability of failure is $1 - p$. We can write the probability function as

$$\Pr(X = x) = 1 - p \quad \text{when} \quad X = 0$$
$$\Pr(X = x) = p \qquad \text{when} \quad X = 1$$

The mean of the distribution is then

$$E(X) = 0(1 - p) + 1p$$
$$= p$$

and the variance is

$$\text{Var}(X) = (0 - p)^2(1 - p) + (1 - p)^2 p$$
$$= p^2(1 - p) + (1 - p)^2 p$$
$$= p(1 - p)$$

We can use these results to obtain the mean and variance of binomial distributions in which n has higher values.

Consider five independent trials in each of which the probability of success is p. For instance we might choose five hand specimens of rock at random for which the probability of finding a fossil in a specimen is p and count the number of the five which are fossiliferous. We can calculate the probability of two successes, for example, as follows.

The probability of obtaining two successes and three failures in a specified order is, from the multiplication rule for independent events,

$$p^2(1 - p)^3$$

Note: the number of ways in which n objects can be arranged when r of them are identical and the remaining $(n - r)$ are identical can be shown to equal

$$\frac{n(n - 1) \ldots (n - r + 1)}{1 \times 2 \times \ldots \times r}$$

which is written $\binom{n}{r}$. Here, there are $\binom{5}{2} = 5 \times 4/(1 \times 2) = 10$ ways.

So there are $\binom{5}{2}$ mutually exclusive orders in which the two successes and three failures may be obtained, some of which are indicated in Fig. 2.17. Each has a probability of $p^2(1 - p)^3$, so the total probability of obtaining two successes and three failures is

$$\binom{5}{2}p^2(1 - p)^3$$

using the addition rule for mutually exclusive events.

We could use this argument for any number n of Bernoulli trials and for any number of successes x from 0 to n so we write the probability function of the number of successes as

$$\Pr(X = x) = \binom{n}{x}p^x(1 - p)^{n-x}, \; x = 0, 1, \ldots, n.$$

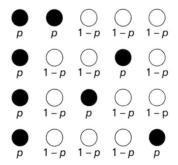

Fig. 2.17 Probabilities of obtaining two successes and three failures in specified orders in a sequence of five independent Bernoulli trials.

We are usually interested in the probability of the number of successes being in some range: for example, we may wish to know the probability that it exceeds a given number or that it is less than or equal to some value. Then we must combine probabilities of obtaining the values in the desired range:

$$\Pr(X > x) = \Pr(X = x + 1) + \Pr(X = x + 2) + \ldots + \Pr(X = n)$$
$$\Pr(X \leqslant x) = \Pr(X = x) + \Pr(X = x - 1) + \ldots + \Pr(X = 0)$$

The second of these sums is the distribution function, $F(x)$.

The calculation is rather laborious and the probabilities can often be obtained from statistical packages or, for a limited number of values of p and n, from tables such as that provided in this book (Appendix 2.1).

The mean of the binomial distribution has immediate interest: it is natural to wish to know how many successes there will be on average in a given number of trials. If we collect 10 hand specimens, how many will, on average, contain fossils? If we make 100 independent point counts on a thin section and the probability of finding hornblende on each occasion is 0.1, how many times, on average, will we find this mineral? If the probability of making a serious error in a certain type of measurement is 0.05, what will be the average number of serious errors made in five measurements? The usefulness of the variance is not obvious at this stage but it will become clearer in Section 2.4, on the normal distribution, and in Section 2.5, on inference. Meanwhile we state the formulae for the mean and variance without derivation:

> *Formulae for mean and variance of binomial distribution*
>
> For a binomial distribution in which the number of trials is n and the probability of success in each trial is p,
>
> Mean $\mu = np$
> Variance $\sigma^2 = np(1 - p)$

The result for the mean is in keeping with intuition: if the probability of a success in each of 100 trials is 0.4, say, then we expect to get successes in about 40 trials. The formula for the variance shows that the variability in numbers of successes is greatest when $p = 0.5$. (Calculate values of $p(1 - p)$ for several values of p including $p = 0.5$ to verify this statement.) This is an important fact because it can give us guidance on the number of trials required to estimate proportions of objects which have some given characteristic: it is necessary to take comparatively large samples when the characteristic of interest is very variable.

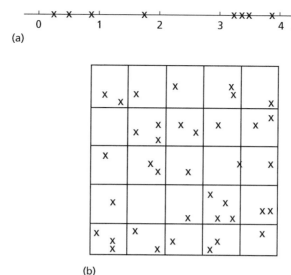

Fig. 2.18 The Poisson distribution applies to discrete, independent random events in (a) time or (b) space.

2.3.7.3   **The Poisson distribution**

Many natural phenomena may be regarded as point events in space or time: the emission of an α-particle occurs in an extremely short time and a major tremor lasts for a brief time in geological terms; similarly, a drumlin or a swallow hole may be regarded as a point, compared with the size of the region in which each is located. The study of such phenomena requires counting of events in given intervals of space or time: we may note the number of eruptions of a certain volcano in each of a number of 25-year periods or the distribution of garnet grains in a thin section of schist (Fig. 2.18). A natural question to ask is whether or not the events occur at random and independently, in some sense, and the model of randomness is provided by the Poisson distribution.

Events will be described as random and independent if the following conditions hold.

1   The probability of a single event occurring in a very short interval of time or space is approximately proportional to the length of the interval. (This is clearly a rough statement, as the probability must remain less than 1, and it must be stressed that we are talking of *small* intervals.)

2   The probability of more than one event occurring in such an interval is virtually zero. (We can make the interval sufficiently small to make this so.) These statements mean that either one event or no event occurs in each small interval.

3   The occurrence or non-occurrence of events in non-overlapping intervals is independent.

When these conditions apply, the number of events occurring in a finite interval t units of time or space has a Poisson distribution with the following probability function.

> *Probability function for Poisson distribution*
>
> $\Pr(X = x) = \exp(-\mu t) \cdot (\mu t)^x / x!$ for $x = 0, 1, \ldots$
>
> where μ is the mean number of events in a unit interval.
>   Worked example: see Box 2.8.

As for the binomial distribution, probabilities in Poisson distributions are readily obtained from many statistics packages or, for limited numbers of values of the mean, from tables (Appendix 2.3). When these aids are not available, we can make use of a simple relationship between probabilities of consecutive values of the random variable. Consider the expression for $\Pr(X = 0)$, $\Pr(X = 1)$, $\Pr(X = 2)$ and so on:

$$\Pr(X = 0) = \exp(-\mu)$$
$$\Pr(X = 1) = \exp(-\mu) \times \mu/1$$
$$\Pr(X = 2) = \exp(-\mu) \times \mu^2/2$$

$$\Pr(X = x) = \exp(-\mu) \times \mu^x/x!$$
$$\Pr(X = x + 1) = \exp(-\mu) \times \mu^{(x+1)}/(x + 1)!$$

Comparison of successive terms shows that the probability of $\Pr(X = x + 1)$ can be obtained from $\Pr(X = x)$ as follows.

$$\Pr(X = x + 1) = \Pr(X = x) \times \mu/(x + 1)$$

Worked example: see Box 2.9.

Mathematically, there is no upper limit to the value of X but if, for example, we were studying major eruptions in 25-year intervals the probabilities of the very high values would be zero for all practical purposes.

If we find that the numbers of intervals in which 0, 1, 2 . . . point events occur are similar to those which would be expected from a Poisson distribution we shall take this as an indication that the events are random and independent in the sense described earlier. (Remember, however, that such a finding cannot be regarded as *proof* of randomness.)

A good example of point events which do satisfy the conditions are emissions of α-particles from radioactive isotopes. On the other hand, tremors and eruptions of volcanoes do not qualify to be called random in the above sense for we know that major tremors are preceded and followed by minor tremors so that events in non-overlapping intervals are not independent. Nevertheless, it is a proper question to ask if *major* tremors appear to occur independently and randomly.

It has been stated that the parameter μ in the probability function is the mean of the distribution and it can be shown that the same parameter is the

Box 2.8 Worked example: Poisson distribution

The number of major floods occurring in 50-year periods in a certain region has a Poisson distribution with a mean of 2.2. What is the probability of the region experiencing:

1 exactly one flood in a 50-year period?
2 exactly one flood in a 25-year period?
3 at least one flood in a 50-year period?
4 not more than two floods in a 25-year period?

Solution

The unit of time here is 50 years, so for parts **1** and **2** the value of t is 1 and we have:

1 The probability of exactly two floods in 50 years is

$$\Pr(X = 2) = \exp(-2.2) \times 2.2^2/2!$$
$$= 0.268$$

2 The probability of at least one major flood is the sum of probabilities of 1, 2, . . . floods and it is obtained by first calculating the probability of the complementary event – no floods – and using the probability rule for complementary events:

$$\Pr(X = 0) = \exp(-2.2) \times 2.2^0/0!$$
$$= \exp(-2.2)$$
$$= 0.1108$$

so the probability of experiencing one or more floods is

$$\Pr(X \geqslant 1) = 1 - 0.1108$$
$$= 0.8892$$

In parts **3** and **4**, we are dealing with 25-year periods, that is, half units of time; then $t = 0.5$ and μt is 1.1.

3 The probability of exactly one flood in 25 years is

$$\Pr(X = 1) = \exp(-1.1) \times 1.1^1/1!$$
$$= 0.366$$

4 The probability of not more than two floods occurring in 25 years is the total of the probabilities of zero, one and two floods:

$$\Pr(X \leqslant 2) = \Pr(X = 0) + \Pr(X = 1) + \Pr(X = 2)$$
$$= 0.900$$

variance. The Poisson distribution is unique in necessarily having a variance equal to the mean, and one test for randomness involves the comparison of the sample mean with the sample variance. Other uses of the variance will be described in Sections 6.2 and 7.1.

Box 2.9 Worked example: Poisson distribution

Pollen grains are dispersed at random in the matrix of a rock and the density on a flat surface is 2.5 per unit area. Find the probabilities of observing zero, one, two, three and four grains in a unit area.

    The solution of the problem on a calculator is set out in the following table.

Value of x	Machine	$Pr(X = x)$	Probability
0	Enter mean 2.5		
	Press $+/-$		
	Press INV LN	$\exp(-2.5)$	0.0821
1	Press $\times$ 2.5	$\exp(-2.5) \times 2.5$	0.2052
2	Press $\times$ 2.5 $\div$ 2	$\exp(-2.5) \times 2.5^2/2!$	0.2565
3	Press $\times$ 2.5 $\div$ 3	$\exp(-2.5) \times 2.5^3/3!$	0.2137
4	etc.	etc.	0.1336

2.3.7.4   **The negative binomial distribution**

This has uses in two completely different contexts, one related to the binomial and the other to the Poisson distribution. Whereas the binomial is the distribution of the number of successes in a fixed number of trials, the negative binomial is the distribution of the minimum number of trials required to produce a fixed number of successes. For example, it might, under suitable conditions, be the distribution of the number of fossils we collect before we find a specimen of a given species or the number of wells drilled to find three exploitable reservoirs.

    It is also used for the distribution of the number of events occurring in an interval when the mean rate is varying in a random manner (more precisely, when the mean has an exponential distribution or a gamma distribution, to be described in Section 2.3.8.2).

    We consider first the case where independent Bernoulli trials are made until the first success is obtained, the probability of success in any trial being p. Denoting successes by S and failures by F, some possible outcomes of the experiment are shown in Table 2.10. Then the probability function for the number of trials up to and including the first success is

$$Pr(X = x) = (1 - p)^{x-1}p \quad \text{for} \quad x = 1, 2, \ldots$$

As the successive values form a geometric progression this, the simplest form of the negative binomial distribution, is called the geometric distribution.

    Now suppose that we wanted to know the probabilities of the numbers of trials which might be required to obtain exactly r successes. The event of interest would happen in this way:

Table 2.10 Geometric distribution

Sequence	Number of trials, x	$\Pr(X = x)$
S	1	p
FS	2	$(1 - p)p$
FFS	3	$(1 - p)^2 p$
FFFS	4	$(1 - p)^3 p$
$x - 1$ failures followed by 1 success	x	$(1 - p)^{x-1} p$

$r - 1$ successes in $x - 1$ trials and the rth success in the final, xth trial

The probability of the first part of the event is given by the binomial distribution in which the number of trials is $x - 1$ and we are interested in the probability of $r - 1$ successes:

$$\binom{r - 1}{x - 1} p^{x-1} (1 - p)^{(r-1)-(x-1)}$$

The probability of the second part is simply p, so the probability of both events is the product of these two terms:

$$\Pr(X = x) = \binom{r - 1}{x - 1} p^x (1 - p)^{r-x}$$

It can be shown that the mean of this distribution is $r(1 - p)/p$.

The justification of this probability function in terms of independent, random, discrete events in time or space is beyond the scope of this book but the following examples are given to bring your attention to possible uses of the negative binomial distribution.

1 Suppose that a region contains volcanoes whose major eruptions can be modelled by Poisson distributions but whose mean rates vary from individual to individual with an exponential distribution. Then the number of major eruptions in time t of a volcano selected at random in the region has a geometric distribution.

2 A large region is divided into blocks for exploration purposes. Suppose that the number of exploitable wells in a block has a Poisson distribution but that the mean of the distribution varies from block to block according to a gamma distribution. Then the number of exploitable wells in a block chosen at random has a negative binomial distribution.

2.3.8 **Continuous random variables and their probability distributions**

When a random variable is measured on a continuous scale, the probability that it takes any particular value exactly is zero, but we can find probabilities

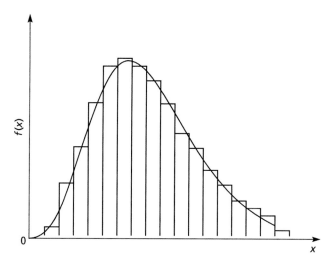

Fig. 2.19 Approximation to the outline of a histogram by a smooth curve when a large number of observations on a random variable is available.

of the value lying in some interval. We stated that the area of a rectangle of a relative frequency histogram represents the proportion of observations which lie in the corresponding interval. If we could make a very large number of observations and use narrow class intervals the outline of the histogram would approximate to a smooth curve which could be described reasonably well by a mathematical function of the random variable X (Fig. 2.19).

The function is called a probability density function (p.d.f.) and it is often denoted by $f(x)$. The probability of obtaining a value of X in some specified interval (x_1, x_2) is represented by the area under the curve between the limits x_1 and x_2. In most practical cases the area must be obtained from tables or computer packages because the calculation is so complex.

Expectations and variances of continuous random variables may be defined using the notation of integral calculus but we shall simply state what these statistical parameters are in terms of the parameters of the p.d.f.

2.3.8.1 The continuous uniform distribution

A random variable with this distribution can be defined on any finite interval, the most useful being (0, 1). The probability of a value of X lying in any interval is proportional to the length of the interval, wherever it may be. Figure 2.20 illustrates the uniform (0, 1) distribution.

The p.d.f. is

$f(x) = 1$ for all values of X in $0 \leqslant X \leqslant 1$
      0 otherwise

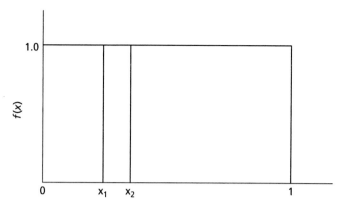

Fig. 2.20 The continuous uniform distribution on the interval from zero to unity.

In this case it is easy to see that

$$\Pr(x_1 < X \leqslant x_2) = x_2 - x_1$$

because the height of the p.d.f. is 1 throughout any interval.

The mean and variance of the uniform $(0, 1)$ distribution are

$$E(X) = 0.5$$
$$\mathrm{Var}(X) = 1/12$$

The distribution is important in statistical theory and for the purposes of computer simulation experiments because it is related to all other probability distributions. We generate data from other distributions by transforming uniform $(0, 1)$ variates.

2.3.8.2 **The exponential distribution**

We have already noted that many geological events may be represented as points in space or time. When discrete events occur randomly and independently in a continuum at mean rate λ per unit interval (so that the number occurring in a unit interval has a Poisson distribution with parameter λ), the intervals between events give rise to relative frequency histograms of the type shown in Fig. 2.21 and the p.d.f. is

$$f(x) = \lambda \exp(-\lambda x), \ \lambda > 0, \ x \geqslant 0$$
$$\quad 0 \text{ otherwise}$$

As the mean rate of occurrence of events per unit interval is λ the mean time between events must be $1/\lambda$:

$$E(X) = 1/\lambda$$

The p.d.f. has only the one parameter λ, so the variance of the random variable X must be a function of it. It is given by

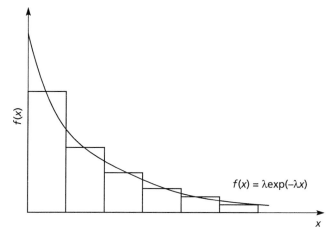

Fig. 2.21 A typical histogram of times between events and its mathematical approximation when the number of events occurring in a unit interval has a Poisson distribution.

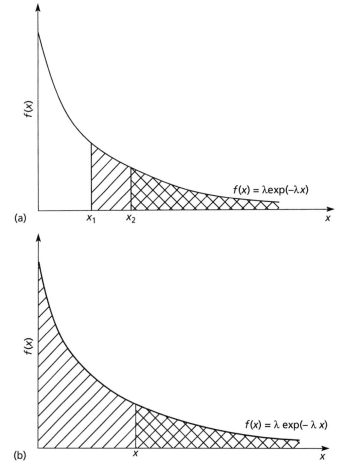

Fig. 2.22 The probability of observing an interval between two given values is represented by the area under the exponential curve between these values. (a) $\Pr(x_1 \leqslant X \leqslant x_2)$; (b) $\Pr(X \leqslant x)$.

$Var(X) = 1/\lambda^2$

Probabilities may be evaluated by using the following result, which is given without proof:

$Pr(X \geqslant x) = \exp(-\lambda x)$

Figure 2.22(a,b) illustrates how it follows that

$Pr(x_1 \leqslant X \leqslant x_2) = \exp(-\lambda x_1) - \exp(-\lambda x_2)$

and

$$Pr(X \leqslant x) = 1 - \exp(-\lambda x)$$

The probabilities may be found with the aid of an electronic calculator.
  Worked example: see Box 2.10.

Box 2.10 Worked example: use of exponential distribution

The number of major tremors occurring in 100-year intervals in a certain region has a Poisson distribution with a mean rate of 2.1. Find the probability that the time between two successive major tremors is
1 more than 25 years;
2 less than 50 years; and
3 between 30 and 40 years.

Solution

The unit of time is 100 years. As the mean rate of major tremors is 2.1 per 100 years, the mean time between them is 1/2.1 of 100 years.
1 25 years is 0.25 unit. Denoting time by T,

$Pr(T > 0.25) = \exp(-2.1 \times 0.25)$
$= 0.59$ to 2 D

2

$Pr(T < 0.5) = 1 - \exp(-2.1 \times 0.5)$
$= 0.65$

3

$Pr(0.3 < T \leqslant 0.4) = Pr(T \geqslant 0.3) - Pr(T \geqslant 0.4)$
$= \exp(-2.1 \times 0.3) - \exp(-2.1 \times 0.4)$
$= 0.10$

2.4 THE NORMAL DISTRIBUTION

2.4.1 Introduction: the importance of the normal distribution

The normal or Gaussian distribution is the most used probability distribution in statistical analysis: most methods depend upon the assumption that the data or some function of the data has a normal distribution. Values tend to come from this distribution when they are the sum of many independent, random contributions. For example:

1   grain sizes in a sandstone will depend on variations of grain composition and current velocity, which will itself be dependent on spatial and temporal variations;

2   errors in analytical results may arise from the sums of effects of friction in bearings of instruments, parallax errors, personal error, etc.; or

3   the size to which a bivalve grows may be determined by environment, genetic factors, age, pathology and life history.

We can imagine a geological (or laboratory) situation being characterised by a preferred value (the mean), but such sources of variation will cause an individual value to deviate from this. As the observed value is the sum of many independent contributions, it is as likely to be above the mean as below it and will more often than not be near the mean. Furthermore, the probability of differing from the mean by more than a given amount will be the same above the mean as below it: in other words, the distribution is symmetrical. A typical histogram is given in Fig. 2.23(a) and superimposed on it is a smooth mathematical curve which approximates to it well. The function has two parameters: one is the mean and is the central value, the other is a scale parameter which describes the spread of the distribution. Figure 2.23(b) shows two normal distributions with different means and equal standard deviations, whilst Fig. 2.23(c) shows distributions with equal means and different standard deviations.

The mathematically perfect normal distribution continues to infinity in each direction (but, of course, with frequency densities becoming vanishingly small), so values from a distribution can take negative as well as positive values. Although this cannot make physical sense when we are dealing with variables such as length, for example, it does not cause difficulties. Mathematical functions are only approximations to what we observe; in the case of Fig. 2.23 the calculated probability of obtaining a negative value will be negligible.

2.4.2 Use of tables of the normal distribution

Probabilities of finding values in specified intervals are given in a variety of · tables. We have chosen to use tables of the distribution function $F(x)$: probabilities of finding values in the interval from minus infinity to a given value. As it is not possible to produce a different table for every possible

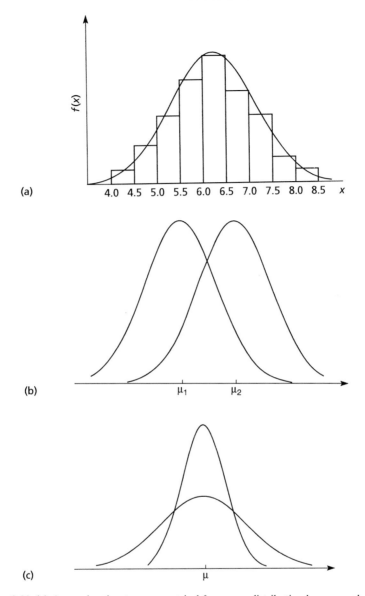

Fig. 2.23 (a) Approximation to a symmetrical frequency distribution by a normal curve. (b) Normal distributions with different means but the same standard deviation. (c) Normal distributions with equal means but different standard deviations.

normal distribution, we standardise values of the variable X so that they come from the standard normal distribution, with a mean equal to zero and standard deviation equal to 1. We have seen that any random variable with mean μ and standard deviation σ can be transformed to one with mean zero and standard deviation 1 as follows:

$$Z = \frac{X - \mu}{\sigma}$$

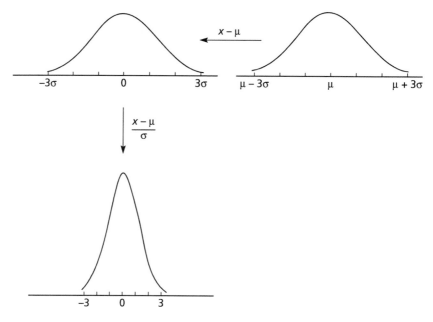

Fig. 2.24 Standardisation of a normal random variable: the effects of subtracting the mean μ from every possible value and dividing by the standard deviation shifts the mean to zero and scales the spread to give a standard deviation of unity.

The effect of the transformation on a variable having a normal distribution is illustrated in Fig. 2.24.

Predictions are made about normally distributed populations by using standard tables such as that presented in Appendix 2.2; for worked examples illustrating the use of these see Box 2.11.

In addition, Box 2.12 introduces the way in which we report the degree of certainty in an estimate of a quantity. More details of such reporting are given in Section 2.5.3, on confidence intervals.

The table in Appendix 2.2(b) provides information in the opposite way of the table in Appendix 2.2(a). It is used to answer the question, 'What is the value which, with probability α will not be exceeded?'. This is known as a percentage points table of the normal distribution. For worked examples of the use of these tables, see Box 2.13.

2.4.3 The distribution of a scaled normal variable

We have stated that if X has a normal distribution with mean μ and standard deviation σ then the scaled variable

$$Z = \frac{X - \mu}{\sigma}$$

also has a normal distribution, with a mean of zero and a standard deviation

Box 2.11 Standardisation and use of tables of the normal distribution

Grades of chip samples from a body of ore have a normal distribution with a mean of 12% and a standard deviation of 1.6%. Find the probability that the grade of a chip sample taken at random will have a grade of:

1　15% or less;
2　14% or more;
3　8% or less; and
4　between 8% and 15%.

Solution

Here, μ is 12 and σ is 1.6 so the standardised values are given by

$$z = \frac{x - 12}{1.6}$$

1　The standardised value of 15 is

$$z = \frac{15 - 12}{1.6}$$
$$= 1.88$$

From Appendix 2.2(a), the probability of obtaining a standardised value of 1.88 or less is 0.9699, i.e.

$$\Pr(X < 15) = \Pr(Z < 1.88) = 0.96995$$

2　Similarly, the standardised value of 14 is 1.25. Now Appendix 2.2(b) shows that the probability of obtaining a grade of 14% or less is 0.89435 and, as the total area under the curve represents 1, the probability of a grade of 14% *or more* is

$$\Pr(X \geq 14) = 1 - \Pr(X < 14)$$
$$= 0.10565$$

3　The standardised value of 8 is -2.5, and as the table only gives values of the distribution function for positive values we use the symmetry of the distribution to solve the problem. The probability of getting a value of Z which is less than or equal to -2.5 is the same as the probability of obtaining one which is greater than or equal to $+2.5$. We can therefore use the method of part **2**.

$$\Pr(Z \leq -2.5) = \Pr(Z > 2.5)$$
$$= 1 - \Pr(Z \leq 2.5)$$
$$= 0.00621$$

continued on p. 78

Box B2.11 *Continued*

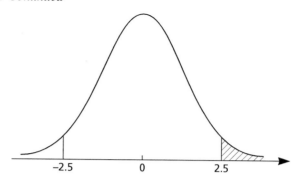

Fig. B2.11.1

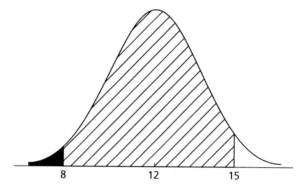

Fig. B2.11.2

4 Figure B2.11.2 illustrates the fact that the probability of obtaining a grade between 8% and 15% is the difference between the probability of the grade being less than or equal to 15% minus the probability of it being less than or equal to 8%.

$$Pr(8 < X < 15) = Pr(X \leqslant 15) - Pr(X \leqslant 8)$$
$$= 0.96995 - 0.00621 \quad \text{(from parts \textbf{1} and \textbf{2})}$$
$$= 0.96374$$

Suppose that it had been claimed that the mean grade in the ore body was 12% and that the standard deviation was 1.6%, how would you react if you found the grade of chip sample was 8%? Part **3** of the example shows that the probabilty of obtaining such a poor sample or worse is only 0.0062. Either a rare event has occurred or the claimed mean is too high (accepting the standard deviation). Although the example is simplified, it demonstrates the kind of thinking we use in assessing claims and evidence, which we discuss in Section 2.5.

Box 2.12 Worked example: use of tables of normal distribution: degree of certainty in estimate

Suppose that it is known that the errors produced by an instrument have a normal distribution with a mean of zero (so it is not biased) and standard deviation of 1.5 units. A single determination of a quantity is made and the reported value is 12.7. What range of values would appear to be reasonable for the quantity measured?

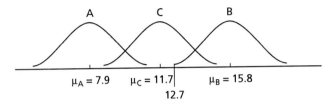

Fig. B2.12.1

Figure B2.12.1 shows three of the infinite number of distributions from which the value may have come. The probabilities of obtaining a value near 12.7 are negligible in the cases of distributions A and B but if C were the correct distribution the probability of obtaining a value of 12.7 or more would be high. Our observation is consistent with the idea that the true value is 11.7, therefore. Where should we draw the boundary line between reasonable values and others?

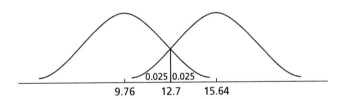

Fig. B2.12.2

Figure B2.12.2 shows one commonly used way of fixing the boundaries. If 9.76 is the true mean then there is a probability of 0.025 of obtaining our value of 12.7 or more; if 15.64 is the correct value there is a probability 0.025 of obtaining 12.7 or less. Then the range from 9.76 to 15.64 is chosen to be reasonable for the value. The probability of the interval containing the true value is 0.95. This idea will be developed in Section 2.5 on statistical inference in connection with estimating values and confidence intervals.

Box 2.13 Worked examples: percentage points table of the normal distribution

1 To find the value of a standard normal variable which, with probability 0.975, will not be exceeded.

Solution

From Appendix 2.2(b), the value of Z is 1.96.

2 Lengths of specimens of a species of fossil have a normal distribution with a mean of 3.5 cm and a standard deviation of 0.1 cm. Find the lengths which will be exceeded in only 5% of specimens and the length which will be exceeded in 97.5% of specimens.

Solution

We find the values of a standard normal variable which would satisfy the given conditions and then find the corresponding values from the distribution whose mean is 3.5 and whose standard deviation is 0.1 cm.

The value of Z which is exceeded with probability 0.05 (5%) is 1.6449 so, using x as the required length,

$$\frac{x - \mu}{\sigma} = 1.6449$$

i.e.

$$\frac{x - 3.5}{0.1} = 1.6449$$

and

$$x = 1.6449 \times 0.1 + 3.5$$
$$= 3.66$$

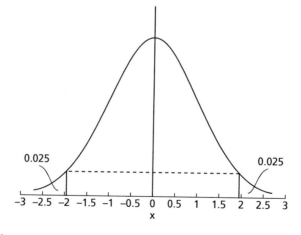

Fig. B2.13.1

> **Box 2.13** *Continued*
>
> For the second part of the problem we must again use the symmetry of the normal distribution. The probability that the value will *not* be exceeded in a specimen chosen at random is 0.025 and Fig. B2.13.1 illustrates the fact that the value of Z will be the reflection in the centre line of the value which *is* exceeded with probability 0.025. Then:
>
> $$\frac{x - 3.5}{0.1} = -1.96$$
>
> so that
>
> $$x = 3.30$$

of 1. This is a particular case of the property that any linear function of the normal variable X, $aX + b$, where a, b are constants, has a normal distribution; the mean value of the new variable is $a\mu + b$ and its standard deviation is $a\sigma$.

2.4.4 Reproductive properties of the normal distribution

In studies involving many variables, such as geochemistry and palaeontology, we shall need to make use of indices formed as linear compounds of the original variables. For example, we may have measurements of the length, height and width of many specimens of brachiopods and wish to form an index of shape by combining these measurements in a compound of the type

$$a_1 \times \text{Length} + a_2 \times \text{Height} + a_3 \times \text{Width}$$

where a_1, a_2 and a_3 are fixed numbers such as those calculated by methods to be described in Chapter 8. Suppose that these dimensions have normal distributions with means μ_1, μ_2, μ_3 and standard deviations σ_1, σ_2, σ_3; then the values of the linear compounds also come from a normal distribution. This is the reproductive property of the normal distribution.

The mean of the new distribution is

$$a_1\mu_1 + a_2\mu_2 + a_3\mu_3$$

The variance and standard deviation will be dealt with later since they involve the covariances between all pairs of variables as well as the individual variances.

The situation is simpler when the variables are uncorrelated because the covariances are zero and the variance of the compound is the total of the variances of the scaled variables. If X_1, X_2, ..., X_p are uncorrelated random variables with means $\mu_1, \mu_2, \ldots, \mu_p$ and standard deviations $\sigma_1, \sigma_2, \ldots, \sigma_p$ then the linear compound

$$y = a_1X_1 + a_2X_2 + \ldots + a_pX_p$$

has a normal distribution with mean

$$\mu_y = a_1\mu_1 + a_2\mu_2 + \ldots + a_p\mu_p$$

2.4.5

Distribution of the sample mean

We have made probability statements about single observations from a normal distribution with a specified mean and variance. In real life we do not know what these parameters are; rather, we obtain a random sample of values and calculate statistics which provide estimates of them. For instance, we shall use sample means to estimate means of populations or to test hypotheses about them.

Each random sample will, in general, produce a different value of a statistic. The statistic is considered to come from a probability distribution called its *sampling distribution* and we need to know its nature in order to draw conclusions about the parameter of interest. In this section we shall deal with the sampling distribution of the sample mean and consider other cases later.

Suppose that 20 palaeontologists each took a random sample of five specimens of a particular species of fossil and recorded the lengths with the

Table 2.11 Twenty hypothetical data samples

Palaeontologist	Lengths of specimens (cm)					Sample mean
1	3.6	3.8	3.9	4.1	3.9	3.8
2	3.6	3.9	3.7	3.8	3.9	3.8
3	4.0	4.1	3.8	4.1	3.7	3.9
4	3.7	3.8	3.9	3.9	3.6	3.8
5	3.7	3.7	3.5	4.1	3.7	3.7
6	3.5	3.8	3.6	3.7	3.5	3.6
7	3.9	3.6	4.0	3.6	4.0	3.8
8	3.3	3.6	3.6	3.6	4.0	3.6
9	4.0	3.9	3.5	3.8	3.5	3.7
10	3.8	4.0	3.5	3.9	3.6	3.8
11	3.8	3.7	4.0	3.8	3.7	3.8
12	3.9	3.8	3.7	3.7	3.7	3.8
13	4.0	3.8	3.9	3.8	4.0	3.9
14	3.8	3.9	3.6	3.8	3.7	3.8
15	3.6	3.8	3.7	3.8	4.1	3.8
16	3.5	3.5	3.7	3.7	3.6	3.6
17	3.8	3.5	3.8	3.9	3.8	3.8
18	3.7	3.9	4.2	3.6	3.6	3.8
19	3.9	3.7	3.8	3.7	4.0	3.8
20	4.2	3.7	4.1	3.9	3.7	3.9

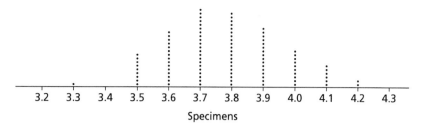

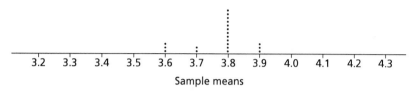

Fig. 2.25 Dot plots for lengths of 100 specimens of a fossil measured by 20 palaeontologists, who each measured 5. The 20 sample means have the same mean as the 100 individual observations but they are less scattered about the overall mean.

objective of estimating the mean length μ of the species. The results might be like those in Table 2.11. As each sample has given a different estimate of the mean μ it is important to assess how much reliance we can place on the estimate obtained from any sample. To do this we must know what probability distribution the sample mean (or other statistic) comes from. Figure 2.25 shows the dot plots for the 500 single observations and for the 20 sample means. The true means are equal, as they must be, because they are means of the same data. The interesting feature is that the sample means are less scattered about the overall mean than are the individual observations: we can place more reliance on the sample mean of several independent, randomly chosen specimens than on a single specimen.

We now place these ideas in the context of probability, assuming for the present that we know the main details of the parent distribution (the distribution from which the single values are taken).

2.4.5.1   **The mean and variance of the sampling distribution of the sample mean**

It can be shown that the distribution of the sample mean has the same mean as that of single values but the variance is lower by a factor which depends on the sample size n. In fact,

$$\mu_{\bar{x}} = \mu_x \qquad \text{and} \qquad \sigma_{\bar{x}}^2 = \sigma^2/n$$

The square root of the variance of this or any other sampling distribution is called the *standard error* (SE) of the statistic. Here,

$$\mathrm{SE}(\bar{x}) = \sqrt{\{\sigma^2/n\}} \qquad \text{or} \qquad \sigma/\sqrt{n}$$

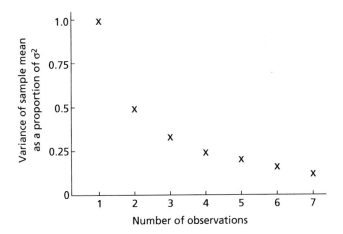

Fig. 2.26 The variances of sample means are in inverse proportion to the size of the sample. The progressive effect becomes smaller as the sample size becomes larger.

Sample means are standardised in the same way as single values, by subtracting the mean of their distribution and dividing by the square root of the variance.

Equation for standardised sample mean Z

$$Z = \frac{\bar{X} - \mu}{\sigma/\sqrt{n}}$$

For a given population or parent distribution whose variance is σ^2, the variance of the sample mean may be made small by using large samples but there are diminishing returns for additional observations, as illustrated in Fig. 2.26. It is important to remember that each observation costs time, effort and money, and the size of a sample should be restricted to the number of observations required to give the desired precision.

2.4.5.2   **The form of the distribution of sample means: the central limit theorem**

When the data come from a normal distribution, the distribution of the sample mean is also normal:

When $X \sim N(\mu, \sigma^2)$, $\bar{X} \sim N(\mu, \sigma^2/n)$

Central limit theorem

Even when the parent distribution is not normal, the sample means of large samples have an approximately normal distribution. The approximation improves with increasing sample size, and its closeness depends also on the symmetry of the parent population; if the parent distribution is symmetrical the sample means of even small samples are well approximated by the normal.

Worked example: see Box 2.14.

The calculations in the worked example show that there is only a probability of about 1 in 500 of obtaining a sample of eight belemnites with a sample mean which is greater than or equal to 3.7 when the population has a mean of 3.5 and standard deviation of 0.2. This probability is so small that

---

Box 2.14 Worked example: central limit theorem

The lengths of a species of belemnite form a normal distribution with mean equal to 3.5 cm and standard deviation equal to 0.2 cm. A random sample of eight belemnites is obtained. What is the standard error of the sample mean? What is the probability that when a belemnite of this species is taken at random, the sample mean length exceeds 3.7 cm?

Compare this with the probability of obtaining a single belemnite whose length exceeds 3.7.

Solution

The standard error of the sample mean is

$$SE(\bar{X}) = 0.2/\sqrt{8}$$
$$= 0.0707 \, \text{cm}$$

The standardised value of 3.7 cm is

$$z = \frac{3.7 - 3.5}{0.0707}$$
$$= 2.82$$

$$Pr(\bar{X} > 3.7) = Pr(Z > 2.82)$$
$$= 0.0024 \quad \text{(from Appendix 2.2)}$$

The standardised value of a single observation of 3.7 is

$$z = \frac{3.7 - 3.5}{0.2}$$
$$= 1.0$$

and

$$Pr(X > 3.7) = 0.1587 \quad \text{(from Appendix 2.2)}$$

The probability that a single value will exceed the true mean by 0.2 cm or more is much greater than the probability of the sample mean doing so.

doubt may be cast on the idea that such a sample came from the stated population. In Section 2.5.4 we shall see that hypotheses are tested in this way: if the observed data have only a small probability of occurring when a hypothesis is true then we shall doubt the hypothesis.

A second question which arises is, 'In the light of the evidence, what values for the mean of the population appear to be reasonable?' Values close to the sample mean would be acceptable but those remote from it would not. Refer to the worked example in Box. 2.15.

2.4.5.3   **Normal approximation to the binomial distribution**

In the point count method of estimating proportions of minerals in rock, it is assumed that the number of times, x, the cross-hairs of the microscope

---

Box 2.15 Worked example: likely values of a population mean

A random sample of five brachiopods had a mean length of 3.2 cm. Assuming that the standard deviation of the population from which they came is 0.2 cm, what range of values would be reasonable for the mean length of the distribution?

Out of all possible normal distributions with a standard deviation of 0.2 cm, some could easily have given rise to the sample mean whilst others could not. Let us say that the extreme values of reasonable means are as follows.

1   The mean μ_L such that the probability of obtaining a sample mean of 3.7 cm or more is 0.025. With this mean, the standardised value of 3.7 would be, from the percentage points of the normal distribution, 1.96:

$$\frac{3.7 - \mu_L}{0.2/\sqrt{5}} = 1.96$$

so that

$$\mu_L = 3.7 - 1.96 \times 0.2/\sqrt{5}$$
$$= 3.52$$

2   The mean μ_U such that the probability of obtaining a sample mean of 3.7 cm or less is 0.025. Then the standardised value of the sample mean would be -1.96 and we would have

$$\mu_U = 3.7 + 1.96 \times 0.2/\sqrt{5}$$
$$= 3.88$$

Then the interval (μ_L, μ_U) includes the values of the mean which we believe could have produced the given sample. The interval is called a 95% confidence interval. This aspect of inference is also discussed more fully in Section 2.5.

coincide with a grain of a particular mineral has a binomial distribution. The proportion is estimated by x/n where n is the total number of stops of the microscope.

By considering a range of values of the true proportion which appear to be reasonable, we could make an assessment of the uncertainty in our estimate. This would involve summing a large number of binomial probabilities, however: for example, if we wanted to know what number of grains of plagioclase would be attained or exceeded with a probability of 0.025 or less when the proportion of plagioclase was 0.25 and the total number of all grains recorded was 200, we would have to find x such that

$$\Pr(X = x) + \Pr(X = x + 1) + \ldots + \Pr(X = 200) \leqslant 0.025$$

where $\Pr(X = x)$ is given by

$$\binom{200}{x} 0.25^x \times 0.75^{200-x}$$

Although many of the terms would be zero to four decimal places, the amount of computation is likely to be enormous.

Normal approximation to the binomial distribution

Provided that the probability of success in each Bernoulli trial is not too extreme, $(0.1 \leqslant p \leqslant 0.9)$, and the number of trials is large enough to make $np \geqslant 5$ and $n(1 - p) \geqslant 5$, the normal distribution with mean np and variance $np(1 - p)$ may be used as an approximation to the binomial. Worked example: see Box 2.16.

This is demonstrated in Fig. 2.27, where normal distribution curves are superimposed on line charts of binomial distributions with the same means and variances. The approximation is better for large values of n than for small ones; but for reasonably symmetrical binomial distributions ($p = 0.5$) it works well, even when the number of trials is small.

2.4.5.4   Normal approximation to the Poisson distribution

The Poisson distribution may also be approximated by the normal distribution with the same mean and variance, provided that the mean is greater than about 30. This may be the case in studies of emissions of α-particles from radioactive sources, for example. As with the binomial distribution, the approximation is improved by a continuity correction.

The variance of a Poisson distribution is equal to its mean μ, so the standardised value of a Poisson variate x will be

$$z = \frac{x \pm 0.5 - \mu}{\sqrt{\mu}}$$

Box 2.16 Worked example: normal approximation to binomial distribution

When a certain procedure is used, the probability that a rock of type A will be wrongly classified as type B is 0.15. What is the probability that more than 10 will be wrongly classified when 50 specimens are examined?

By summing 40 binomial probabilities $Pr(X = 11) + \ldots + Pr(X = 50)$ or, for simple numbers such as these by using tables, we obtain the result

$$Pr(X \geqslant 11) = 0.1199$$

Now calculate the probability of this event, using a normal distribution with mean

$$\mu = 50 \times 0.15$$
$$= 7.5$$

and variance

$$\sigma^2 = 50 \times 0.15 \times 0.85$$
$$= 6.375$$

The standardised value of 11 is

$$z = \frac{11 - 7.5}{\sqrt{6.375}}$$
$$= 1.39$$

Then

$$Pr(X \geqslant 11) = Pr(Z \geqslant 1.39)$$
$$= 0.0823$$

The approximation is low, compared with the exact result given by the binomial distribution. This always happens when a continuous distribution is used as an approximation to a discrete distribution. As an extreme example, the probability $Pr(X = 11)$ would be zero for any continuous distribution; we should use $Pr(10.5 \leqslant X \leqslant 11.5)$ as an approximation and this provides the solution of our problem in general: it is called a continuity correction. For the purposes of calculating normal probabilities we replace

$$Pr(X \geqslant x) \quad \text{with} \quad Pr(X \geqslant x - 0.5)$$
$$Pr(X \leqslant x) \quad \text{with} \quad Pr(X \leqslant x + 0.5)$$
$$Pr(x_1 \leqslant X \leqslant x_2) \quad \text{with} \quad Pr(x_1 - 0.5 \leqslant X \leqslant x_2 + 0.5)$$

to increase the range of values of the binomial variable.

Applying this to our last example, we now obtain the approximation from

$$z = \frac{11 - 0.5 - 7.5}{\sqrt{6.375}}$$
$$= 1.19$$

$$Pr(X \geqslant 11) = Pr(Z \geqslant 1.19)$$
$$= 0.1170$$

which is much closer to the exact answer.

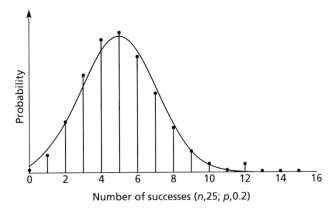

Number of successes (n,25; p,0.2)

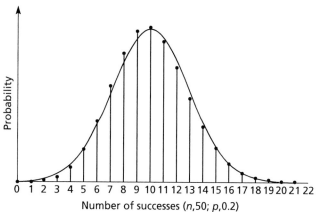

Number of successes (n,50; p,0.2)

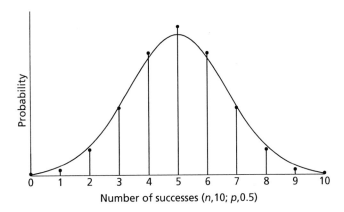

Number of successes (n,10; p,0.5)

Fig. 2.27 For large numbers of trials and probabilitites of success which are not too extreme, the probability of a value of a binomial variate which lies in a given interval may be obtained approximately by treating the distribution as if it were normal, with the same mean and variance as the true binomial distribution. Here, normal curves are superimposed on examples of line charts of binomial variates.

Box 2.17 Worked examples: normal approximation to Poisson distribution

The number of α-particles emitted from a radioactive isotope in 30 seconds has a Poisson distribution with a mean of 800. Find the probability that more than 850 will be emitted in a 30-second interval.

Solution

The distribution may be approximated by a normal distribution with mean and variance both equal to 800. Using the continuity correction, the standardised value of 850 is

$$z = \frac{850 - 0.5 - 800}{\sqrt{800}}$$
$$= 1.75$$

Then

$$Pr(X \geqslant 850) = Pr(Z \geqslant 1.75)$$
$$= 0.0401$$

This approximation will be useful in calculating confidence intervals for means of Poisson distributions.

---

where the sign of 0.5 is chosen to increase the value of the approximated probability.

Normal approximation to the Poisson distribution

When a Poisson distribution has a mean μ which is greater than 30, it may be approximated by a normal distribution with mean μ and variance μ.
    Worked example: see Box 2.17.

2.5   INFERENCE

Given a specified frequency distribution or a particular model of a geological process, we have seen that it is possible to make probability statements about the properties of samples of data which may be obtained from it. Now we consider the reverse problem: what are the characteristics of the geological population or model from which the sample has been obtained? This is the problem of statistical inference.

    Consider the following geological problems.

1   The data are numbers of major earthquakes experienced in a certain region during 20 consecutive decades:

1 2 1 2 1 5 1 2 1 5 2 5 5 1 3
1 3 1 1 3

It is important to know if the major earthquakes occur independently and are randomly scattered through time, or exhibit a pattern or trend. If they are independent, then the data should be compatible with a sample of observations from a Poisson distribution.

2 Suppose that an igneous rock classification scheme is being used in which a rock must contain on average more than 20% quartz in order to be classified as a granite. Quartz percentage was obtained from eight thin sections from an unidentified rock:

23.5   16.6   25.4   19.3   19.1   22.4   20.9   24.9

Does the evidence suggest that this rock is a granite? We need to assess the characteristics of the source population: what is the mean quartz percentage and what degree of certainty can we have in our estimate?

3 Brachiopods were collected from two horizons and their lengths (cm) were measured, with the following results:

Horizon A: 3.2   3.1   3.1   3.3   2.9   2.9   3.5   3.0
Horizon B: 3.1   3.1   2.8   3.1   3.0   2.6   3.0   3.0   3.1   2.8

Is there strong evidence for the idea that brachiopods from horizon A are longer on average than those from horizon B? We need to assess whether or not the characteristics of the two samples are compatible with having been derived from the same parent population.

In each of these cases there is a geological question to be answered which can be rephrased in statistical terms: problem **1** is a statistical problem about the form of the distribution from which the data come; problem **2** involves statements about the mean of a probability distribution; and problem **3** requires the comparison of the means of two populations.

Statements about the shape or parameters of a population or probability distribution are called *statistical hypotheses*, and the ways in which these hypotheses are tested and parameters are estimated constitute the subject of statistical inference. This chapter will deal with cases where the data are from distributions which are normal or can be transformed to be normal, and inferences will be made about the means and variances of normal distributions.

2.5.1 **Methods of estimating parameters**

In a random sample, each value is a representative of the population from which it was obtained and provides some information about that population. Even a single observation tells us something about the mean but at least two would be needed to give minimal information about the variability. We now seek methods of combining such data in order to estimate the parameters of the parent population. The principles most commonly used in formulating estimators for calculating optimal estimates of population parameters are those of least squares (LS) and maximum likelihood (ML).

Least squares methods

In the LS approach, we regard all observations from a common source as consisting of a fixed value, for example μ, plus a random term e which comes from a distribution with a mean of zero and unknown variance denoted by σ^2. Then, if the data are $x_1, x_2, \ldots, x_n$, we have:

$$x_1 = \mu + e_1$$
$$x_2 = \mu + e_2$$
$$\vdots$$
$$x_n = \mu + e_n$$

so that the random terms are:

$$x_1 - \mu = e_1$$
$$x_2 - \mu = e_2$$
$$\vdots$$
$$x_n - \mu = e_n$$

Suppose we now choose a certain value of K as an arbitrary estimate of μ. We can compute the differences between each item of data and this estimate; these differences are called residuals and are denoted by $R_1, R_2, \ldots, R_n$:

$$R_1 = x_1 - K$$
$$R_2 = x_2 - K$$
$$\vdots$$
$$R_n = x_n - K$$

(see Fig. 2.28a).

If K is a good estimate of μ then the residuals will be small. In fact, we choose K to make the sum of squares of the residuals as small as possible. In principle, we could plot values of the sum of squared residuals against values of K (Fig. 2.28b), but in practice calculus is used to derive an appropriate formula. In the simplest case where there is a single source of data it is found that the required function of the data for estimating μ is:

$$\bar{x} = \frac{1}{n} \sum_{i=1}^{n} x_i$$

Maximum likelihood methods

If we know the distribution from which the data are derived, we can use the ML approach. This addition of information about the distribution results in

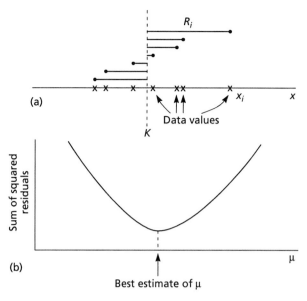

Fig. 2.28 If we convey the data in the form of the residuals R_i from an estimate of the mean K (a), then the sum of squared residuals could, in principle, be calculated for each value of K (b) and the minimum point of K gives the best estimate of the mean.

lower variance of sampling distributions, so in these circumstances ML gives better estimators.

In ML methods we treat the joint probability function or p.d.f. for the model as a function of the unknown parameter rather than of the data. It is then called a joint likelihood function. For example, if the earthquakes in the problem **1** above (p. 90) can be regarded as independent, the data will be from a Poisson distribution with probability function:

$$Pr(X = x) = \frac{e^{-\mu}\mu^{x}}{x!}$$

(see Section 2.4).

Using the data, the joint likelihood function is:

$$\frac{e^{-\mu}\mu^{1}}{1!} \times \ldots \times \frac{e^{-\mu}\mu^{3}}{3!} = \frac{e^{-20}\mu^{6}}{1!\ldots 3!}$$

where μ is the mean rate of occurrence. The principle of ML tells us to choose for our estimate the value of μ which makes the value of this likelihood function as large as possible.

In principle, we could do this by calculating values of the likelihood function for several values of μ and plotting them, as in Fig. 2.29. Then, of all the values of μ, we choose the one, $\hat{\mu}$, which maximises the function. It is called the ML estimate of the mean μ. In practice, the natural logarithm of the joint likelihood is used for simplicity and calculus is used to find a

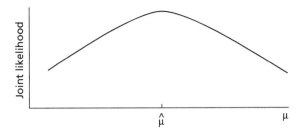

Fig. 2.29 The maximum likelihood method finds the value µ which maximises the ML function. µ is then taken as the best estimate of the mean.

Table 2.12 Maximum likelihood estimators of parameters of distributions

Distribution	Probability function	Parameter	Data	Estimates of parameters
Binomial	$\binom{n}{x}p^x(1-p)^{n-x}$	p	x	$\hat{p} = x/n$
Poisson	$\dfrac{e^{-\mu}\mu^x}{x!}$	μ	$x_1,\ldots,x_n$	$\hat{\mu} = \bar{x}$
Exponential	$\lambda e^{-\lambda x}$	λ	$x_1,\ldots,x_2$	$\hat{\lambda} = 1/\bar{x}$
Normal	$\dfrac{1}{\sigma\sqrt{2\pi}}\exp\left\{-\dfrac{1}{2}((x-\mu)/\sigma)^2\right\}$	$\begin{matrix}\mu\\\sigma^2\end{matrix}$	$x_1,\ldots,x_2$	$\begin{matrix}\hat{\mu} = \bar{x}\\\hat{\sigma}^2 = \text{CSS}(x)/n\end{matrix}$

mathematical expression for the estimate. This expression, a function of the data, is called an ML estimator of the parameter.

In the case of the Poisson distribution, the ML estimator of the mean µ is the sample mean $\bar{x}$. Table 2.12 shows ML estimators of parameters of some distributions which are useful in geological models.

ML estimators have some useful attributes. Their sampling distributions have lower variances than those of other types of estimator and, when large samples are used, they are approximately normal. In order to obtain ML estimators, however, we must know the distribution from which the data come.

2.5.2 Sampling distributions of estimators of parameters

The estimates obtained by ML and LS are single (point) values and give no idea of the degree of uncertainty which attaches to them, and yet we can be sure that if we took a number of similar samples then each one would give us a slightly different estimate. The estimator is considered to have a probability distribution, called its sampling distribution, and this can be used to find *intervals* within which is likely to occur the true value of the quantity in which we are interested. For instance, if we estimate the alumina content of a rock to be 15% we might find that an interval from 14.7 to 15.3% includes the true value with some specified probability such as 0.95 (19 times out of 20).

The idea of a sampling distribution was raised in Section 2.4.5, where we discussed the variation of the sample mean, assuming the variance to be known. At this stage we shall introduce the sampling distribution of:

1 a quantity which is required for making probability statements about sample means of data taken from a normal distribution when the variance is not known (Student's t); and

2 the ratio of two independent estimates of the same variance when the data come from normal distributions (the F distribution).

Some of the applications of these sampling distributions will be presented in the two sections — on confidence intervals and tests of hypotheses — which follow this. You may find it convenient to look at the applications of Student's t before working on the F distribution.

2.5.2.1   **Student's t distribution**

Our discussion of the distribution of the sample mean has required a knowledge of the value of the standard deviation in the quantity

$$\frac{\bar{x} - \mu}{\sigma/\sqrt{n}}$$

The population standard deviation σ is usually unknown, however. We must estimate it with the statistic s, which gives the following.

Student's t statistic

$$t = \frac{\bar{x} - \mu}{s/\sqrt{n}}$$

The distribution of this quantity (investigated by W.S. Gossett, who used the pen-name 'Student' for his research papers) is not normal, though it is again bell-shaped and centred on zero. It is found to be independent of the unknown standard deviation and to have one parameter (compared with two for the normal distribution) which depends only on the number of observations used to estimate σ^2. In fact, it is equal to the divisor used in calculating an unbiased estimator of the variance. The parameter, which is called the degrees of freedom (d.f.) and denoted by v (nu), is equal to $(n - 1)$ in the present case since the estimate of the variance based on a single sample of observations is

$$s^2 = \frac{\Sigma(x_i - \bar{x})^2}{n - 1}$$

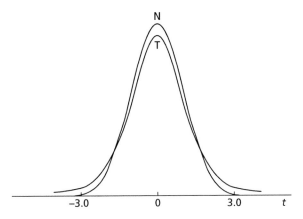

Fig. 2.30 The t distribution with four degrees of freedom compared with the standard normal distribution. Distributions of t have more probability in the tails than the standard normal distribution, reflecting our lack of knowledge of the variance of the parent distribution. Large samples give us more information about the variance and the corresponding t distributions are more like the normal. T, t distribution with parameter equal to 4; N, standard normal density.

(Other values will be given as applications are dealt with in the sections on inference and regression.)

In Fig. 2.30, the t distribution with 5 is compared with a standard normal curve. The t distribution becomes increasingly close to the standard normal as the number of d.f. increases: this is to be expected because, as we gather more data (provided that they are obtained by a suitable random procedure), the more information we have about the variance. Tables of the t distribution are given in Appendix 2.4.

Worked example: see Box 2.18.

2.5.2.2   The *F* distribution

We shall need to test pairs of quantities to see if they can be regarded as independent estimates of the same variance. For example, we may have data on a variable from rock specimens, obtained from two methods of determination. The comparison is made by finding the ratio of one to the other because the resulting statistic has a known probability distribution, called Fisher's F (after Sir R.A. Fisher, who discovered it).

F has two parameters, which we shall denote by v_1 and v_2, and they are equal to the numbers of d.f. associated with the independent estimates s_1^2 and s_2^2 of the variance. Suppose that these were obtained from samples of n_1 and n_2 observations respectively: then the d.f. for these estimates are $(n_1 - 1)$ and $(n_2 - 1)$ and the ratio

$$s_1^2/s_2^2$$

comes from the F distribution, in which the parameters are

$$v_1 = n_1 - 1 \qquad \text{and} \qquad v_2 = n_2 - 1$$

Box 2.18 Worked examples: *t* distribution

1 A random sample of 12 observations is obtained from a normal distribution. What value of the t-statistic will be exceeded with probability 0.025?

Solution

The number of degrees of freedom (d.f.) for the *t* distribution is 11 and
from Appendix 2.4 we see that the 2.5% point is 2.201. For a standard
normal distribution it is 1.96. Note how the percentage points in the 0.025
column – and in the others – become smaller as the d.f. increase until, for
infinite d.f., they are the same as those of the standard normal distribution.

   A second example will help to show one way in which the distribution
can be useful in assessing conjectures about mean values of normal
distributions.

2 A random sample of eight hand specimens of rock was analysed for
organic material; the sample mean was found to be 5.8% and the sample
standard deviation was 2.3. Do you think it reasonable to suppose that the
organic content of the rock is 5.0%?

Solution

If the mean organic content is 5.0% then the quantity

$$t = \frac{\bar{x} - 5.0}{s/\sqrt{n}} = \frac{5.8 - 5.0}{2.3/\sqrt{8}}$$

or 0.98 comes from a *t* distribution with 7 d.f. The percentage points table
of the *t* distribution shows that even a value of *t* as large as 1.4 would not be
unusual if the mean was 5.0 (the 10% point is 1.415) so 0.98 could easily be
attained or exceeded by chance. There is nothing here to cast doubt on the
conjecture that the organic content is 5.0.

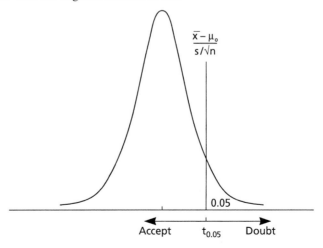

Fig. B2.18.1

continued on p. 98

Box 2.18 *Continued*

What kind of value would have cast doubt on it? The table shows that, in a *t* distribution with 7 d.f., 1.895 is only exceeded with probability 0.05 (1 in 20) and 2.998 has a very low probability (0.01 or 1 in 100) of being exceeded. These would make us hesitate to accept that the true organic content was 5.0, and this is the subject of Section 2.5.4.

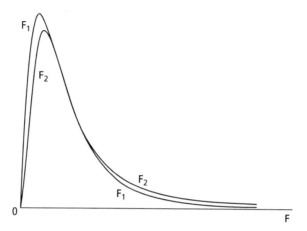

Fig. 2.31 *F* distributions have two parameters, v_1 and v_2. The diagram compares the distributions $F_{4,12}$ and $F_{12,4}$. F1, density for F with $v_1 = 4$ and $v_2 = 12$; F2, density for F with $v_1 = 12$ and $v_2 = 4$.

The distribution is written $F_{v1,v2}$ and values which are exceeded with probability α are denoted by $F_{\alpha;v1,v2}$.

The order of the parameters is important: the ratio s_2^2/s_1^2 comes from a different *F* distribution from that just described: the parameters are now

$$v_1 = n_2 - 1 \quad \text{and} \quad v_2 = n_1 - 1$$

The $F_{4,12}$ and $F_{12,4}$ distributions are compared in Fig. 2.31.

In the version of the *F* table given in Appendix 2.5 there is a separate page for each percentage point.

Worked example: see Box 2.19.

2.5.3   Confidence intervals for parameters of distributions

The estimates obtained by ML and LS are single values (called point estimates) and give no idea of the degree of confidence we can have in them. It is important to be able to quote an interval of values which we can say includes the true value of the parameter with a high level of probability. For instance, with reference to the problem **2** above (p. 91) concerning quartz

---

Box 2.19 Worked example: F distribution

What value of a statistic from the $F_{4, 10}$ distribution will be exceeded with probability 0.01?

Answer

5.99 (from Appendix 2.5)

Note that for an $F_{10, 4}$ distribution the answer would be 14.5.

Published tables of F contain information on the probabilities in the upper tail only, in the interests of economy, but lower percentage points can be calculated by using the relation

$$F_{1-\alpha;\, v1,\, v2} = 1/F_{\alpha;\, v2,\, v1}$$

so that, for example, the value of an F variable with parameters $v_1 = 3$ and $v_2 = 7$ which is exceeded with probability 0.95 is given by

$$F_{0.95;\, 3,\, 7} = 1/F_{0.05;\, 7,\, 3}$$
$$= 1/8.89$$
$$= 0.11$$

---

percentage in a rock, we may wish to give an interval which, with a probability of 0.95 or 95%, includes the true mean quartz content of the population. Such an interval is called a 95% confidence interval and the ends of the interval are called 95% confidence limits. In order to quantify them, we need to know something about the sampling distribution of the estimator or of some function of it. In this section, we shall deal with cases in which data come from normal distributions, so that the sampling distributions required will be t and F.

Confidence intervals for the mean of a normal distribution

In general, we do not know the variance of the normal distribution from which the data come, but have to estimate it with s^2. Then we may use the fact that the quantity

$$t = \frac{\bar{x} - \mu_0}{\sqrt{\dfrac{s^2}{n}}}$$

has a t distribution with $(n - 1)$ d.f., and this does not depend upon the variance of the normal parent distribution (see Section 2.5.2).

It is easy to find, from the percentage points table of the t distribution, an interval in which values of t will lie with a specified probability. For example, 95% of values will lie between $-t_{0.025;\, v}$ and $t_{0.025;\, v}$: see Fig. 2.32.

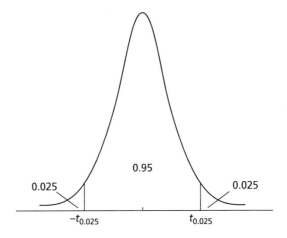

Fig. 2.32 Ninety-five percent of values lie between $-t_{0.025}$ and $t_{0.025}$.

We now ask, what range of values for μ would produce the given interval of values of t? For the highest value in the t interval we have:

$$\frac{\bar{X} - \mu_U}{\sqrt{\dfrac{s^2}{n}}} = -t_{0.025;n-1}$$

(The negative sign is used with t here because the upper limit of μ will exceed $\bar{x}$.)

Confidence interval for mean of a normal distribution

Confidence limit at highest value of t:

$$\mu_U = \bar{x} + t_{0.025;n-1}\sqrt{(s^2/n)}$$

Confidence limit at lowest value of t:

$$\mu_L = \bar{x} - t_{0.025;n-1}\sqrt{(s^2/n)}$$

The values μ_U, μ_L are the required 95% confidence limits for the mean and (μ_U, μ_L) is the 95% confidence interval.
  Worked example: see Box 2.20.

We could also see this graphically by plotting values of $(\bar{x} - \mu)/\sqrt{(s^2/n)}$ against μ: see Fig. 2.33(a).

Interpretation of the confidence interval

Suppose that we could take a large number of independent samples of data from a given distribution with mean μ and that we calculated a 95% con-

---

Box 2.20 Worked example: confidence intervals

Given data on the percentage of quartz in thin sections from an igneous rock, what is the confidence interval around the estimated mean quartz percentage in the rock?

Data: 23.5   16.6   25.4   19.1   19.3   22.4   20.9   24.9

From the data and the table of the t distribution we have:

Sample mean $\bar{x}$	Sample variance s^2	Sample size n	$\sqrt{s^2/n}$	$t_{0.025;\,n-1}$
21.512	9.507	8	1.09	2.365

The lower confidence limit is:

$$\bar{x} - t_{0.025;\,n-1}\sqrt{s^2/n} = 21.512 - 2.365 \times 1.09 = 21.512 - 2.578 = 18.93$$

and the upper limit is:

$$\bar{x} + t_{0.025;\,n-1}\sqrt{s^2/n} = 21.512 + 2.365 \times 1.09 = 21.512 + 2.78 = 24.09.$$

so that a 95% confidence interval is (18.93, 24.09).

---

fidence interval from each sample. Then, in the long run, about 95% of the intervals would include the true mean (but, of course, 5% would not!): see Fig. 2.33(b).

Size of the confidence interval

So far we have only considered 95% confidence intervals but we may require to have more than 95% confidence that our interval includes the true mean, or we may be prepared to accept less certainty. For a given sample, greater confidence may be obtained by using a longer interval (we will be 100% confident that the true mean ppm of gold in geochemically surveyed stream sediments is in the interval 0 to 1000 000!), whereas the use of a shorter interval decreases our confidence. This is reflected in the percentage points table of the t distribution (Appendix 2.4). Consider 90%, 95% and 99% confidence intervals for the mean quartz percentage in a rock (Fig. 2.34). To obtain these we need the 5%, 2.5% and 0.5% points respectively of the t distribution. In general, $100(1 - \alpha)\%$ confidence intervals require the $100\alpha/2\%$ point (Fig. 2.35).

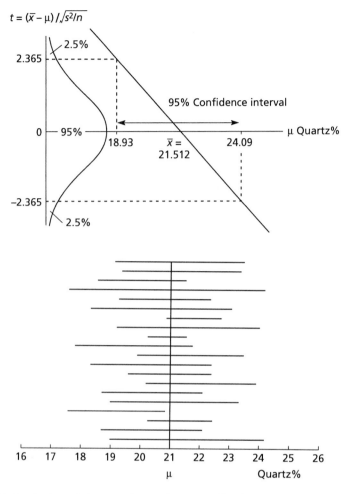

$t = (\bar{x} - \mu)/\sqrt{s^2/n}$

Fig. 2.33 (a) Graph of t statistic vs. μ, showing how the 95% point on the t axis translates into the 95% confidence interval on the axis. (b) Suppose (although such knowledge is unattainable!) that the true mean quartz percentage for the rock in problem number **2** (p. 91 and Box 2.20) is 21%. The lines constructed on this diagram show 95% confidence intervals for a series of samples from the same rock: the lowest line conveys the result gained in Box 2.20. Of 20 samples, at 95% confidence level, we may expect one confidence interval not to include the true mean.

Estimation of the sample size required

It is possible to increase the precision of an estimate and so decrease the size of a confidence interval by obtaining more information, that is to say a larger sample. The half-width of an interval is $t_{\alpha/2;\,n-1}\sqrt{(s^2/n)}$.

Now s^2 will vary to some extent from sample to sample but larger values of n will definitely produce smaller values of $\sqrt{(s^2/n)}$ and so give shorter confidence intervals. In fact, we can control the length of an interval at a given level of confidence by choosing an appropriate sample size.

The statement that a certain interval $(\bar{x} - K, \bar{x} + K)$ includes the mean

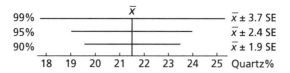

Fig. 2.34 Confidence intervals at various percentage levels for quartz percentage in a rock (data from p. 91, Box 2.20).

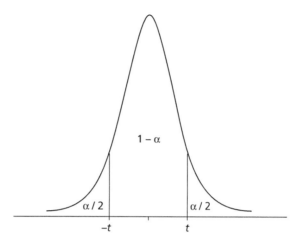

Fig. 2.35 The $100(1 - \alpha)\%$ confidence limits are given by the $100\alpha/2$ percentage points of the sampling distribution.

with probability 95% is equivalent to saying that we are 95% certain that the sample mean does not differ from the true mean by more than an amount K. We are 95% certain, for example, that the sample mean quartz percentage does not differ from the true mean value by more than 2.578% (see Box 2.20). Suppose we wish to be 95% certain that the sample and true means do not differ by a smaller amount, say 1%. How many thin sections will be needed? We could estimate the number by saying that we require $t_{0.025;n-1}\sqrt{(s^2/n)}$ to be not more than γ:

$$t_{0.025;n-1}\sqrt{(s^2/n)} < \gamma$$

from which we find:

$$n > t_{0.025;n-1}s^2/\gamma$$

Unfortunately, $t_{0.025;n-1}$ depends upon the value n which we want to estimate and s^2 cannot be known in advance. In practice we use a pilot sample of n_1 observations, calculate a provisional estimate s_1^2 of the variance, base t on the size of the pilot sample and use:

$$n > t_{0.025;n1-1}s_1^2/\gamma$$

This is not usually an integer so we use the next highest integer as the minimum number of observations needed. As n_1 values have already been obtained only $(n - n_1)$ more are required. Worked example: see Box 2.21.

Box 2.21 Worked example: choice of sample size

What is the required sample size (i.e. number of thin sections) in order to be 95% sure of obtaining a sample mean within 1% of the true mean for the quartz percentage problem (p. 91 and Box 2.20)?
Using

$$n > \frac{t_{0.025;\, n1-1} s_1^2}{y}$$

from the data $n_1 = 8$, $s_1 = 3.083$, $y = 1\%$:

$$n > (t_{0.025;\, 7} \times 3.083/1)^2 = 53.17$$

so we require $n - n_1 = 53.17 - 8$ which, rounded up, means 46 more thin sections!

Confidence interval for differences between two means

In the third problem cited at the beginning of Section 2.5, data were collected on two assemblages of brachiopods in order to compare the mean lengths. A point estimate of the difference in means is provided by the difference in sample means but once again we should compute a confidence interval for the difference to indicate the precision of the estimate. Now, if the data come from normal distributions with means μ_A and μ_B, the difference in sample means also comes from a normal distribution: its mean is $\mu_A - \mu_B$. By analogy with the case of a single sample where the quantity

$$\frac{\bar{x} - \mu}{\text{(estimated SE of } \bar{x})}$$

has a t distribution, we may expect the quantity

$$\frac{(\bar{x}_A - \bar{x}_B) - (\mu_A - \mu_B)}{\text{(estimated SE of } (\bar{x}_A - \bar{x}_B))}$$

to have a t distribution, and this is so: provided that the two parent normal populations have equal variances.

The best estimate of the common variance is obtained by combining the two CSS of the samples and also combining the d.f.:

$$s^2 = \frac{\text{CSS}(x_A) + \text{CSS}(x_B)}{(n_A - 1) + (n_B - 1)} = \frac{\text{CSS}(x_A) + \text{CSS}(x_B)}{n_A + n_B - 2}$$

If a calculator is used it is more efficient to calculate s^2 in the form:

$$s^2 = \frac{(n_A - 1)s_A^2 + (n_B - 1)s_B^2}{n_A + n_B - 2}$$

Box 2.22 Worked example: confidence intervals for differences between two samples

Is there any evidence that the two brachiopod samples (p. 91) could have been derived from populations having the same mean?

Statistics derived from data:

Horizon	$\bar{x}$	s	s^2	n	CSS(x)
A	3.125	0.205	0.042	8	0.294
B	2.96	0.171	0.029	10	0.264

The separate sample variances appear to be compatible (a formal test for equality of variances is given in Section 2.5.5) so we calculate an estimate of the common variance:

$$s^2 = \frac{(n_A - 1)s_A^2 + (n_B - 1)s_B^2}{n_A + n_B - 2} = \frac{(7 \times 0.042) + (9 \times 0.029)}{16} = 0.0348$$

The appropriate t distribution has 16 degrees of freedom and we find from the t tables that:

$$t_{0.025;\ 16} = 2.12$$

Then

$$t_{0.025;\ 16}\sqrt{s^2(1/n_A + 1/n_B)} = 2.12\sqrt{0.0348(1/8 + 1/10)}$$
$$= 0.1876$$

and 95% confidence limits for the difference in mean lengths of the brachiopods found in horizons A and B are:

$$3.125 - 2.96 - 0.1876 = -0.0226$$

and

$$3.125 - 2.96 + 0.1876 = 0.3526$$

Because these values straddle zero, the two samples may be regarded as coming from populations with equal means, as the observed difference could have arisen by chance more than one time out of 20.

where s_A^2, s_B^2 are the separate sample estimates of the variance. Then, the estimated standard error of the difference in sample means is:

$$\sqrt{s^2(1/n_A + 1/n_B)}$$

and

$$\frac{\bar{x}_A - \bar{x}_B - (\mu_A - \mu_B)}{\sqrt{s^2(1/n_A + 1/n_B)}}$$

has a t distribution with $(n_A + n_B - 2)$ d.f.

For the confidence interval, see the following.

Confidence interval for difference between two means
Using

$$L = t_{\alpha/2; nA+nB-2}\sqrt{s^2(1/n_A + 1/n_B)}$$

the limits of a $100(1 - \alpha)\%$ confidence interval are then given by

$$\bar{x}_A - \bar{x}_B - L \quad \text{and} \quad \bar{x}_A - \bar{x}_B + L$$

Assumptions

1 the data are drawn without bias from normally distributed populations; and

2 the variances of the two populations are equal.

Worked example: see Box 2.22.

Assumption of equal variances

If it is found that the variances cannot be regarded as equal, then the standard error of the estimate of the difference in sample means is estimated by

$$\sqrt{(s_A^2/n_A + s_B^2/n_B)}$$

The quantity (called the Welch statistic)

$$\frac{\bar{x}_A - \bar{x}_B - (\mu_A - \mu_B)}{\sqrt{s_A^2/n_A + s_B^2/n_B}}$$

has a distribution which is approximately like t but the d.f. ν lie between the smaller of $(n_A - 1)$ and $(n_B - 1)$ and the total d.f. and it is not in general an integer:

$$\nu = \frac{(a + b)^2}{(a^2/(n_A - 1)) + (b^2/(n_B - 1))}$$

where $a = s_A^2/n_A$ and $b = s_B^2/n_B$.

Statistical experiments show, however, that, if the sample sizes are at least approximately equal, the simpler form of t works sufficiently well and the Welch statistic is unnecessary.

2.5.4 **Testing hypotheses using the *t* distribution**

A good scientific hypothesis suggests an experiment or survey which can be performed to test it. The hypothesis predicts the outcome of the experiment or trial and if the prediction is correct we do not reject the hypothesis. For example, if it is thought that a sandstone was deposited under certain conditions and we can create these conditions in a flume tank, we may compare flume tank results with the natural sediment. If there is a marked difference between them, we reject the hypothesis about the conditions of sedimentation; if they are similar, we can provisionally accept the hypothesis, until further evidence, if any, discredits it. The hypothesis cannot be proved to be correct; it just becomes more likely after each supportive result. This is in the nature of scientific activity.

Suppose that the data on quartz percentage in an igneous rock (p. 91) had been collected to test an assertion (the hypothesis) that the rock had in excess of 20% quartz, and hence, according to some classification, was a granite. We would require firm evidence of the rock containing more than 20% quartz before believing the identification. Once again, we are looking at the difference between evidence and hypothesis:

$$\bar{x} - \mu$$

Evidence $\qquad$ Hypothesis

The difference can only be judged against background variation so it is measured in (estimated) standard errors of the sample mean:

$$\frac{\bar{x} - \mu}{\sqrt{s^2/n}} = \frac{21.5 - 20}{\sqrt{(s^2/n)}}$$

If 20% is the true mean quartz percentage, this quantity comes from a *t* distribution with $(n - 1)$ d.f. (Notice that the *statistical* hypothesis implied in this calculation is that the granite content is 20%, but the geological hypothesis would be that the content is less than or equal to 20%. The calculation of the test statistic requires us to use a single value, so we choose the geological dividing line between types of rock.) In the present case $n = 8$ so we can identify a range of values in which the quantity is likely to lie, and the complementary range in which values will rarely occur, given that the sample mean is 21.5. We may choose, as an example, a probability of 0.05 (5%) as the dividing point between rare and common events. That is to say, a value which would be exceeded only with a probability of 0.05 is to be considered unlikely to occur: see Fig. 2.36. Such values will therefore cast doubt on our first hypothesis that the mean quartz was 20% and favour the other hypothesis that the mean exceeds 20%. Note that our first hypothesis may be correct and yet, by chance, we may obtain a sample with unusually high quartz: this will lead us to reject the hypothesis even though it is true. Another possibility is that the mean quartz percentage really exceeds 20% but we obtain a sample unusually poor in quartz, which causes us to accept

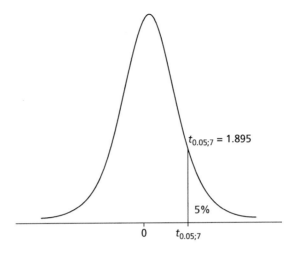

Fig. 2.36 t distribution for data on p. 91: values greater than 1.895 can be regarded as unlikely to occur.

the first hypothesis. We must recognise that either type of error may arise in random sampling and accept that our hypothesis cannot be proved or disproved by statistical means. We can, however, attach probabilities to the errors.

This discussion illustrates the ideas used in the theory of testing statistical hypotheses about parameters of probability distributions. The components of the theory are:

1 two (converse) hypotheses about the parameter or parameters of interest;
2 sample data;
3 statistics which provide information about parameters;
4 a function (called the test statistic) of the statistics which has a known probability distribution when the parameter of interest is known;
5 a level of probability (called a level of significance) for distinguishing between rare and common events;
6 the use of the probability level to identify sets of values of the test statistic which would cast doubt on the hypothesis; and
7 a recognition of the possibility of rejecting a true hypothesis or of accepting a false one.

2.5.4.1   Null and alternative hypotheses

Two hypotheses are considered.
1 A null hypothesis, usually denoted by H_0. The word 'null' indicates 'no difference' and examples of the form of H_0 are:

$H_0: \mu = \mu_0$
$H_0: \mu_1 = \mu_2$ i.e. $\mu_1 - \mu_2 = 0$
$H_0: \sigma_1^2 = \sigma_2^2$ i.e. $\sigma_1^2/\sigma_2^2 = 1$

It is the assertion in H_0 which is used in the test statistic.

2 An alternative hypothesis, H_1. The form of the alternative depends upon the object of the investigation. Examples are:

$H_1: \mu \neq \mu_0$ (two-sided alternative)
$H_1: \mu > \mu_0$ (one-sided alternative)
$H_1: \sigma_1^2 \neq \sigma_2^2$ (two-sided alternative)

The alternative hypothesis is the statement whose truth we would like to establish; it is often the more interesting hypothesis, contrasting with the boring 'no difference' of the null hypothesis. If we believe that the grade of an ore exceeds some economically viable value, say 15%, we must remain sceptical: the test is performed on the assumption that the grade is 15%, not >15%:

$H_0: \mu = 15$
$H_1: \mu > 15$

If supplier A of an analytical instrument tells us that one of its instruments has a higher precision (producing a lower variance) than the corresponding instruments supplied by B, we take the view that they are as good or bad as each other in the null hypothesis, and use supplier A's statement as the alternative:

$H_0: \sigma_A^2 = \sigma_B^2$
$H_1: \sigma_A^2 < \sigma_B^2$

where σ_A^2, σ_B^2 are variances of errors produced by the two instruments.

2.5.4.2 Choice of estimators of parameters and test statistics

The choice of estimators was discussed earlier (Section 2.5.1). When the parent populations are not normal, the fact that ML estimates come from approximately normal distributions when the sample size n is large can be exploited.

The sampling distributions of sample means and variances calculated from data which come from normal distributions were described in Sections 2.4 and 2.5.2 and used in the latter section for calculating confidence intervals for parameters and functions of parameters. For convenience, a summary of test statistics and their uses is given in Table 2.13.

2.5.4.3 Significance levels and rejection or critical regions

Two levels of probability which are commonly regarded as describing unusual samples are 5% and 1%. The level chosen is called the *level of significance* of the test. It should be chosen in advance – before data are obtained – to avoid unintentional (or intentional!) bias, to assist in the

Table 2.13 Summary of test statistics

Parameters	H_0	H_1	Estimates	Test statistic	Sampling distribution
μ	$\mu = \mu_0$	$\mu \begin{Bmatrix} \neq \\ > \\ < \end{Bmatrix} \mu_0$	$\bar{x}$	$\dfrac{\bar{x} - \mu}{\sqrt{s^2/n}}$	t_{n-1}
μ_1, μ_2	$\mu_1 = \mu_2$	$\mu_1 \begin{Bmatrix} \neq \\ > \\ < \end{Bmatrix} \mu_2$	$\bar{x}_1, \bar{x}_2$	$\dfrac{(\bar{x}_1 - \bar{x}_2) - (\mu_1 - \mu_2)}{\sqrt{s^2(1/n_1 + 1/n_2)}}$	$t_{n_1+n_2-2}$
σ_1^2, σ_2^2	$\sigma_1^2 = \sigma_2^2$	$\sigma_1^2 \begin{Bmatrix} \neq \\ > \\ < \end{Bmatrix} \sigma_2^2$	s_1^2, s_2^2	s_1^2/s_2^2	F_{n_1-1, n_2-1}

choice of number of observations and in response to the degree of confidence required of the result in the particular geological context.

Many computer packages display the probability level at which the value of the test statistic would be statistically significant. A high level of probability (e.g. 0.5 or 0.2) means very little significance: it shows that there is a high probability of attaining the value (or a more extreme value) when the null hypothesis is true, so that there is no need to doubt the null hypothesis. A low level of probability (e.g. 0.01) is significant for it implies that there is little chance of attaining (or exceeding) the value of the test statistic when the null hypothesis is true.

Samples which cast doubt on the null hypothesis constitute a critical or rejection region for the test: a lower value of the level of significance is associated with a smaller rejection region and we will less often reject a null hypothesis at such a level. The level chosen may be decided after considering the consequences of rejecting a null hypothesis when it is true. If they are costly or serious in other respects (e.g. where the null hypothesis is that a major ore prospect is economical, or where a major geological hypothesis is at stake) we shall try to avoid the error by using a low level of significance, say 1% or less; but if the consequences are not so serious (e.g. an incorrect fossil or rock identification) we may choose the 5% level. We must also take into account the consequences of accepting a null hypothesis which is incorrect.

2.5.4.4 Types of error and their probabilities

Errors which are made in testing hypotheses and the conventional symbols for their probabilities are shown in Table 2.14.

The probability of rejecting the null hypothesis when the alternative is

Table 2.14

Error	Probability
Type I: reject a true null hypothesis	α
Type II: accept H_1 when H_0 is true	β

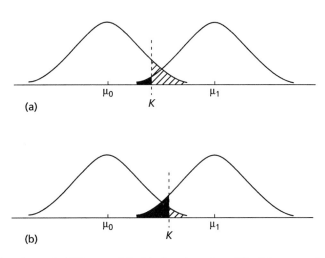

Fig. 2.37 Changing probabilities α and β with changing values (K) of the test statistic. Hatched shading: α; solid shading: β. See text for explanation.

true is $1 - \beta$ and it is called the power of the test. It is natural to wish to use tests which have a high power.

The relationship between α and β will be understood by considering a test of H_0: $\mu = \mu_0$ against H_1: $\mu = \mu_1$ in the case of normal distributions, assuming, for simplicity, that the variance is known. The distribution of the sample mean under the null and alternative hypotheses can be represented as in Fig. 2.37(a).

We may choose K as the value of the test statistic

$$\frac{\bar{x} - \mu_0}{\sqrt{(\sigma^2/n)}}$$

so that lower sample values lead to the acceptance of H_0: $\mu = \mu_0$, higher ones to the acceptance of H_1: $\mu = \mu_1$ and the probability of a type I error is α. If we reduce α by moving the dividing point to the right, we increase β, the probability of a type II error: see Fig. 2.37(b).

2.5.4.5   **Number of observations required for a test**

We shall consider tests on the mean of a single normal distribution. It is possible to reduce the probability of a type I error and yet to control the probability of a type II error by using large enough samples. Recall that the

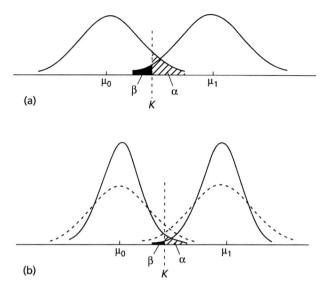

μ_0 β α μ_1

(a)

K

μ_0 β α μ_1

(b)

K

Fig. 2.38 The effect of increased sample size (shown in (b)) on the probabilities of error α and β. Symbols and shading as for Fig. 2.37. See text for explanation.

variance of the distribution of a sample mean is σ^2/n so that it decreases as the sample size increases. The effect is illustrated in Fig. 2.38. Although the value of α has been reduced by moving the boundary to the right, the value of β is small because of the reduced variance of the distribution of sample means.

Worked examples: the principles of inference applied to t tests are exemplified in Boxes 2.23 and 2.24.

2.5.5

F-test for the equality of the variances of two independent normal distributions

The F distribution was described in Section 2.5.2.2. Now, when two independent samples are taken from normal distributions whose variances are σ_1^2 and σ_2^2, and estimates s_1^2, s_2^2 of these variances are calculated, the ratio

$$s_1^2/\sigma_1^2$$
$$\vdots$$
$$s_2^2/\sigma_2^2$$

comes from an F distribution whose parameters are

$$v_1 = (n_1 - 1) \quad \text{and} \quad v_2 = (n_2 - 1)$$

where n_1, n_2 are the numbers of observations in the first and second samples. If the variances of the two normal distributions are the same, the ratio reduces to s_1^2/s_2^2, and this provides a test for the equality of two variances. The null and alternative hypotheses are usually

Box 2.23 Worked example: inference applied to *t*-test for equivalence of population mean to a hypothetical mean

Is there any evidence that the igneous rock from which the eight measurements of quartz were taken (see p. 91) has a mean quartz percentage greater than 20%?

Data:

Percentage quartz:

23.5   16.6   25.4   19.1   19.3   22.4   20.9   24.9

H_0: $\mu = 20$
H_1: $\mu > 20$

Test statistic:

$$t = \frac{\bar{x} - \mu_0}{\sqrt{s^2/n}}$$

Level of significance required:

$\alpha = 0.05$ (5%)

From the data:

$\bar{x}$	s^2	n	$\sqrt{s^2/n}$
21.5	9.51	8	1.09

Degrees of freedom:

$\nu = 7$

Critical value of *t*:

$t_{0.05;\,7} = 2.36$

Calculation:

$t = (21.5 - 20)/1.09 = 1.38$

Calculated *t* does not exceed critical *t*.

Conclusion: there is no strong evidence against the hypothesis that the rock quartz content is 20%, as opposed to the hypothesis that the quartz content exceeds 20%. It is not safe to assume that the rock has >20% quartz.

Box 2.24 Worked example: inference applied to t test for equivalence of two population means

Using the data on brachiopods introduced in Box 2.22, there are two alternative questions that might need to be answered.

1 Is there any evidence that the brachiopods from horizon A are longer than those from horizon B?

Denote the mean lengths of the two populations by μ_A, μ_B.

H_0: $\mu_A = \mu_B$ i.e. $\mu_A - \mu_B = 0$
H_1: $\mu_A > \mu_B$ i.e. $\mu_A - \mu_B > 0$

Test statistic:

$$t = \frac{\bar{x}_A - \bar{x}_B - (\mu_A - \mu_B)}{\sqrt{s^2(1/n_A + 1/n_B)}}$$

$$= \frac{\bar{x}_A - \bar{x}_B}{\sqrt{s^2(1/n_A + 1/n_B)}} \quad \text{according to } H_0$$

Level of significance required:

$\alpha = 0.05$ (5%)

From the data:

Horizon	$\bar{x}$	s	s^2	CSS(x)	n
A	3.125	0.2053	0.0421	0.295	8
B	2.960	0.1713	0.0293	0.264	10

Degrees of freedom:

$\nu = n_A + n_B - 2 = 10 + 8 - 2 = 16$

Critical value of t:

$t_{0.05;\,16} = 1.746$

Pooled estimate of variance (assuming a common variance):

$s^2 = (0.295 + 0.264)/16 = 0.0349$

Calculation:

$t = (3.125 - 2.96)/\sqrt{0.0349 \times (1/8 + 1/10)} = 1.86$

Calculated t exceeds critical t.

Conclusion: we reject the null hypothesis that the means are equal: we favour the hypothesis that the brachiopods from horizon A are longer.

2 Is there evidence of difference in lengths between A and B?

Box 2.24 *Continued*

H_0: $\mu_A = \mu_B$

H_1: $\mu_A \neq \mu_B$ i.e. $\mu_A - \mu_B > 0$ or $\mu_A - \mu_B < 0$

Data as above.

Level of significance:

$\alpha = 0.05$ (5%)

Critical value of t:

$t_{0.025;\ 16} = 2.12$

Note: the 2.5% point (0.025) from the t tables is used because both positive and negative values may lead us to doubt the null hypothesis (so we need 2.5% at the positive end and 2.5% at the negative end to allow $\alpha = 5\%$): it is the magnitude which is of importance, not the sign, so we use:

$$t = \frac{\bar{x}_A - \bar{x}_B}{\sqrt{s^2(1/n_A + 1/n_B)}} = 1.86$$

Calculated t does not exceed critical t.

   Conclusion: there is no statistical evidence against the hypothesis that the brachiopods in the two horizons are of equal length on average. We do not accept that there is a difference.

These two approaches and results illustrate the importance of the question asked: the different results are correct statistical responses to different questions. The analyst has to ensure that the null and alternative hypotheses chosen are appropriate for the solution of the pertaining geological problem.

H_0: $\sigma_1^2 = \sigma_2^2$

against

H_1: $\sigma_1^2 \neq \sigma_2^2$

and a two-sided test is required. In order to avoid the difficulty of working out the values of F for the lower tail, however, the ratio is always written with the larger estimate of variance in the numerator and a one-sided test of the null hypothesis is performed. If we are testing at the 5% level of significance, for example, we compare the ratio with the 2.5% point of the appropriate F distribution (Appendix 2.5).

2.5.6 Testing for distributions

We have described the logic and methodology of inference specifically as applied to t-tests on the population mean above, but similar procedures are used in all statistical tests. In this section, we include applications of inference directed towards solving a different type of univariate problem. We have so far assumed that we know the form of the distribution required to describe the random variation in our data. We now consider tests of those assumptions. There is no point in estimating parameters or testing hypotheses about them if they are parameters of the wrong distribution. Furthermore, the form of the distribution may give clues to the nature of geological processes. We have said that when discrete events occur randomly and independently in time or space then the Poisson distribution is useful for calculating probabilities of 0, 1, ... events occurring in any unit interval. Conversely, if the Poisson distribution is found to be useful for describing the random variation this is at least consistent with the idea that the events occur randomly and independently. (As always we cannot *prove* that they occur in this manner.)

We describe four methods:

1 the chi-squared test, which is better used for discrete than continuous distributions and which does not require a knowledge of the parameters;
2 the Kolmogorov–Smirnov test, which, at least in its simple form, requires that the parameters be specified;
3 a test for the normal distribution, using order statistics; and
4 an assessment of normality using probability paper.

2.5.6.1   The chi-squared test

Chi-squared (χ^2) tests involve a comparison of observed frequencies of events or values with those which would be expected, on average, when a given null hypothesis is true. The name of the test refers to the sampling distribution of the test statistic required. Consider this simple example as an introduction to the tests.

Suppose that some theory leads us to believe that proportions of four minerals in a granite are in the proportions $4:1:2:3$, and that we have a random sample of 100 grains in a thin section consisting of 35, 12, 22 and 31 of these species respectively. We would like to see if such a sample offers any evidence against the theory. A natural procedure is to compare the observed frequencies with those expected in samples of 100, on average,

Table 2.15 Comparison of observed and expected frequencies of four minerals

Mineral	*A*	*B*	*C*	*D*	Total
Observed frequency	35	12	22	31	100
Expected frequency	40	10	20	30	100
Observed − Expected	−5	2	2	1	0
$\dfrac{(\text{Observed} - \text{Expected})^2}{\text{Expected}}$	25/40	4/10	4/20	1/30	1.258

when the theory is true. This is done in Table 2.15. The so-called expected frequencies were obtained as follows:

Number of 1st mineral expected on average = 4/10 × 100 = 40
Number of 2nd mineral expected on average = 1/10 × 100 = 10

and so on.

Large differences between observed and expected frequencies will cast doubt on the theory. An overall measure of the departure of observed from expected values is provided by calculating

$$\frac{(\text{Observed frequency} - \text{Expected frequency})^2}{\text{Expected frequency}}$$

for each mineral, then adding the four values together. In mathematical symbols the statistic is

$$\sum \frac{(O_j - E_j)^2}{E_j}$$

The square in the numerator removes the difficulty that negative and positive differences annul each other. Division by the expected frequency compensates for the effect of the overall magnitude of the numbers.

For the example with the four minerals the statistic obtained is 1.258. The null hypothesis here is that the data were obtained from a population with the specified proportions of minerals. A large value casts doubt on the null hypothesis – and 'large' means a value such that there is only a small probability (0.05, say) of it being exceeded when the null hypothesis is true. To make the judgement we need to know what the sampling distribution of the statistic is when the null hypothesis is true. It is, if the expected frequencies are not too small, approximately the same as the χ^2 distribution.

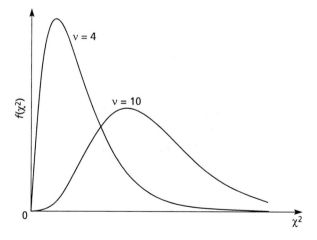

Fig. 2.39 Densities of χ^2 distributions with parameters (degrees of freedom) equal to 4 and 10. The distribution becomes less skew to the right as the parameter increases.

This distribution, which is used in many other contexts, is skew to the right, the degree of skewness depending upon a parameter v, the d.f. A general idea of the shape of such a distribution is given in Fig. 2.39. The parameter v is related to the number of frequencies compared (four, in the present case) and to the number of parameters of the distribution under test which must be estimated in order to carry out the test (none in this case). In fact,

v = (Number of frequencies compared − 1 −
    Number of parameters estimated)

Given four items to compare, only three independent comparisons are possible: if the first item is compared with the second and then with the third we cannot find anything new by comparing the second with the third, for example. Also, each estimate of a parameter imposes a further restriction.

The number of d.f. in the present example is, then, three. We can find from tables that, for χ^2 distributions with v = 3, 95% of randomly obtained values are less than or equal to 7.815 (Appendix 2.6). Testing at the 5% level of significance a greater value would fail to support the theory. In our case, the value is 1.258 and the data do not cause us to doubt the theory that abundances of the minerals are in the ratio 4:1:2:3.

χ^2 statistic

$$\chi^2 = \sum_{j=1}^{k} \frac{(O_j - E_j)^2}{E_j}$$

where k = number of categories, O_j = frequency observed in jth category, and E_j = frequency expected in jth category if null hypothesis is true.

H_0: Data drawn from population having specified proportions in each of a number k of categories.

H_1: Data drawn from population not having specified proportions.
Degrees of freedom = $k - 1 -$ (number of parameters estimated).
Worked examples: see Box 2.25.

Warnings

1 There should be at least five observations expected in each class. This condition can usually be satisfied by decreasing the number of classes.
2 The result is very susceptible to the number of classes used: it is more likely that a null hypothesis will be rejected if a large number of classes are used. With a small number of classes, failure to reject H_0 may simply be due to very coarse and unimportant similarities between data and model.

The χ^2 test is very versatile: it can be used to compare any data presented as frequencies in categories with any model described in the same form. As the examples show, it is probably most often used to assess the fit of data to the standard types of frequency distributions. In this case, the categories are

Box 2.25 Worked examples: chi-squared test for a normal distribution

1 The frequency distribution of data on uranium content (in log ppm) in lake sediments from 71 sites in Saskatchewan, Canada (Appendix 3.14), is as follows:

Midpoint	Count	
−1.0	2	**
−0.5	3	***
0.0	8	********
0.5	17	*****************
1.0	14	**************
1.5	12	************
2.0	6	******
2.5	5	*****
3.0	1	*
3.5	1	*
4.0	1	*
4.5	0	
5.0	1	*

Are the data drawn from a normally distributed population?
  We will use the χ^2 test with the hypotheses:

H_0: data drawn from a normally distributed population
H_1: data drawn from a non-normally distributed population

First, standardise the data and obtain frequencies in appropriate classes:

Midpoint	Count	
−2.250	0	
−1.750	3	***
−1.250	4	****
−0.750	17	*****************
−0.250	18	******************
0.250	9	*********
0.750	11	***********
1.250	4	****
1.750	2	**
2.250	1	*
2.750	1	*
3.250	0	
3.750	1	*

continued on p. 120

classes such as those of a histogram and the model is the idealised frequency distribution, such as normal or Poisson. Tabulated relative probability distributions are converted so as to be directly comparable with the data.

Box 2.25 *Continued*

Using $\chi^2 = \Sigma(O_j - E_j)^2/E_j$, the calculation (using $E_j\%$ from tables of the normal distribution, Appendix 2.2a) is:

Class	O_j	$E_j\%$	E_j	$(O_j - E_j)^2/E_j$
$-\infty--1.5$	3	6.68	4.74	0.64
$-1.5--1.0$	4	9.19	6.52	0.97
$-1.0--0.5$	17	14.98	10.63	3.82
$-0.5-0.0$	18	19.15	13.60	1.42
$0.0-0.5$	9	19.15	13.60	1.56
$0.5-1.0$	11	14.98	10.63	0.01
$1.0-1.5$	4	9.19	6.52	0.97
$1.5-\infty$	5	6.68	4.74	0.01
			$\chi^2 = 8.134$	

Degrees of freedom:

$v = k - 3 = 5$ (k is no. of classes)

Level of significance:

$\alpha = 0.05$

Critical $\chi^2 = 11.07$ (from tables, Appendix 2.6)

Calculated χ^2 does not exceed critical χ^2.

Fail to reject H_0: the data do not differ significantly from normal distribution. In this case, as the data used were log-transformed, we can proceed on the assumption that the raw data were drawn from a log-normal distribution.

2   Unfortunately, the outcome of the test may depend upon the choice of subintervals; the best choice appears to be that one which would lead to the expected frequencies in the subintervals being approximately equal. Another difficulty is that the amount of calculation needed to perform the test rigorously is large, but we shall describe a simplified version and note the effect the modification has on the distribution of the test statistic.

The following data are values of the organic content of 60 rock specimens:

11.8	7.0	5.3	5.5	5.9	2.3	11.6	5.1	8.3	8.0
6.5	7.8	3.5	4.9	15.2	13.9	5.9	12.6	12.2	3.6
7.7	6.3	8.6	17.1	11.7	10.2	5.4	3.0	9.6	7.1
22.8	3.2	2.3	16.2	7.6	7.7	7.6	14.1	2.4	6.5
5.9	11.0	7.9	7.1	10.7	13.4	7.1	1.6	4.9	7.4
5.4	7.2	2.0	11.5	10.1	8.0	11.7	9.3	8.3	5.4

Can the data be regarded as coming from a normal distribution?

Box 2.25 *Continued*

The mean and standard deviation of the distribution are estimated from all the available data. As we would like to obtain classes with reasonably high expected frequencies it is sensible to divide the line into 10 subintervals in each of which the expected frequency is about 6.

The sample mean and standard deviation of these data are 8.16 and 4.13 respectively. We may use the percentage points table of the normal distribution (Appendix 2.2b) to calculate nine class boundaries which should divide the frequency distribution into 10 equal parts. The percentage points are:

−1.2816
−0.8416
−0.5284
−0.2533
0
0.2533
0.5284
0.8416
1.2816

and the class boundaries needed are obtained from:

$$\frac{1\text{st boundary} - 8.16}{4.13} = -1.2816$$

so that

$$1\text{st boundary} = 8.16 - 4.13 \times 1.2816$$
$$= 2.87$$

and so on.

The numbers of values lying in the computed intervals are compared with the average frequency, 6, expected if the frequency distribution is normal, using the test statistic described above:

$$\frac{(5 - 6)^2}{6} + \frac{(4 - 6)^2}{6} + \ldots + \frac{(5 - 6)^2}{6}$$

We find

$$\{(-1)^2 + (-2)^2 + \ldots + (-1)^2\}/6 = 9.5$$

The number of degrees of freedom is $(10 - 1 - 2)$, as two parameters (the mean and standard deviation of a normal distribution) have been estimated. Comparing the value of the test statistic with the 5% point of the chi-squared distribution having 7 degrees of freedom (Appendix 2.6), we find that the evidence does not conflict with the notion that the organic content of the rock has a normal distribution.

2.5.6.2 **Kolmogorov–Smirnov test**

In this method, we compare the cumulative relative frequencies of the sample data with values of the distribution function $F(x)$ of the distribution from which we think the data come.

Suppose the data in our sample are, in increasing order, $x_{(1)}, \ldots, x_{(n)}$ and that, for simplicity, they are all different. Then the cumulative relative frequencies (proportions of the data which are less than or equal to these values) are

$$1/n, 2/n, \ldots, 1$$

If the data come from the distribution specified in the null hypothesis the values of $F(x_{(1)}), \ldots, F(x_{(n)})$ should be close to $1/n, \ldots, 1$ respectively. When the variable has no upper limit the cumulative frequencies are divided by $(n + 1)$ rather than n to prevent the 'cumulative relative frequency' reaching a value of 1. The test consists of calculating the magnitude of the greatest difference and comparing it with a tabulated critical value. Provided that the null hypothesis is true the table holds good, whatever the hypothesised distribution may have been, and in this sense the test is nonparametric. (Some other nonparametric procedures are described in Chapter 4.)

Worked examples: see Box 2.26.

2.5.6.3 **Normal scores test**

Suppose that we took a random sample of values from a normal distribution, standardised them and arranged them in increasing order and let us denote the resulting order statistics by

$$z_{(1)}, z_{(2)}, \ldots, z_{(n)}$$

where $z_{(1)}$ is the lowest value and $z_{(n)}$ is the highest.

They come from distributions whose means depend upon the number of observations, n. These mean values are known as normal scores; they give the mean value of z for each data point that would be expected if the data come from a normal distribution. It is usual to use statistics software to find these values for the specified value of n.

Data are plotted against these normal scores and an approximately straight line indicates that the data come from a normal distribution. The plot obtained from the data in Table 2.16 is reproduced in Fig. 2.40. It may be difficult at first to judge whether lines can be described as approximately straight, and we concentrate on the middle portion to see if the curvature is marked. The straightness of the line can be quantified by calculating the correlation coefficient (see Section 2.2) between the normal scores and the data. It is inevitable that normal scores increase as data values increase, so we expect a fairly high correlation coefficient and usual tests of significance cannot be used. There are, however, special tables of critical values of corre-

Box 2.26 Worked examples: Kolmogorov–Smirnov test for normal distribution

1 The data are errors in eight observations and it is suspected that they come from a normal distribution with mean equal to zero and standard deviation equal to 0.1.

0.07   0.12   −0.06   −0.04   −0.05   0.08   0.04   0.00

The hypotheses are:

H_0: the distribution is normal with $\mu = 0$, $\sigma = 0.1$
H_1: the data come from some other distribution

The data are ordered and values of the sample cumulative relative frequencies and theoretical values of $F(x_1), \ldots, F(x_n)$ are calculated, using the table of the normal distribution. These values have been tabulated and plotted (Fig. B2.26.1). Note that the cumulative relative frequency of the data is plotted as a step function.

Table B2.26

Error	Cumulative frequency	C.f./$(n + 1)$	z	$\Pr(Z \leqslant z)$
−0.06	1	0.11	−0.6	0.2743
−0.05	2	0.22	−0.5	0.3085
−0.04	3	0.33	−0.4	0.3446
0.00	4	0.44	0.0	0.5000
0.04	5	0.56	0.4	0.6554
0.07	6	0.67	0.7	0.7580
0.08	7	0.78	0.8	0.7881
0.12	8	0.89	1.2	0.8449

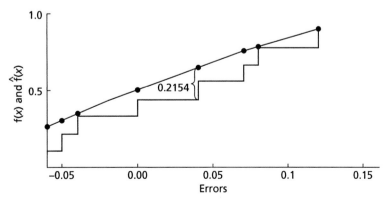

Fig. B2.26.1

continued on p. 124

Box 2.26 *Continued*

The magnitude of the greatest difference is

$D_8 = 0.6554 - 0.44$
$\quad\;\; = 0.2154$

From Appendix 2.7, the value which would be exceeded with probability 0.05 when the hypothesis is true is 0.457. At the 5% level of significance we conclude that the data are consistent with the hypothesis that the errors come from a normal distribution with mean equal to zero and standard deviation equal to 0.1.

2 The data are numbers of failures of an instrument in each of 12 consecutive months and we wish to test the hypothesis that the failures occur randomly and independently with a mean rate of two per month.

1 0 2 1 4 3 2 1 2 2 5 3

H_0: the distribution is Poisson with mean equal to 2.0
H_1: the data come from some other distribution

The cumulative relative frequencies of the values 0, 1, ..., 5 and the associated values of the distribution function $F(x)$ for a Poisson distribution with mean equal to 2 are given in Appendix 2.3. In this case both are step functions because the variable is discrete and it is easy to find, without a diagram, that the magnitude of the greatest difference is

$0.7500 - 0.6767 = 0.0733$

This is less than the tabulated value for a test at the 5% level of significance and we can accept that the failures occur randomly and independently with a mean rate of two per month.

lation coefficient for the normal scores test – see Appendix 2.9. The null hypothesis is that the data are drawn from a normal distribution; if the calculated value is *lower* than the critical value, we reject this null hypothesis.

2.5.6.4   **Normal probability graphs**

Graph paper for this procedure is prepared commercially and it is constructed in such a way that a plot of the distribution function of a normal distribution would produce a straight line. In practice the values of the distribution function are plotted along the horizontal axis and data on the vertical axis – the opposite to the usual way. Data are usually arranged into classes for which frequencies are counted; these are converted to cumulative frequency and then a percentage cumulative frequency distribution is calculated from

Table 2.16 Data with normal scores

Value (ordered)	6.17	6.66	6.92	7.12	7.36	7.37	7.48	7.49	7.71
Normal score	−1.87	−1.40	−1.13	−0.92	−0.74	−0.59	−0.45	−0.31	−0.19

Value (ordered)	7.73	7.80	7.81	8.08	8.53	8.91	8.94	9.19	9.64	9.67	10.14
Normal score	−0.06	0.06	0.19	0.31	0.45	0.59	0.74	0.92	1.13	1.40	1.87

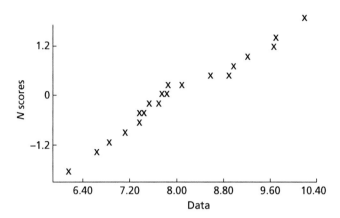

Fig. 2.40 Normal scores test. If the data come from a normal distribution a plot of them against the average values obtained in a sample of the same size from a standard normal distribution will be approximately straight. Data from Table 2.16.

$100 \times$ Cumulative frequency/$(n + 1)$

This method yields a result that can only be assessed subjectively. However, data plotted in this form are also traditionally used for grain size analysis, and many workers become very practised at accurate assessment of such plots. Grain size parameters are also routinely derived by this method: see Box 2.27.

Box 2.27 Use of normal probability paper and Folk grain size parameters

Grain size is usually quantified on the phi (ϕ) scale. Values can be calculated from grain diameters (d) in mm by:

$$\phi = -\log_2(d) = -\ln(d)/\ln(2)$$

Notice that this involves a log transformation, and that the negative sign means that higher positive values are associated with finer grain sizes.

The following percentages of two clastic rocks A and B were recorded in 0.5ϕ classes:

continued on p. 126

Box 2.27 *Continued*

	A%	A cum. %	B %	B cum. %
−4 to −3.5	0	0	12	12
−3.5 to −3	0	0	46	58
−3 to −2.5	2	2	22	80
−2.5 to −2	0	2	10	90
−2 to −1.5	4	6	2	92
−1.5 to −1	8	14	4	96
−1 to −0.5	10	24	2	98
−0.5 to 0	12	36	0	98
0 to 0.5	22	58	0	98
0.5 to 1	12	70	0	98
1 to 1.5	14	84	2	100
1.5 to 2	8	92		
2 to 2.5	6	98		
2.5 to 3	0	98		
3 to 3.5	2	100		

The cumulative percentages have been calculated for plotting on normal probability graph paper: a normal distribution will plot as a straight line.

Plotting the points, we obtain a graph as in Fig. B2.27.1. We can observe that sediment A seems to be close to a normal distribution, but the convexity of the sediment B curve indicates a positive skew. At the top and the bottom of the graph, the curves are affected by minor percentages, so there are likely to be insignificant irregularities. There is no 0% or 100% on the scale because the perfect normal distribution only reaches these values at infinity. The calculated sample 100% values cannot be plotted.

The horizontal position of the curves is related to the average grain size; the slope gives the dispersion (sorting): steeper curves are less dispersed along the scale, i.e. better sorted. Sedimentologists use such curves to estimate parameters of the frequency distribution. Calculations are based on the ϕ values at a series of specific percentages. To find, for example, ϕ_{50} for sediment A, we find the 50% point on the vertical scale, follow the line across to the grain size curve, then read down to the ϕ scale: this gives a value of $\phi = 0.33$. Other values used are ϕ_{95}, ϕ_{84}, ϕ_{75}, ϕ_{25}, ϕ_{16}, ϕ_5; these are read off in a similar way. The equations used were first proposed by Folk and are called the Folk parameters. They are:

$$\text{Median} = \phi_{50}$$
$$\text{Mean} = (\phi_{84} + \phi_{50} + \phi_{16})/3$$
$$\text{Standard deviation (sorting)} = (\phi_{84} - \phi_{16})/4 + (\phi_{95} - \phi_5)/6.6$$
$$\text{Skewness} = (\phi_{84} + \phi_{16} - 2\phi_{50})/2(\phi_{84} - \phi_{16})$$
$$- (\phi_{95} + \phi_5 - 2\phi_{50})/2(\phi_{95} - \phi_5)$$
$$\text{Kurtosis} = (\phi_{95} - \phi_5)/2.44(\phi_{75} - \phi_{25})$$

Box 2.27 *Continued*

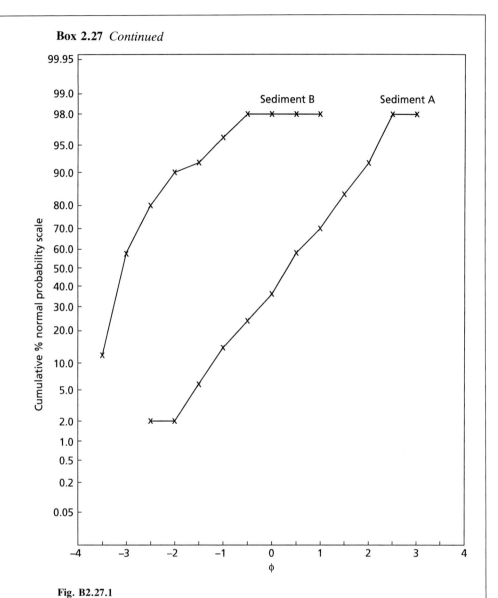

Fig. B2.27.1

For the two sediments we have:

	A	B
ϕ_{95}	2.18	−1.16
ϕ_{84}	1.5	−2.32
ϕ_{75}	1.16	−2.62
ϕ_{50}	0.33	−3.07
ϕ_{25}	−0.45	−3.32
ϕ_{16}	−0.88	−3.43
ϕ_{5}	−1.6	−3.6 (requires extrapolation)

continued on p. 128

Box 2.27 *Continued*

and from this we can calculate:

	A	B
Median	0.33	−3.07
Mean	0.32	−2.94
St. dev.	1.17	0.20
Skewness	−0.019	0.458
Kurtosis	0.962	1.43

These figures convey the coarser, skewed but better-sorted characteristics of sediment *B*, as compared with *A*.

Table 2.17 Rates of extraction of an ore under differing conditions of temperature and solvent

Temperature (°C)	Solvent		
	A	B	C
50	59	58	54
	61	59	57
80	65	62	58
	67	60	62

2.6 ANALYSIS OF VARIANCE

2.6.1 Introduction

2.6.1.1 Objectives

The method required for a data analytical job depends upon the way in which the data are collected. Too many geologists make observations unthinkingly and then ask what to do with them. Some even suggest hypotheses on the basis of properties of exploratory data and then test their ideas on the same data. In the present section we shall consider ways of obtaining information which leads to efficient methods of analysis and scientific tests of hypotheses, and to economic use of time and material. An example will help to introduce the most important ideas.

It was required to find how, if at all, the rate of extraction of an ore was affected by temperature and the solvent used. The results of one experiment are given in Table 2.17.

The extraction process has been run twice with each combination of temperature and solvent, producing 12 observations altogether. There are said to be two replications per cell. Temperature and solvent are called *factors* and different temperatures and solvents are called *levels* of those factors. Each combination of levels is a *treatment*. (If a study were being carried out in the field and the factors were, say, depth and location, each combination of levels would still be called a treatment in statistical terms.) Six treatments are tried in the experiment.

The data would be examined for variation which can be assigned to differences in temperature and solvent; any variation remaining after allowing for these effects (residual variation) is ascribed to chance.

In the first cell, where the solvent is A and the temperature 50°C, the difference is considered to be random, for the conditions have been kept constant as far as possible. Any difference between the means in the first cell and the one in which the solvent is A and the temperature 80°C, however, might be caused by the difference in temperature. Similarly, differences between cell means for different solvents but the same temperature could be the effect of changes in solvent.

Another possibility is that the effect of temperature may depend upon the solvent: this variation in dependence is called the interaction between factors.

In analysing the data we would test hypotheses that

1 temperature has no effect;

2 the solvent does not affect the yield; and

3 there is no interaction.

If the data fail to support any of the hypotheses then we shall accept that yield appears to be affected by the factor or factors concerned and try to estimate the effect. This is the type of problem that analysis of variance is designed to solve.

2.6.1.2   **Randomisation of experimental material**

There is potentially a further source of variation which, if not controlled, could lead to bias: it lies in the experimental material. If the ore assigned to treatments at 80°C is inherently richer than that used at 50°C, for example, the results will be biased in favour of 80°C. In order to remove any such risk, we not only try to make the experimental material homogeneous but also assign the units of material to treatments at random, so that effects of remaining homogeneity are averaged out between treatments.

Here, we could divide the ore into 12 parts and label them 1, 2, . . . , 12. In order to choose the specimens for the treatment at 50°C and solvent A we could draw two numbers between 1 and 12, using tickets or a table of random numbers, and assign the corresponding specimens to that treatment. The rest of the material would be assigned in the same way. The essential point is that the observations can be regarded as a random sample from some distribution.

2.6.1.3   **Analysis of variance**

The objective is to split the overall variation in the data (in terms of the total CSS) into parts which may be ascribed to:

1   the effects of factors; and
2   random variation (Fig. 2.41).

The sums of squares (SS) for factors are then tested to see if they indicate real differences between effects, or are no more than can be expected to arise by chance. (For a reminder on the meaning of SS, see Section 2.1.4.3.)

Usually, statistical packages will be used to analyse the data from experiments and therefore the most important consideration here is to understand the output. Nevertheless, it is worthwhile carrying out the tests described with the aid of a calculator in order to increase awareness of what is involved. In any case, if the recommended procedures are used, the calculations can be completed in a very short time.

The splitting of the SS requires the following conditions for the random term in our model.

1   It comes from a distribution with a mean of zero and a variance σ^2 which is the same, whatever treatment is being applied.
2   The random terms are not correlated with each other.

The condition of at least approximately constant variance is vital and in some cases the data are transformed in order to achieve it. The presence of correlation will also undermine the analysis.

Formal statistical tests require, further, that the data are independent observations from *normal* distributions, but departures from this condition are not very serious provided that the distribution is reasonably symmetric.

A description of possible sources of variation is called a model, and two kinds must be distinguished. In the first, called fixed effects, model I or parametric models, levels of factors are chosen deliberately and there is a particular interest in them and in their means. In the second, model II or random effects models, the levels are random samples from all possible levels; there is no interest in them as individuals, only as representatives of the population, and here it is the variance of the distribution of effects which is of concern – for example, in reliability trials of equipment.

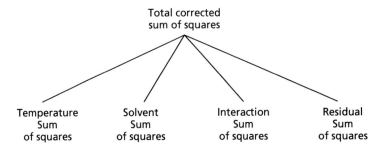

Fig. 2.41 Decomposition of the overall variation into effects of factors and random variation.

2.6.2   **Fixed effects models**

2.6.2.1   **Single-factor experiments**

A number of independent observations are made on each level of the factor and a test is carried out of:

H_0: no factor effects, on average

against

H_1: at least one level produces a different outcome from the rest

When there are exactly two levels, their effect may be examined by the two-sample t-test (Box 2.24); but, when three or more levels are used, that method fails. Suppose there were 10 levels: 45 pairwise comparisons would be possible, though not all independent. We know that when performing one of the tests at the 5% level of significance there is a 1 in 20 chance of finding a statistically significant result even when there is no real difference. The probability of finding at least one significant result in 45 tests is greater than 5%, so the level of significance of our test is not the one required: it is, in fact, higher than 5%.

The method used will be described with the aid of an example.

Three methods were used to determine the grade of an ore in order to see whether they gave the same or different results, on average. All the determinations were made at the same laboratory, so there were no differences associated with this aspect of the trial. For reasons of time and cost, different numbers of determinations were made with the three methods, but the experimental material was nevertheless assigned by a random procedure. The results are given in Table 2.18.

We may write down a model for the variation in the data and state the hypotheses which are to be tested. The observations on the first, second and third levels of the factor (i.e. methods of determining the grade) are considered to come from distributions with the same variance, σ^2, and means of μ_1, μ_2 and μ_3 respectively.

Table 2.18 Measured grades (%) of specimens of ore according to three different methods of analysis

Method of determination		
A	B	C
42.7	44.9	41.9
45.6	48.3	44.2
43.1	46.2	40.5
41.6		43.7
		41.0

The model may be written

$$x_{ij} = \mu_j + \varepsilon_{ij} \qquad \text{for} \qquad j = 1, 2, 3$$

where x_{ij} is the ith observation on level j and ε_{ij} is the random term in it. Frequently it is written in terms of deviations from an overall mean μ:

$$x_{ij} = \mu + \alpha_j + \varepsilon_{ij}$$

where α_j is the difference from the overall mean caused by using the jth treatment. Then the sum of the terms α_1, α_2 and α_3 is zero.

The null hypothesis to be tested is

H_0: all methods give the same result, on average

i.e.

$$\mu_1 = \mu_2 = \mu_3$$

(equivalently, all the effects α_1, α_2, α_3 are equal to zero); and the alternative is

H_1: at least one method gives different results on average from the others

First, we should examine the data graphically, using dot plots or box-and-whisker plots. Figure 2.42 contains nothing to suggest that the spread differs among the three sets of values (though there are admittedly few observations on the second method with which to detect a difference in spread). The individual sets are also reasonably symmetrical. The first and third sets look as if they have approximately equal means, but the values in the second are comparatively high.

The next task is to calculate the sample means and variances, and these statistics are shown in Table 2.19, with the numbers of observations made on each method. The sample variances will be used to produce a pooled estimate of the variance of the random term in the model as follows.

Denoting the numbers of observations on each treatment by n_1, n_2, n_3, and the three sample variances by s_1^2, s_2^2, s_3^2 the respective CSS are $(n_1 - 1)s_1^2$ and so on. The best estimate of the variance σ^2 can be shown to be the total value of these CSS divided by the total of the d.f.:

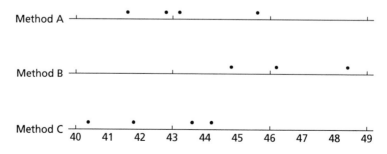

Fig. 2.42 Graphical comparison of data on three methods of determining the grade of an ore.

Table 2.19 Summary statistics for determinations of the grade of an ore

	Method		
	A	B	C
Sample mean $\bar{x}_i$	43.25	46.47	42.26
Sample s.d. s_i	1.69	1.72	1.63
Number of values n_i	4	3	5

$$s^2 = \frac{(n_1 - 1)s_1^2 + (n_2 - 1)s_2^2 + (n_3 - 1)s_3^2}{(n_1 - 1) + (n_2 - 1) + (n_2 - 1)}$$

and the denominator is written more simply as $(n_1 + n_2 + n_3 - 3)$.

This is the most convenient way of calculating the estimate of the variance if a calculator is used because of the ready availability of the standard deviations, but, of course, a computer program would produce the CSS directly.

The total CSS of the 12 observations can be expressed as the sum of two parts.
1 The CSS associated with differences between the levels of the factor.
2 The remainder, called the residual SS, which is associated with random variation. Although it is called the residual SS it is, in practice, easier to calculate this portion first, as described above, and to find the treatment SS as the difference from the total:

Total CSS = Treatment SS + Residual SS

so:

Treatment SS = Total SS − Residual SS

We would now like to know if the treatment SS represents significant variation or is no more than that expected from random terms.

The results of the calculations performed so far, and their use in testing, are best set out in an analysis of variance table (Table 2.20). Tabulation in this form is the standard way of presenting analysis of variance results, so we will explain the table, column by column.

Table 2.20 Analysis of variance for comparison of methods of determining the grade of an ore

Source of variation	d.f.	Sum of squares	Mean square	Mean square ratio, F
Between methods	2	34.1005	17.0503	6.11
Residual	9	25.1087	2.7899	
Total	11	59.2092		

Degrees of freedom

Total d.f.: the number of observations minus 1, which is the number of independent squared deviations from the mean. This is familiar from our work on the estimation of variance from a single sample of n observations.

Treatment d.f.: the number of independent comparisons which can be made between treatments; it is one less than the number of treatments.

Residual d.f.: the divisor in the estimate of σ^2. Here, it can be calculated as the sum of the d.f. associated with each level of the factor. It is also the difference between the total and treatment d.f. Sum of squares

Total SS: the CSS for all observations, with the deviations measured from the grand mean.

Residual SS: calculated either as the difference between the total and treatment SS or, when using a calculator, from

$$(n_1 - 1)s_1^2 + (n_2 - 1)s_1^2 + (n_3 - 1)s_3^2$$

Treatment SS: although it can be obtained from the total and residual SS, it is instructive to notice that it is a measure of the scatter of the treatment means $\bar{x}_1$, $\bar{x}_2$, $\bar{x}_3$ about the grand mean, with each squared deviation weighted by the number of observations on the treatment:

$$\text{Treatment SS} = n_1(\bar{x}_1 - \bar{x})^2 + n_2(\bar{x}_2 - \bar{x})^2 + n_3(\bar{x}_3 - \bar{x})^2$$

Mean squares

These are obtained by dividing the treatment and residual SS by their respective d.f.

The residual mean square (MS) always gives an unbiased estimate of the variance, σ^2, just as, in the two-sample t-test, the sum of the CSS of the two samples was divided by the sum of the d.f. to form a pooled estimate of the common variance.

The treatment SS has been shown to provide an estimate of the quantity

σ^2 + a positive term involving the squares of the α terms, caused by systematic differences between the means for the three levels

Then, if there are no differences in these means (i.e. the α terms are all zero), as the null hypothesis states, the treatment MS gives another unbiased estimate of the variance σ^2 and should have a value 'near' that of the residual MS. If the values differ greatly, there is reason to suppose that the treatment MS includes a further positive term and that there are some differences between the effects of the levels of the factor.

From Section 2.5.2.2 we know that the ratio of two independent estimates of a variance σ^2 comes from an F distribution, provided that the data are independent values from a normal distribution – and it is here that the assumption is required for the first time. The parameters v_1, v_2 are the d.f. for the treatments and residual respectively. Here,

$$F = 17.0503/2.7899$$
$$= 6.11$$

The 5% point of the appropriate F distribution is $F_{0.05;2,9} = 4.26$ (from Appendix 2.5).

The MS ratio is significantly large and the evidence does not support the hypothesis that all the methods produce the same results on average. The test indicates only that at least one mean differs from the rest, not which mean or means. It remains to determine which ones differ and to estimate parameters.

2.6.2.2 **Examination of residuals**

The examination of residuals has an important part in testing the assumptions in models, and statistical packages should include a facility for computing them in their analysis of variance routines. For the single-factor design a residual is defined to be the difference between an observation and the mean of the values obtained on the level of the factor:

$$R_{ij} = x_{ij} - \bar{x}_j$$

The residuals for these results are given in Table 2.21.

If the data have normal distributions with the same variance in each case then the residuals will come from a normal distribution with a mean equal to zero. Furthermore, there should be no pattern in plots of residuals against levels of treatment or against the treatment means. Figure 2.43(a) is a plot of normal order statistics against the residuals of this experiment; the residuals are plotted against levels of the factor in Fig. 2.43(b) and against treatment means in Fig. 2.43(c).

The results of this examination indicate that some of our assumptions may not be satisfied. Although the sum of the residuals is necessarily zero, two-thirds of them are negative, balanced by a few positive values, and the normal scores plot does not support the idea that the data come from

Table 2.21 Residuals from the comparison of methods of determining the grade of an ore

Method of determination		
A	B	C
−0.55	−1.57	−0.36
2.35	1.83	1.94
−0.15	−0.27	−1.76
−1.65		1.44
		−1.26

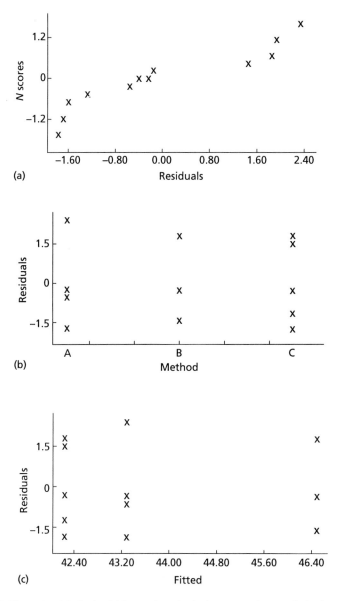

(a)

(b)

(c)

Fig. 2.43 Plots of residuals should be used to check the assumptions made in the analysis of variance. In (a) they are plotted against normal order statistics to justify the use of the F test. The plot in (b) tests the assumption of equal variances for all treatments and the plot against fitted values in (c) should have no trend.

normal distributions. There is no suggestion that the condition of constant variance has been broken, however, and the test remains useful as a guide. There is one particularly large residual (2.35) and perhaps the recorded determination which produced it should be inspected.

2.6.2.3   **Randomised blocks design**

Suppose it was required to test three different methods of porosity measurement four times but there was insufficient material from one borehole for 12 observations. One solution might be to obtain batches of material from four boreholes. There would now be an additional source of variation which could make it more difficult to detect any differences between methods.

The problem is solved by dividing the material from each borehole into three parts and assigning them to methods at random, as indicated in Fig. 2.44. Then differences between results from a given batch are ascribed to methods and error only, since the material is relatively homogeneous.

In statistical terms, the batches constitute *blocks* and the type of experiment is called a randomised blocks design. It is used to reduce errors arising from sources which are not of interest in their own right but which are important in that they provide an extra possible source of variation.

Suppose that three methods of porosity measurement were compared, using the rock from four sources as four blocks, as we have described. The data in Table 2.22 were obtained.

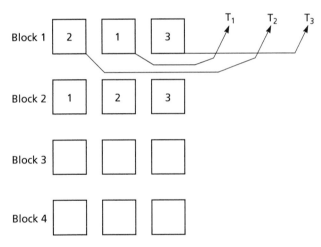

Fig. 2.44 Assignment of experimental material to treatments in a randomised blocks experiment.

Table 2.22 Results of randomised blocks experiment with row and column totals. Data give porosity (in % × 10)

Source	Method M_1	M_2	M_3	Total
S_1	83	80	75	238
S_2	79	76	77	232
S_3	88	82	79	249
S_4	85	84	82	251
Total	335	322	313	990

Model

Denoting the observation on the ith block and jth level of the factor by x_{ij},

$$x_{ij} = \mu + \alpha_i + \beta_j + \varepsilon_{ij}$$

where μ is an overall mean, α_i is the effect of the ith method, β_j is the effect of the jth block and ε_{ij} is the error term.

Another situation which could arise is that there is time for only three observations to be made during one shift of technicians, so the work is spread out over four shifts; in this case the blocks would be shifts.

Using this design, then, we are able to examine the factor of real interest after removing the effects of another source of variation, making our statistical test more sensitive. We now consider the statistics involved and the analysis of variance.

Null hypotheses

$$\alpha_1 = \alpha_2 = \alpha_3 = 0$$

and

$$\beta_1 = \beta_2 = \beta_3 = \beta_4 = 0$$

against alternatives that at least one method and at least one block differ from the rest.

Notation

Sample means of observations on treatments: $\bar{x}_{1.}, \bar{x}_{2.}, \bar{x}_{3.}$
Sample means of observations on blocks: $\bar{x}_{.1}, \bar{x}_{.2}, \ldots, \bar{x}_{.4}$
Grand mean: $\bar{x}_{..}$

Decomposition of the total corrected sum of squares

The total SS is now partitioned into

SS between treatments
SS between blocks
Residual SS = Total SS − Treatment SS − Blocks SS

The SS between treatments is calculated as the SS of the treatment means about the grand mean, weighted by the number of values entering into each mean:

$$4\{(\bar{x}_{1.} - \bar{x}_{..})^2 + (\bar{x}_{2.} - \bar{x}_{..})^2 + (\bar{x}_{3.} - \bar{x}_{..})^2\}$$

and the SS between blocks is calculated as

$$3\{(\bar{x}_{.1} - \bar{x}_{..})^2 + \ldots + (\bar{x}_{.4} - \bar{x}_{..})^2\}$$

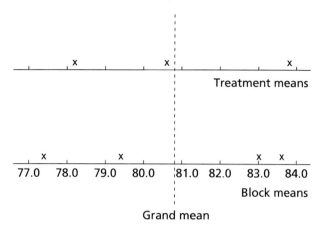

Fig. 2.45 Scatter of treatment and block means about the grand mean in a randomised blocks experiment.

The scatter of the treatment and block means is illustrated in Fig. 2.45. Again, the SS for treatments and blocks are SS of deviations from the grand mean. They will usually be obtained by means of a statistical package, but Box 2.28 describes how to obtain them with a calculator.

Box 2.28 Computation of sums of squares for the analysis of variance

Calculators enable us to obtain sums of squares in the analysis of variance easily from totals of rows and columns. Table 2.22 shows the data from a randomised blocks experiment, with these totals.

The corrected sum of squares for rows (blocks, in this experiment) is calculated in the following steps, and the column (method) sum of squares similarly.

1 Enter the 4 row totals in SD mode and press the standard deviation button.

2 Square the result.

3 Multiply by (number of rows −1)

4 Divide by the number of values, here 3, making up each row.

The total sum of squares is calculated from the 12 individual values in the usual way (Box 2.3) and the residual sum of squares is then calculated as

Total SS − (Row SS + Column SS)

in the analysis of variance table.

The same method can be applied to treatment totals in one-factor experiments when the same number of observations is available on each treatment.

Table 2.23 Analysis of variance for a randomised blocks experiment

Source of variation	d.f.	Sum of squares	Mean square	Mean square ratio, F
Between methods	2	61.1667	30.58	8.03
Between sources	3	81.6667	27.22	7.14
Residual	6	22.8333	3.81	
Total	11	165.6667		

Results

The results are set out in the analysis of variance table (Table 2.23). The d.f. are calculated on the same basis as those in Table 2.20 for the single-factor experiments; for example, there are four blocks, so these are 3 d.f. for them.

Again, the residual MS gives an unbiased estimate of the variance σ^2. Also, if there are treatment effects, the treatment MS gives an unbiased estimate of:

σ^2 + a positive term caused by differences between treatment effects, involving squares of α terms

and the blocks MS gives an estimate of

σ^2 + a positive term caused by differences between blocks involving squares of β terms

As in the case of single-factor experiments, we can argue that if there are no differences between treatments (all α terms are zero) then the treatment MS gives a second, independent estimate of σ^2 and we test it against the residual MS using the ratio

$$F = \frac{\text{Treatment MS}}{\text{Residual MS}}$$

which comes from the F distribution with parameters $v_1 = 2$ and $v_2 = 6$. Here, we find

$F = 30.58/3.81 = 8.03$

and as the 5% point of the F distribution is $F_{0.05;2,6} = 5.14$ (Appendix 2.5) the evidence does not support the hypothesis that the three methods give the same yield on average.

Further, if there are no differences between blocks (all β terms are zero), the blocks MS gives another independent estimate of σ^2, so we test the blocks MS against the residual. Here, the MS ratio is

$F = 27.22/3.81 = 7.14$

and $F_{0.05;3,6} = 4.76$. There appear to be differences between the blocks, so the effort required to organise the experiment as a randomised blocks design was worth while.

Residuals

These are computed from differences of observations, treatment and block means about the grand mean:

$$(\text{Observation} - \bar{x}_{..}) - (\text{Block mean} - \bar{x}_{..}) - (\text{Treatment mean} - \bar{x}_{..})$$

which simplifies to

Observation − Block mean − Treatment mean + Grand mean

but it is recommended that the calculations are left to the computer. The examination of the residuals from the data in Table 2.22 is left as an exercise.

2.6.3 Random effects models: components of variance

In the experiments we have considered so far, the levels of factors were fixed because there was a particular interest in them and in differences between their mean results. In this section we introduce random effects models for variation where the levels are random samples from all possibilities for a given factor and then the point of interest is the variance of the distribution of results. The total variance is now divided into components of variance: a part due to random errror and others due to treatment effects. The computations required for sums of squares are exactly the same as those used for fixed effects models; but there is a difference in the way the mean squares are tested in two factor experiments, and in the interpretation of the results of all experiments. We shall also introduce another design, one which finds most of its applications when levels are chosen at random.

2.6.3.1 Single-factor experiments

A company was considering the purchase of instruments for measuring resistivity from a supplier and wished to investigate the random variation between instruments as well as that between observations made with a given instrument. A random sample of five instruments was made available and four observations performed with each on specimens of a standard material. One observation was lost by the technician.

The actual instruments used in the trial are not of interest in themselves but only as representatives of a population and we can regard the contributions of the instruments to the observations as values of a random variable.

Model

$$x_{ij} = A + u_j + \varepsilon_{ij}$$

where n_j is the number of observations made with the jth instrument, x_{ij} is the ith observation on instrument j, ε_{ij} is the usual error term and u_j is the value of a random variable associated with values obtained from the jth instrument.

The random terms u_j are considered to come from a distribution whose mean is zero and whose variance is σ_u^2, and they must be uncorrelated with each other and with the error terms ε_{ij}.

Hypotheses

H_0: The variance σ_u^2 is zero

against

H_1: σ_u^2 is not zero

Decomposition of the total corrected sum of squares

When the variance σ_u^2 is not zero the treatment MS gives an estimate of

σ^2 + a positive term involving σ_u^2

The residual MS gives an unbiased estimate of σ^2, as before, so a large value of the ratio

$$F = \frac{\text{Treatment MS}}{\text{Residual MS}}$$

indicates that σ_u^2 is not zero whereas a low value suggests that there is no significant component of variance caused by instruments.

The ratio is assessed in the way described for the fixed effects model, referring to the F distribution, whose parameters are the treatment and residual d.f.

Result

The analysis of variance for a realisation of the trial of instruments is given in Table 2.24. When the null hypothesis is true the MS ratio comes from the F distribution with parameters $v_1 = 4$ and $v_2 = 14$ (recall that one result was lost by the technician). Now $F_{0.05; 4, 14} = 3.11$ so the evidence suggests that the component of variance between instruments is significant. As the null hypothesis is not supported we shall want to estimate the component of variance due to instruments. When the same number of observations n is made on each instrument the treatment MS gives an estimate of

Table 2.24 Analysis of variance for a one-factor random effects model

Source of variation	d.f.	Sum of squares	Mean square	Mean square ratio, F
Between instruments	4	32.9904	8.2476	4.7
Residual	14	24.5672	1.7548	
Total	18	57.5576		

$$\sigma^2 + n\sigma_u^2/(k - 1)$$

where k is the number of levels of the factor instruments (that is, the number of instruments used).

An estimate of σ_u^2 is calculated by treating the estimates of this quantity and of σ^2 as if they were the correct values of the parameters:

$$\text{Treatment MS} = \sigma^2 + n\sigma_u^2/(k - 1)$$
$$\text{Residual MS} \;\;= \sigma^2$$

and then as

$$\text{Treatment MS} - \text{Residual MS} = n\sigma_u^2/(k - 1)$$

the component σ_u^2 is estimated by

$$(\text{Treatment MS} - \text{Residual MS}) \, (k - 1)/n$$

When the numbers of observations on the k treatments are not equal, n must be replaced by

$$N - (\Sigma n_j^2)/N$$

where N is the total number of observations made in the experiment. In this example, $N = 19$ and $\Sigma n_j^2 = 4^2 + 4^2 + 3^2 + 4^2 + 4^2 = 73$, so

$$N - (\Sigma n_j^2)/N = 19 - 73/19$$
$$= 15.16$$

and the estimate of the component of variance due to instruments is

$$s_u^2 = 4(8.2476 - 1.7548)/15.16$$
$$= 1.71$$

2.6.3.2   Nested classification

In the experiments we have analysed so far, there has been a single factor of intrinsic interest. In nested or hierarchical designs, each level of a first (main) factor is divided according to a second criterion into different levels of a second (subgroup) factor. This would be the case where random samples of material are themselves divided into subsamples for examination:

the main factor is sample and the subgroup factor is subsample (Fig. 2.46). In another context, the main factor might be region and the subgroup factor be areas within regions.

Consider this example.

Three cores are examined from the same two depths down the borehole and four analyses of the material are made at each depth. The coded data are in Table 2.25. The types of variation which are of interest are:

1 variation between readings within depths, i.e. at the same depth in a given borehole (which will be used to estimate the variance due to random variation);

2 variation between depths within cores; and

3 variation between cores.

Model

$$x_{ijk} = \mu + u_i + v_{j(i)} + \varepsilon_{ijk}$$

where u_i is a random term for the ith core, variance σ_u^2; $v_{j(i)}$ is a random term for depth j down borehole i, variance σ_v^2; and ε_{ijk} is a random term associated with the kth observation at the jth depth down the ith borehole, variance σ^2.

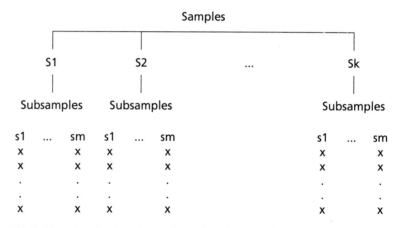

Fig. 2.46 Nesting of levels of a subgroup factor in main group levels of a nested design.

Table 2.25 Data from a study of cores (C) at two depths (D)

C_1		C_2		C_3	
D_1	D_2	D_1	D_2	D_1	D_2
5.3	4.8	5.1	4.6	5.9	5.2
5.2	4.4	4.9	4.6	5.7	5.1
5.3	4.3	5.2	4.4	5.6	5.3
5.1	4.7	4.7	4.3	5.9	5.5

Null hypotheses

H_0: $\sigma_u^2 = 0$; H_0: $\sigma_v^2 = 0$

against the alternatives that they are not zero.

Decomposition of the total corrected sum of squares

This is represented in Fig. 2.47 and it can be considered to occur in two stages.

1 The data are treated as the results of a one-factor experiment in which levels are the subgroups (depths), as in Table 2.26. The breakdown of the SS is given in Table 2.27.

2 Subgroup means are used as data in another one-factor experiment, in which the levels are main groups, as in Table 2.28. Table 2.29 contains the breakdown of the SS between depths.

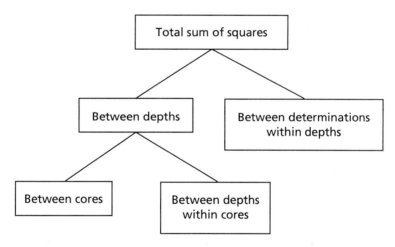

Fig. 2.47 The decomposition of the total sum of squares in a nested classification, considered in two parts: (a) scatter of observations about subgroup (depth) means and that between subgroup means; (b) scatter of the main group (core) means about the grand mean.

Table 2.26 Subgroup levels treated as the levels in a one-factor experiment

	Depth					
	D_1	D_2	D_1	D_2	D_1	D_2
	5.3	4.8	5.1	4.6	5.9	5.2
	5.2	4.8	4.9	4.6	5.7	5.1
	5.3	4.3	5.2	4.4	5.6	5.3
	5.1	4.7	4.7	4.3	5.9	5.5
Total	20.9	18.2	19.9	17.9	23.1	21.1

Table 2.27 Decomposition of total corrected SS

Source of variation	d.f.	Sum of squares
Between depths	5	4.7721
Between observations	18	0.5675
Total	23	5.3396

Table 2.28 Main group levels treated as the levels of a one-factor experiment

	Core		
	C_1	C_2	C_3
	20.9	19.9	23.1
	18.2	17.9	21.1
Total	40.1	37.8	44.2

Table 2.29 Decomposition of the between-depths SS

Source of variation	d.f.	Sum of squares
Between cores	2	2.8608
Between depths within cores	3	1.9113
Total	5	4.7721

Table 2.30 Final analysis of variance for the nested model

Source of variation	d.f.	Sum of squares	Mean square	Mean square ratio, F
Between cores	2	2.8608	1.4304	2.245
Between depths within cores	3	1.9113	0.6371	20.225
Between readings within depths	18	0.5675	0.0315	
Total	23	5.3396		

For the purposes of obtaining the SS with a calculator, the totals are used rather than means, as they were in the analyses described earlier. The computations can be performed by following the description of one-factor experiments (see Box 2.28). (Not all statistical packages have a facility for dealing with nested designs.)

We deal here only with experiments in which there are equal numbers of subgroup factors in each main group and equal numbers of observations within each subgroup. These conditions can be relaxed, but at the cost of more complicated calculations for testing. Furthermore, if the numbers within subgroup levels are not equal, we cannot test easily the null hypothesis that $\sigma_u^2 = 0$.

In the simple case, then, where there are s subgroup factors (depths) in each main group (cores), r observations on each subgroup level and m main group levels so that the total number of observations N is equal to mrs, the expected values of the MS are as follows:

Between main group levels: $\sigma^2 + r\sigma_u^2 + rs\sigma_v^2$
Between subgroups within main groups: $\sigma^2 + r\sigma_u^2$
Within subgroups: σ^2

Results

The results are combined in the final analysis of variance table (Table 2.30). The hypothesis that $\sigma_v^2 = 0$ is tested by comparing

$$F = \frac{\text{Between main group MS}}{\text{Between subgroups MS}}$$

with $F_{0.05; m-1, m(s-1)}$ and

$$F = \frac{\text{Between subgroups MS}}{\text{Within subgroups MS}}$$

with $F_{0.05; m(s-1), ms(r-1)}$.

The evidence suggests that, while there is no significant component of variance between cores, there is one between depths within cores.

FURTHER READING

The methods described in this chapter are generally undertaken as routine low-level analyses and are consequently not usually documented in research papers. Textbooks are the main source of examples, but see also:

Ashley G.M. (1978) Interpretation of polymodal sediments. *Journal of Geology*, **86**, 411–421.
Folk R.L. (1966) A review of grain size parameters. *Sedimentology*, **6**, 73–93.

for analysis of grain size frequency distributions.

Flinn D. (1959) An application of statistical analysis to petrochemical data. *Geochimica et Cosmochimica Acta*, **17**, 161–175.

a lucid application of *t*-tests.

3 Statistics with Two Variables

INTRODUCTION: THE BIVARIATE SCATTER

In Chapter 3 we will consider the data analytical methods used when we need to consider, simultaneously, the variation of two variables, where both are measured on each object in a sample. In addition to providing extra information about the frequency distribution of a sample, these methods give us information about the relationship between variables. The relationships between, say, gold and tungsten in exploration geochemistry, or between the sonic wireline log and porosity in hydrocarbon reservoir appraisal, are of prime importance.

With univariate methods, we symbolise the data in the form:

$x_1, x_2, x_3, \ldots, x_n$

In this chapter, we need a further variable: we will be calling the two variables x and y, so that the data are:

$x_1 \quad y_1$
$x_2 \quad y_2$
$x_3 \quad y_3$
$\vdots$
$x_n \quad y_n$

Such data are conventionally stored with one column for each variable and one row for each object (= rock specimen, fossil, soil sample, etc.). It is crucial to be sure that a pair of measurements, x_i and y_i, relates to the same object.

It was noted in Section 2.1 that the simplest graphical concept of univariate data was a series of points along a scaled line. With one variable, this array of points is one-dimensional; with two variables, we can produce a scatter of points in a two-dimensional space, with the (x_i, y_i) data pair giving the coordinates with respect to two perpendicular scaled axes (Fig. 3.1). Bivariate plots and analyses are, no doubt, particularly popular simply because the dimensionality of the plot matches the dimensionality of the sheet of paper or computer screen!

All the techniques of bivariate statistics can be regarded as ways of describing and analysing the shape of the bivariate scatter.

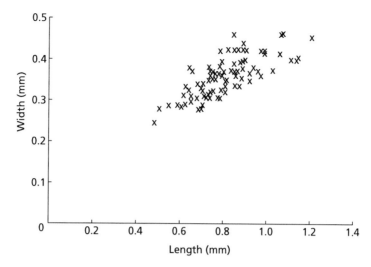

Fig. 3.1 Example of a simple bivariate scatter of length vs. width of specimens of the echinoid *Micraster*.

THE CORRELATION COEFFICIENT

Covariance and correlation

Two variables in a data set are said to be correlated if a simple bivariate scatter plot of the data has a significantly rectilinear (straight line) trend. This can be of great importance as it may indicate that the variables are linked, directly or indirectly, in the underlying causative geological process: for example, an observed correlation in a carbonate rock between porosity and magnesium content would probably reflect the well-known phenomenon of volume reduction associated with the process of dolomitisation. The correlation coefficient ρ is a measure of the degree of linear correlation. In practice, we will need to obtain estimates of ρ from sample data: this estimate is known as the sample correlation coefficient and is symbolised by r. Correlation coefficients are extensively used at the exploratory stage of data analysis in geology as an initial rough measure of intervariable relationship, but can also be applied rigorously where appropriate.

The correlation coefficient is based on a measure known as the corrected sum of products (CSP):

$$CSP_{xy} = \sum_{i=1}^{n}(x_i - \bar{x})(y_i - \bar{y})$$

This measure has a clear graphical interpretation which enables us to understand the properties of the data that the correlation coefficient conveys: see Box 3.1.

CSP will clearly depend on the sample size (n). This effect can be

Box 3.1 Graphical interpretation of the corrected sum of products

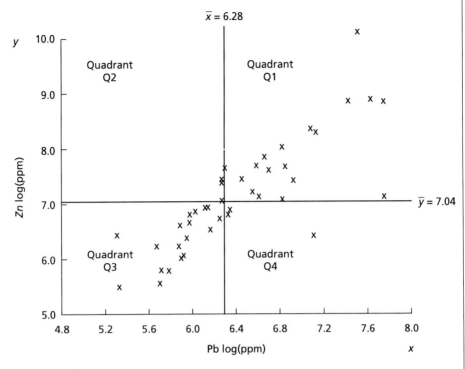

Fig. B3.1.1

The scatter of points is divisible into four quadrants by the means of x and y. The value of $(x_i - \bar{x})(y_i - \bar{y})$ for a data point will be positive or negative according to the signs of each of $(x_i - \bar{x})$ and $(y_i - \bar{y})$; and signs of these obviously depend on whether the value is smaller or larger than the mean. Therefore, the sign of $(x_i - \bar{x})(y_i - \bar{y})$ will depend on which quadrant the point is in.

Quadrant	Sign of $(x_i - \bar{x})$	Sign of $(y_i - \bar{y})$	Sign of $(x_i - \bar{x})(y_i - \bar{y})$
Q1	+	+	+
Q2	−	+	−
Q3	−	−	+
Q4	+	−	−

The corrected sum of products (CSP) is obtained by adding together the values of $(x_i - \bar{x})(y_i - \bar{y})$ for all points, so it will have a value related to the distribution of the points among the quadrants. In the scatter shown, the data points fall mostly in quadrants 1 and 3, so the resulting CSP will be large and positive. A data scatter with points largely in quadrants 2 and 4 will have a large negative CSP value; if the data points are evenly distributed between the four quadrants the positive and negative $(x_i - \bar{x})(y_i - \bar{y})$ values will tend to cancel out and the value of CSP will be near zero.

removed by division by $(n - 1)$. The resulting measure is the covariance (COV) between the two variables:

$$COV_{xy} = \frac{CSP_{xy}}{n - 1}$$

The justification for using $(n - 1)$ and not just n is the same as for the variance (see Section 2.1.4): division by n tends to produce an underestimate of the population value. The covariance is used extensively in many multivariate methods (see Chapter 8) but is of limited use here because it is clearly unit-dependent: covariance between length and width of a brachiopod species will yield a higher value if measured in millimetres rather than centimetres. The result is even less satisfactory if the two variables are measured in different units altogether. The effects of units are removed by standardising the data: $(x_i - \bar{x})$ is divided by the standard deviation of x (s_x) and $(y_i - \bar{y})$ by s_y. This is equivalent to the division of the covariance by the product of the standard deviation $(s_x s_y)$. The result is a dimensionless measure of correlation, the full name of which is Pearson's product-moment correlation coefficient (r), usually known simply as the correlation coefficient.

Pearson's product-moment correlation coefficient (r)

$$r_{xy} = \frac{COV_{xy}}{s_x s_y} = \frac{\sum\limits_{i=1}^{n}(x_i - \bar{x})(y_i - \bar{y})}{(n - 1)s_x s_y}$$

(There are various other equivalent forms of this equation which allow more efficient manual calculation.)

Warnings

1 This correlation coefficient is not a general measure of relationship between two variables, but just specifically of the degree of a straight line tendency: see Fig. 3.2. The bivariate scatter should therefore always be viewed in conjunction with the correlation coefficient.
2 The correlation coefficient is liable to spurious high values if outliers are present: see Fig. 3.3.
3 Correlation coefficients of two log-transformed variables should be interpreted with caution: see Fig. 3.4.
4 Correlation coefficients of closed data (e.g. data expressed as percentages) are likely to be spurious: see Section 3.2.2.

The correlation coefficient ranges from -1 (straight line, negative slope) to 1 (straight line, positive slope; see Fig. 3.2); these two extremes indicate equally strong relationships between variables, but with different implications for interpretation. Values near zero obviously indicate a lack of a rectilinear

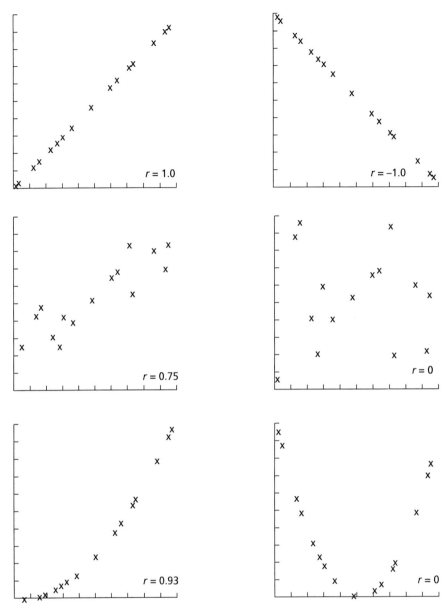

Fig. 3.2 Examples of correlation coefficients from various types of hypothetical data scatter. Note that perfect curvilinear relationships are not well described by r.

trend, but if the data plot as a vertical or horizontal linear scatter r will also be near zero (or undefined if $s_x = 0$ or $s_y = 0$).

It is important to establish a threshold for the significance of the correlation coefficient, so that we can infer the probability of an r value having arisen due to random sampling from a population having $\rho = 0$. For this, we must consider the sample size: a correlation coefficient of 0.4 may be

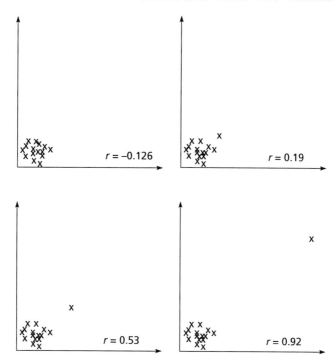

Fig. 3.3 Correlation coefficients from a series of hypothetical bivariate scatters, showing the effect of outliers. The values of the correlation coefficient become more misleading for data with more extreme outliers.

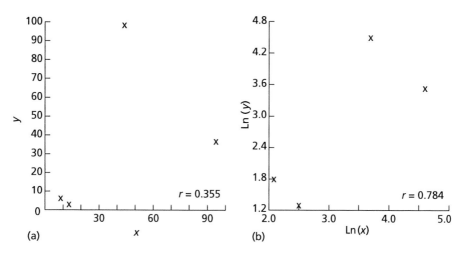

Fig. 3.4 The effect of the log transform on correlation coefficients: the shape of data scatter shown in (a) produces much higher correlation coefficients after log transform (b).

significant if $n = 100$ but not if $n = 10$. The significance of the correlation coefficient is estimated using a t statistic:

Test for significance of correlation coefficient

$$t = r\sqrt{\frac{n-2}{1-r^2}}$$

Null hypothesis (H_0): $\rho = 0$
Alternative hypothesis (H_1): $\rho \neq 0$
Degrees of freedom $= n - 2$

Worked example: see Box 3.2.

Warning

This test is only valid if the data are taken without bias from a population which is normally distributed with respect to both variables. This is not often the case in geology: alternative approaches are non-parametric coefficients (see Chapter 4) or linear regression (see Section 3.3).

The significance of the correlation coefficient can also be assessed by direct comparison with statistical tables of r, if these are available.

Box 3.2 Worked example: test for significance of correlation coefficient

Using lead–zinc geochemical data from Derbyshire, illustrated in Box 3.1, is there a significant correlation between $\ln(Pb)$ and $\ln(Zn)$?

H_0: $\rho = 0$
H_1: $\rho \neq 0$

Test statistic:

$$t = r\sqrt{\frac{n-2}{1-r^2}}$$

$r = 0.765$
$n = 44$
$t = 0.765 \times \sqrt{[(44 - 2)/(1 - 0.765^2)]} = 7.7$
Degrees of freedom $= n - 2 = 42$
$\alpha = 0.05$

From tables (Appendix 2.4):

Critical $t = 1.68$

Calculated t exceeds critical t, so we reject the null hypothesis.
    Conclusion: there is a significant correlation between $\ln(Pb)$ and $\ln(Zn)$.

3.2.2

Closed data and induced correlations

Special problems arise in geology where data are expressed as percentages, or any other means of conveying a fraction of a fixed total. This situation is extremely common in petrology: rocks are often described in terms of the percentages of constituent minerals or oxides. The problem in bivariate statistics is that the values of any two such variables are not, in principle, independent: regardless of the underlying geological process, the value of one variable will automatically tend to affect the values of the other variable. This is best illustrated by an extreme example: if we are studying a suite of felsic intrusions consisting entirely of quartz and feldspar, we might record, for each intrusion, the percentage of quartz and the percentage of feldspar. It is obvious in this situation that percentage quartz + percentage feldspar = 100%, so that a scatter of such data lies along a straight line and has a correlation of −1 (Fig. 3.5a). This analytical result is mundane and tells us nothing about the geological situation. This is an extreme example of closed data, but the same problem occurs in more realistic situations. Suppose samples of igneous rocks are divided into three components, quartz, feldspar and other constituents, and that 'other constituents' never constitute more than 50%. This still constrains the quartz–feldspar data scatter to a band with a negative slope which will almost certainly give a spurious induced negative correlation coefficient (Fig. 3.5b). The problem here is that there is a poor corresondence between the observations and the inferred processes: if we add quartz to the system, keeping all else constant, the feldspar

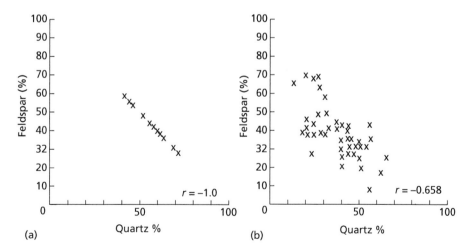

Fig. 3.5 Spurious negative correlations in closed data. (a) Scatter of points resulting from random amounts of two components (quartz and feldspar) of a two-component system, with amounts expressed as percentages. The scatter is obviously constrained to a line where quartz + feldspar = 100%. (b) Scatter of points of two components of a three-component system, where the amounts of all three components are random except that the third component is constrained to be less than 50%. The result appears to show a negative correlation.

percentage declines, and the negative correlation indicates to the naïve interpreter a process in which quartz is substituted for feldspar. In fact, of course, feldspar is constant but the total volume changes. Such a total volume change would be completely undetectable in real rocks.

A graphical representation of the closure problem in igneous petrology is the Harker diagram, where the percentage of an oxide is plotted against percentage of SiO_2. Trends on such diagrams tend to be dominated purely by the variation in SiO_2, with the other oxide showing an induced negative correlation.

In contrast, a palaeoecological study on the constituents of a benthic community might involve recording the frequency of a species as the percentage of the total number of fossils found in a $1\,m^2$ quadrat. In this case we may still get spurious induced negative correlations, but here we have the option, not available in petrology, of rejecting the use of percentage data and using the raw absolute counts to convey frequencies. Unless the fossils are so abundant that it would be difficult to cram any more into a square metre, the raw data are not closed: the addition of more specimens of one species will not necessarily reduce the value ascribed to any other species.

Spurious correlations with closed data are not exclusively negative. Compositional data expressed in small percentages (or ppm), for example in trace element geochemistry, are likely to yield induced positive correlations. For example, if we halve the absolute amount of a major constituent such as SiO_2, while keeping absolute amounts of all other constituents constant, then the percentage or ppm of all other constituents will inevitably increase: any pair of minor constituents, then, will tend to appear to be positively correlated in a sample with varying SiO_2.

Box 3.3 shows the effect of closure on some simulated open data. The data were generated with various inherent correlations; recalculation as percentages corrupts these in various ways. If we imagine the data in the context of stream sediment geochemistry, the original correlation between a pair of variables could be due to their association in specific source rocks, with this provenance being variably represented in different parts of the area.

In petrology and many other fields of geology, it is impossible to escape dealing with closed data, so guidelines are needed on how to detect real bivariate relationships. Various approaches have been used and some have become standard procedures.

Ratios

Any ratio between two variables is, in principle, open. Petrologists have suggested using ratios such as x/SiO_2 or x/K, but these are clearly not adequate solutions because the ratio will be dependent on variation in the variable chosen as the denominator: positive correlations are likely if the

Box 3.3 Closed data: demonstration using simulated data

An open data set of 100 items and 13 random variables was simulated by, for each variable, specification of the means and standard deviations and correlations with other variables. The following is an extract:

Variables

1	2	3	4	5	6	7	8	9	10	11	12	13
46.9	5.11	4.55	10.78	4.25	0.07	0.07	0.02	0.00	0.08	0.15	0.06	0.02
97.1	3.00	3.93	16.96	10.18	0.20	0.04	0.01	0.03	0.06	0.14	0.11	0.09
97.2	5.73	2.71	16.71	10.09	0.15	0.04	0.03	0.03	0.07	0.17	0.07	0.04
21.2	6.24	2.92	10.38	7.54	0.04	0.06	0.02	0.03	0.07	0.15	0.05	0.05
144.1	6.28	2.12	19.47	4.14	0.19	0.06	0.02	0.02	0.06	0.14	0.06	0.03
70.5	3.30	3.22	15.51	3.33	0.12	0.04	0.02	0.03	0.01	0.15	0.00	0.01
42.0	6.63	3.16	5.50	5.53	0.05	0.04	0.03	0.03	0.07	0.15	0.09	0.04
136.7	6.75	4.30	21.02	2.35	0.17	0.05	0.02	0.03	0.10	0.16	0.12	0.01
111.4	3.57	1.84	20.99	8.66	0.16	0.04	0.02	0.03	0.07	0.17	0.07	0.06
108.3	6.17	0.35	19.01	11.29	0.14	0.02	0.02	0.03	0.03	0.18	0.05	0.05

The first column is the major component and columns 6–13 simulate minor or trace components. The units are the same for each column but are otherwise irrelevant. Each row can be imagined as being a stream sample site and each column a mineral, oxide or element. These open data could represent the quantities of each component being deposited at each site in a given time interval – information not normally available to geologists in the real world! The varying amounts and correlations represent different sedimentation rates and different sediment source areas. The correlations are as follows:

	1	2	3	4	5	6	7	8	9	10	11	12
2	0.036											
3	0.014	−0.083										
4	0.851	0.065	−0.064									
5	−0.009	−0.413	−0.044	−0.018								
6	0.764	0.015	−0.010	0.693	0.046							
7	−0.078	−0.043	0.043	−0.032	0.093	−0.135						
8	−0.121	0.144	0.055	−0.065	−0.133	−0.192	−0.121					
9	0.069	0.067	−0.037	0.042	0.060	0.217	−0.620	0.045				
10	0.074	0.462	0.053	0.151	−0.264	0.177	−0.014	−0.014	−0.052			
11	0.098	0.555	−0.231	0.125	−0.229	0.017	−0.046	0.028	0.069	0.142		
12	0.106	0.271	0.096	0.087	−0.183	0.208	0.011	−0.177	−0.072	0.733	−0.007	
13	0.159	−0.347	−0.035	0.138	0.729	0.178	0.101	−0.118	0.039	−0.054	−0.265	−0.044

If such a situation were to be sampled by a geologist, the data for each variable would be recorded as closed data – each variable being recorded as a proportion of the rock sample, perhaps with the total adding to 100%.

continued on p. 158

Box 3.3 *Continued*

Although the geological situation is reflected in the absolute magnitudes of components in the open data, the geologist collecting the sediment inevitably loses this information and is left with only relative quantities of components.

If we calculate percentages for the simulated open data, assuming there are no other components, we obtain the following simulated closed data (extract comparable to above):

Variables

1	2	3	4	5	6	7	8	9	10	11	12	13
65.1	7.1	6.3	15.0	5.9	0.09	0.09	0.02	0.01	0.11	0.20	0.08	0.03
73.7	2.3	3.0	12.9	7.7	0.15	0.03	0.01	0.02	0.04	0.10	0.08	0.07
73.1	4.3	2.0	12.6	7.6	0.11	0.03	0.02	0.02	0.06	0.13	0.05	0.03
43.5	12.8	6.0	21.3	15.5	0.08	0.13	0.04	0.06	0.14	0.31	0.10	0.10
81.6	3.6	1.2	11.0	2.3	0.11	0.03	0.01	0.01	0.03	0.08	0.03	0.02
73.2	3.4	3.3	16.1	3.5	0.13	0.04	0.02	0.03	0.01	0.16	0.00	0.01
66.4	10.5	5.0	8.7	8.7	0.08	0.06	0.04	0.04	0.11	0.23	0.14	0.07
79.6	3.9	2.5	12.2	1.4	0.10	0.03	0.01	0.02	0.06	0.09	0.07	0.01
75.7	2.4	1.3	14.3	5.9	0.11	0.03	0.01	0.02	0.05	0.11	0.05	0.04

The closed data have the following correlations:

	1	2	3	4	5	6	7	8	9	10	11	12
2	−0.690											
3	−0.683	0.488										
4	−0.595	0.272	0.153									
5	−0.807	0.292	0.485	0.259								
6	−0.011	−0.081	−0.053	0.127	0.003							
7	−0.759	0.537	0.629	0.343	0.628	−0.118						
8	−0.687	0.607	0.541	0.381	0.454	−0.079	0.465					
9	−0.553	0.449	0.379	0.291	0.421	0.274	0.189	0.504				
10	−0.649	0.736	0.449	0.399	0.291	0.154	0.472	0.525	0.407			
11	−0.817	0.826	0.603	0.351	0.552	−0.067	0.732	0.649	0.540	0.643		
12	−0.464	0.597	0.461	0.155	0.197	0.183	0.362	0.377	0.300	0.837	0.481	
13	−0.632	0.134	0.327	0.272	0.819	0.050	0.477	0.395	0.297	0.256	0.335	0.161

Comparing this with the open data correlation matrix, it is clear that closure has corrupted the original correlation information. The major component (1) now appears to be inversely related to most variables, simply because more of it leaves less room for anything else! Other variables appear more correlated with each other, as the major component has a similar diluting effect on all of them.

Box 3.3 *Continued*

The log-ratio transform applied to the closed data, using the major constituent as the divisor, gives the following data:

2	3	4	5	6	7	8	9	10	11	12	13
−2.2	−2.3	−1.5	−2.4	−6.5	−6.6	−8.0	−9.2	−6.4	−5.8	−6.7	−7.7
−3.5	−3.2	−1.7	−2.3	−6.2	−7.8	−9.1	−8.1	−7.4	−6.6	−6.8	−7.0
−2.8	−3.6	−1.8	−2.3	−6.5	−7.7	−8.1	−8.2	−7.2	−6.4	−7.2	−7.7
−1.2	−2.0	−0.7	−1.0	−6.4	−5.8	−7.0	−6.5	−5.8	−5.0	−6.1	−6.0
−3.1	−4.2	−2.0	−3.5	−6.6	−7.8	−8.8	−8.9	−7.9	−6.9	+7.8	−8.5
−3.1	−3.1	−1.5	−3.1	−6.4	−7.5	−8.0	−7.7	−8.9	−6.2	−11.2	−9.0
−1.8	−2.6	−2.0	−2.0	−6.7	−7.0	−7.4	−7.3	−6.4	−5.7	−6.2	−6.8
−3.0	−3.5	−1.9	−4.1	−6.7	−7.9	−8.7	−8.5	−7.2	−6.7	−7.0	−9.2
−3.4	−4.1	−1.7	−2.6	−6.6	−7.8	−8.4	−8.4	−7.4	−6.5	−7.4	−7.5
−2.9	−5.7	−1.7	−2.3	−6.6	−8.4	−8.5	−8.1	−8.1	−6.4	−7.8	−7.6

This has the following correlation matrix:

	2	3	4	5	6	7	8	9	10	11	12
3	0.498										
4	0.545	0.397									
5	0.223	0.385	0.410								
6	0.186	0.189	0.317	0.134							
7	0.579	0.551	0.564	0.456	0.154						
8	0.661	0.533	0.537	0.372	0.136	0.539					
9	0.427	0.268	0.373	0.272	0.359	0.089	0.399				
10	0.483	0.266	0.381	0.034	0.239	0.262	0.272	0.159			
11	0.847	0.586	0.640	0.440	0.222	0.712	0.710	0.473	0.334		
12	0.434	0.316	0.278	0.063	0.246	0.278	0.193	0.088	0.535	0.371	
13	0.198	0.322	0.344	0.618	0.133	0.448	0.283	0.273	0.151	0.377	0.185

Although this has not retrieved the original open data structure, the transformation has been shown to allow tests on the correlations and is generally useful: see Sections 2.1.4.4 and 8.5.1.

denominator is a major component (like SiO_2). Probably the most common approach is to use as the denominator a variable which is independent of both the major 'diluting' components and the geological process being investigated. This attempts to reduce the situation to a geological, rather than a statistical, problem. There remains, though, the serious problem of circularity: needing to make assumptions about the geological processes in advance, when these assumptions will affect the results, and the results are intended to elucidate the geological processes.

Remaining space

Niggli numbers are ratios of (component percentage)/(100 − SiO_2 percentage). These clearly only remove the 'dilution' effect of SiO_2; there are likely to be other components which have a diluting effect on the remaining space after SiO_2 is removed.

A similar approach is to compare two components x and y by comparing x' and y' where $x' = x/(100 − y)$ and $y' = y/(100 − x)$. This removes the direct effect that one component has on the other due to closure, but does not remove the 'dilution' effect of a third component.

Principal components (PCs)

Principal components analysis (PCA) is a multivariate technique (Section 8.5) which finds the dominant directions of variation in a data set on the basis of the covariance matrix. It might be hoped that PCs can be identified which correspond to the intercorrelations induced by closure, and that these can be separated from other variation. However, there are complications: the trends produced by closure are not linear, and the various real and spurious sources of correlation are intimately mixed.

Hypothetical open arrays

We can conceptualise some original open data (the *basis*), with correlations generated by the geological process, which, when closed, becomes the composition that we observe. Some authors have thought it possible to reconstruct some properties of the basis from the compositions, but this has since been discredited.

Log-ratio transformation

Aitchison (1984, 1986) has demonstrated that the transformation $x' = \ln(x/y)$, where y is a component (e.g. SiO_2) used as the denominator for all variables, yields an open data array which can be tested against the null hypothesis of no correlations in the basis (complete subcompositional independence). If this is rejected, individual correlations can be tested. Unfortunately, the open data array does not reconstruct the hypothetical original open data, and the statistical test is consequently complex and requires special software! Nevertheless, the log-ratio transformation is recommended for routine investigations of closed data.

3.3   BIVARIATE REGRESSION

In bivariate statistics, regression allows the shape of the data scatter to be described and tested by the process of fitting a line through the data points.

Although the correlation coefficient is a useful convenient measure of linear relationship between two variables, regression is often to be preferred because: (a) rigorous statistical testing can be obtained with non-normally distributed data (though there are other constraints); (b) a variety of curvilinear, as well as rectilinear, relationships can be tested; and (c) the nature of the bivariate relationship can be more precisely defined in the form of an equation.

The latter point is probably the most important in geology: the derived equation can be used to describe and aid understanding of the geological process, and permits predictions to be made. Prediction can be done by

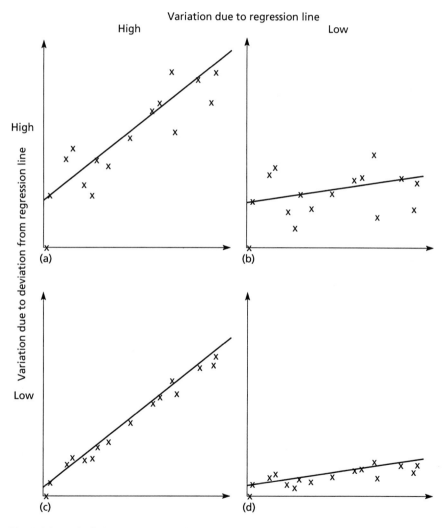

Fig. 3.6 Hypothetical regression situations resulting from combinations of high and low variation due to the regression line and due to deviation or error. In these examples, (a), (c) and (d) are significant regressions (especially (c)) but in (b) the deviation overwhelms the slight gradient so that the regression is not significant.

interpolation and extrapolation, or by applying the equation to new situations where unknown values of one variable are needed from known values of another. It is inherent to the subject of geology that most data are literally buried and difficult to access, so prediction on the basis of partial information is of great importance. In regression, the best-fit line is fitted in such a way as to minimise the deviation of the data points from the line. In general, the total variation in the data can be divided into the variation due to the line, and the variation due to the deviation of the data points from the line, or error (see Fig. 3.6). Although the process of regression always results in the error being minimised, there are a variety of ways in which the error can be defined.

1 The error can be regarded as residing in only one of the two variables – this is classical regression: see Section 3.3.1.

2 The error can be regarded as inherent in both variables – this is structural regression: see Section 3.3.2.

3 The error in one or both variables may be known in advance by replication of measurements – this is the weighted least-squares method: see Section 3.3.3.

Note that regression may also be used to fit surfaces to three-dimensional scatters of spatial data (trend surface analysis: see Section 7.3) and to investigate relationships between a larger number of variables (multiple regression: see Section 8.3).

3.3.1 **Classical regression**

In classical regression, the two variables are conceptually very different.

1 The independent or regressor or predictor variable. This is conventionally symbolised by x and placed on the horizontal axis (abscissa) of a graph. Normally, this variable is accurately measured and is not regarded as incorporating any error: sometimes its values are specified by the user, for example a geologist may decide the distances (x) along a traverse at which measurements of a variable (y) will be taken.

2 The dependent or regressed or predicted variable. This is conventionally symbolised by y and is placed on the vertical axis (ordinate). It is usually not so accurately measurable and all of the deviation of the data points from the regression line, the error, may be regarded as being in this direction. In geology, the x variable is often an easily measurable property such as distance or size and the y variable is often a more specifically geological measurement such as grain size or the amount of an element or oxide. The error in y may relate to the accuracy of the instrumentation used in the measurement process or error due to sampling from the geological population. However, where there is error in x, classical regression is still sometimes used if y can be regarded as a function of x (e.g. carbonate porosity as a function of magnesium percentage due to dolomitisation) or if we wish to predict y from x (e.g. clay percentage from gamma log). The y

variable, for example sandstone permeability, may be difficult and expensive to measure in comparison with the x variable, for example an electric well log measurement: in such a case the objective may be to derive an equation for y in terms of x so that y might be predicted just on the basis of the cheaper x measurement in future.

3.3.1.1 **Classical linear regression**

In classical linear regression, a best-fit straight line is sought. The general equation for the relationship in the source population is:

$$y_i = \beta_0 + \beta_1 x_i + \varepsilon_i$$

β_0 and β_1 are known as the coefficients of the equation: β_0 gives the intercept of the line on the y axis (where $x = 0$) and β_1 gives the gradient of the line. ε_i is the deviation of the point from the line.

The equation of the best-fit regression line is:

$$y = b_0 + b_1 x$$

b_0 and b_1 are the sample-based estimates of β_0 and β_1; the problem is to find b_0 and b_1 so that the total error $\Sigma \varepsilon_i^2$ is minimised. The error measure to be minimised is called the sum of squares due to deviation (SS_D):

$$SS_D = \sum_{i=1}^{n} (\hat{y}_i - y_i)^2$$

where $\hat{y}_i$ is the estimate of y: the value of y where a vertical line through the point hits the regression line.

It can be shown that the problem resolves into the solution of simultaneous equations:

$$\sum_{i=1}^{n} y_i = b_0 n + b_1 \sum_{i=1}^{n} x_i$$

$$\sum_{i=1}^{n} x_i y_i = b_0 \sum_{i=1}^{n} x_i + b_1 \sum_{i=1}^{n} x_i^2$$

This is a problem which is readily solved using standard computer software. The results of a worked example are shown in Box 3.4.

However, calculating b_0 and b_1, and hence finding the regression line, is not an adequate end in itself: clearly, these numbers could be calculated for any scatter of points, even a totally random scatter with no linear trend. We still need to know, then, whether or not the data scatter has a trend which is significantly different from what could be expected in a random scatter, or, in statistical terms, whether or not the data could have been drawn without bias from a population with $\beta_1 = 0$.

As mentioned at the beginning of Section 3.3, regression can be regarded as partitioning the total variation in the data into two sources: variation due

Box 3.4 Worked example: regression equation

The percentage of copper in samples of vein material varies along a mine adit (Appendix 3.1). What is the best-fit straight line through the data?

Distance is the obvious choice for the independent (x) variable and copper for the dependent (y) variable: we wish to ascertain how the copper percentage changes as a function of distance.

A standard statistics package produced the following coefficients:

$b_0 = 0.336 \quad b_1 = 0.0303$

so the equation of the line is:

$y = 0.336 + 0.0303x$

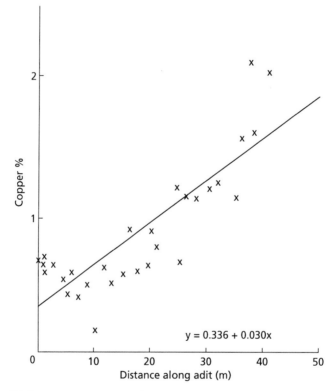

Fig. B3.4.1

The line is shown superimposed on the scatter graph: note that b_0 is the intercept on the y axis and b_1 is the gradient. Regardless of the quality of the fit (see Box 3.5) the line should not be used for prediction: extrapolation of the line will clearly take it to values of copper greater than 100% and less than 0%!

to the regression line and variation due to deviation from the regression line. The former can be regarded as due to the causative geological process, the latter as due to randomness. These two types of variation can be readily quantified from the raw data and calculated regression line.

Sum of squares due to regression:

$$SS_R = \sum_{i=1}^{n} (\hat{y}_i - \bar{y}_i)^2$$

Sum of squares due to deviation:

$$SS_D = \sum_{i=1}^{n} (\hat{y}_i - y_i)^2$$

This is normally calculated by $SS_D = SS_T - SS_R$ where the total sum of squares:

$$SS_T = \sum_{i=1}^{n} (y_i - \bar{y}_i)^2$$

The goodness-of-fit statistic R^2 is often used to convey the quality of a regression:

$$R^2 = SS_R/SS_T$$

The significance of the regression line depends on the ratio of the variances from the two sources: the significance is greater with higher variance due to the regression and lower variance due to error (Fig. 3.6). This is the same situation as that encountered in multisample univariate statistics (Section 2.6) and the same technique, analysis of variance (ANOVA), is used here. These calculations are standard on statistics software packages.

ANOVA for linear regression

Source	d.f.	Sums of squares	Mean squares	F-test
Regression	1	SS_R	$MS_R = SS_R$	MS_R/MS_D
Error	$n - 2$	SS_D	$MS_D = SS_D/(n - 2)$	
Total	$n - 1$	SS_T		

H_0: $\beta_1 = 0$ (no gradient on line)
H_1: $\beta_1 \neq 0$ (line has gradient)

As usual, MS values are found by dividing SS values by the degrees of freedom. (All values are normally provided by statistical software).

If the calculated F exceeds the critical F (from tables, Appendix 2.5) with degrees of freedom $= 1, n - 2$, we reject the null hypothesis.

Worked example: see Box 3.5.

Warnings

1 Points which are outliers (particularly with respect to the independent variable, x) can have an undue effect on the regression. This is similar to the effect of outliers on the correlation coefficient (Fig. 3.3). Many statistics packages draw the user's attention to such outliers, and their influence should be ascertained by removing them and repeating the regression (see Fig. 3.7).

2 The difference in the y direction between a point and the regression line is known as the residual. The validity of the regression statistics depends on the distribution of the residuals: (a) the residuals must be normally distributed; (b) the residuals must be homoscedastic: this means that there is no trend in the distribution of variance along the line; and (c) the residuals must not be autocorrelated (see Section 6.3). It appears to be the case in geology that these criteria are seldom checked. However, it is only where there are large departures in these respects, or where there is a result of borderline significance, that these become important. Nevertheless, it is always advisable to investigate residuals to check that the straight line model, rather than curvilinear, is appropriate (see Box 3.5).

3 Regardless of how significant the regression, care should be taken in using the resulting equation for prediction. In particular, extrapolation beyond the range of the original data values is not advisable. To justify this, it must be arguable that the equation is inherent in the causative geological process.

3.3.1.2 **Curvilinear regression**

Curves may be fitted to data by two methods:

1 transforming variables and then using linear regression; or
2 polynomial regression.

Transformation of variables

Certain types of curved data trend can be readily modelled by simple transformations applied to the original data, for example log, reciprocal or power transformation, followed by simple linear regression. There are a variety of transformations available but it is not always clear in advance which are likely to prove useful. A trial-and-error approach can be used, though the choice of trial transformation should be moderated by the geologist's understanding of the processes involved. It is often the case that the best transformation is the same one that is required to normalise the univariate data. For example, the log transform should be a first choice if trace element data are involved (see Section 2.1.4.4). Some transformations, however, should be avoided if they are meaningless when applied to the type of data involved; for example, measurements of temperature in Celsius

Box 3.5 Worked example: significance of linear regression using analysis of variance

Using the same data as in Box 3.4, is there a significant linear relationship between distance along adit and copper?

H_0: $\beta_1 = 0$
H_1: $\beta_1 \neq 0$

The results from a statistics package are as follows.

ANOVA table

Source	d.f.	SS	MS	F
Regression	1	4.4452	4.4452	64.67
Error	28	1.9245	0.0687	
Total	29	6.3697		

$R^2 = 0.698$
Degrees of freedom: 1, 28
$\alpha = 0.05$

From tables (Appendix 2.5),

Critical $F = 4.2$

Calculated F exceeds critical F so we reject the null hypothesis: the slope is significantly different from zero.
    Conclusion: there is a significant linear relationship between distance and copper.

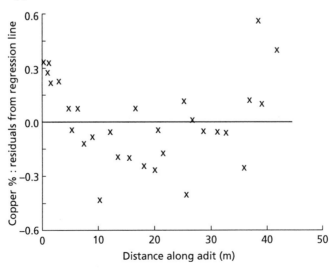

Fig. B3.5.1

continued on p. 168

> **Box 3.5** *Continued*
>
> However, the scatter plot (Box 3.4) and the residuals plot (Fig. B3.5.1) show that a curvilinear fit is likely to be more appropriate: this example is pursued further in Box 3.8 where a curved-line fit is obtained using polynomial regression. Also, the residuals fail the criterion of homoscedasticity: there is greater deviation to the right of the plot.

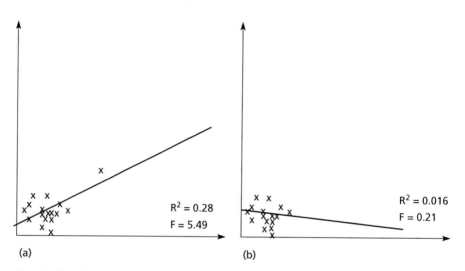

(a)   (b)

Fig. 3.7 The effect of outliers on regression: the outlying point in (a) is solely responsible for the apparently significant regression. After removal, (b), the significance vanishes. Critical $F = 4.49$.

should not be subject to the log transformation (and many others) because it is not a ratio scale measurement: the zero point is arbitrary (see Section 1.3). Using the trial-and-error approach, the best transformation is chosen by finding the highest goodness-of-fit or highest F value in the ANOVA.

Worked example: see Box 3.6.

Polynomial regression

Polynomial regression may be appropriate if the data scatter shows certain types of more complex curvature, for example parabolic. The equations fitted to the data in this method are polynomial equations with terms for x, x^2, x^3, etc. Each power of x has an associated coefficient, and the regression must be solved for all coefficients. Regression problems such as this with more than one independent variable must be solved using multiple regression

Box 3.6 Worked example: curvilinear regression using transformations

Using data on temperature and age of major oilfield source rocks (Appendix 3.2), what equation best describes temperature as a function of age? Of the enormous number of possibilities, four are shown in Fig. B3.6.1: temperature vs. age, temperature vs. ln(age), temperature vs. age^2 and temperature vs. 1/age. For each, the F value resulting from ANOVA (as in Box 3.5) are shown. The critical F ($\alpha = 0.05$, degrees of freedom = 1, 16) is 4.49, so all regressions are significant. However, the F values are directly comparable so the equation involving ln(age) is the best fit, having the largest F and hence the greatest significance.

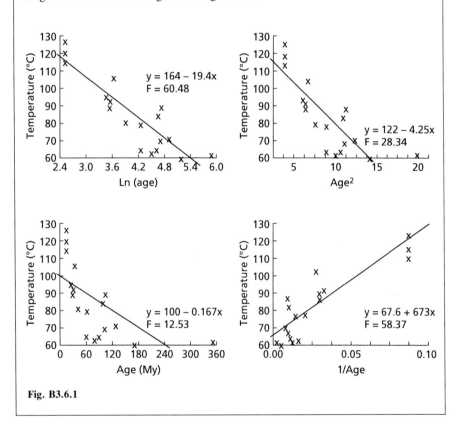

Fig. B3.6.1

techniques (see Section 8.3). Polynomial regression is just a special case of multiple regression where the independent variables are powers of x.

After the linear regression, the simplest polynomial is the quadratic, which allows parabolic curvature and has the general equation

$$y = b_0 + b_1 x + b_2 x^2$$

This clearly just involves the addition of the x^2 term to the linear equation. If we test a quadratic equation for significance, the hypotheses are:

H_0: $\beta_1 = 0$ and $\beta_2 = 0$
H_1: β_1 and/or $\beta_2 \neq 0$

If a linear regression with the same data has already been established as significant, then we will have already rejected H_0: $\beta_1 = 0$, so we will inevitably reject the quadratic regression H_0 cited above. In this case, we only need to test the β_2 coefficient.

    Increasing the number of terms in the polynomial gives us the cubic equation:

$$y = b_0 + b_1 x + b_2 x^2 + b_3 x^3$$

and the quartic:

$$y = b_0 + b_1 x + b_2 x^2 + b_3 x^3 + b_4 x^4$$

and so on. Each extra term gives us an extra sense of curvature: quadratic curves have no inflection points, cubic curves permit one inflection, quartic two inflections, etc.

ANOVA for quadratic regression

Source	d.f.	Sums of squares	Mean squares	F-test
Quad. regr.	2	SS_{R2}	$MS_{R2} = SS_{R2}/2$	(1) MS_{R2}/MS_{D2}
Linear regr.	1	SS_{R1}		
Increase	1	$SS_{R2-1} = SS_{R2} - SS_{R1}$	$MS_{R2-1} = SS_{R2-1}$	(2) MS_{R2-1}/MS_{D2}
Quad. error	$n - 3$	SS_{D2}	$MS_{D2} = SS_{D2}/(n - 3)$	
Total	$n - 1$	SS_T		

Test 1:

H_0: $\beta_1 = 0$ and $\beta_2 = 0$
H_1: β_1 and/or $\beta_2 \neq 0$
Degrees of freedom = 2, $n - 3$

Test 2:

H_0: $\beta_2 = 0$
H_1: $\beta_2 \neq 0$
Degrees of freedom = 1, $n - 3$

Calculations normally need to be done from SS values provided by statistical software. H_0 is rejected if the calculated F exceeds the critical F (from tables, Appendix 2.5) using the appropriate degrees of freedom.

This ANOVA can be generalised for other polynomials, as follows.

ANOVA for polynomial regression (For regression of order k, where k is the number of terms in the equation, excluding the constant b_0 term.)

Source	d.f.	Sums of squares	Mean squares	F-test
kth Regr.	k	SS_{Rk}	$MS_{Rk} = SS_{Rk}/k$	(1) MS_{Rk}/MS_{Dk}
$(k-1)$th Regr.	$k-1$	SS_{Rk-1}		
Increase	1	$SS_{RI} = SS_{Rk} - SS_{Rk-1}$	$MS_{RI} = SS_{RI}$	(2) MS_{RI}/MS_{Dk}
kth Error	$n-k-1$	SS_{Dk}	$MS_{Dk} = SS_{Dk}/(n-k-1)$	
Total	$n-1$	SS_T		

Test 1:

H_0: all of $\beta_1, \beta_2, \beta_3, \ldots, \beta_k = 0$
H_1: at least one of the above $\neq 0$
Degrees of freedom $= k, n - k - 1$

Test 2:

H_0: $\beta_k = 0$
H_1: $\beta_k \neq 0$
Degrees of freedom $= 1, n - k - 1$

Worked example: see Box 3.7.

Warnings (in addition to those listed for linear regression)

1 Whereas linear and quadratic terms are inherent in many physical processes, higher orders of polynomial are less likely to be natural models for geological situations, so good fits may be coincidental.
2 High orders of polynomial can form extreme gradients which will yield implausible estimates if extrapolated.

If we try each equation successively, we only need to assess the significance of the additional term. Usually, we rapidly come to the point where the addition of new terms adds nothing useful to the equation.

3.3.2   Structural regression

In structural regression, the two variables are treated equally: error is regarded as residing in both variables. This is often the case in geology, particularly in geochemistry and palaeontology: if a relationship is sought between, say, tungsten and gold in a geochemical soil survey or between length and width in a trilobite population, it is clear that the variables

Box 3.7 Worked example: polynomial regression

Using the same data as Boxes 3.4 and 3.5, can we improve the fit of the regression line by using quadratic and cubic models?

Quadratic regression

The regression equation is found from a statistics package to be:

$$y = 0.624 - 0.0225x + 0.00136x^2$$

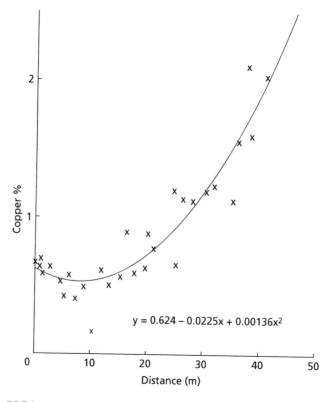

Fig. B3.7.1

ANOVA table

Source	d.f.	SS	MS	F
Quad. regr.	2	5.5193	2.7596	87.61
Lin. regr.	1	4.4452		
Increase	1	1.0741	1.0741	34.098
Quad. error	27	0.8505	0.0315	
Total	29	6.3697		

Box 3.7 *Continued*

The information for the linear regression comes from Box 3.5. Test for significance of increase in fit for quadratic over linear:

H_0: $\beta_2 = 0$
H_1: $\beta_2 \neq 0$
Degrees of freedom $\nu = 1, 27$
$\alpha = 0.05$

From tables (Appendix 2.5):

Critical $F = 4.21$

Calculated F exceeds critical F so we reject H_0 and conclude that the quadratic equation gives a significantly better fit than the linear.

Cubic regression

The regression equation is found from a statistics package to be:

$$y = 0.656 - 0.0351x + 0.00220x^2 - 0.000014x^3$$

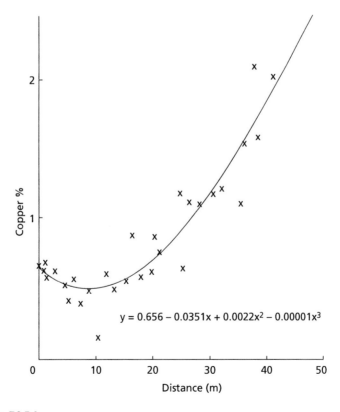

Fig. B3.7.2

continued on p. 174

Box 3.7 *Continued*

ANOVA table

Source	d.f.	SS	MS	F
Cubic regr.	3	5.5311	1.8437	57.16
Quad. regr.	2	5.5193		
Increase	1	0.0118	0.0118	0.365
Cubic error	26	0.8386	0.0323	
Total	29	6.3697		

Test for significance of increase in fit for cubic over quadratic:

H_0: $\beta_3 = 0$
H_1: $\beta_3 \neq 0$
Degrees of freedom $v = 1, 26$
$\alpha = 0.05$

From tables (Appendix 2.5):

Critical $F = 4.23$

Calculated F does not exceed critical F so we fail to reject H_0 and conclude that the cubic term does not give a significant improvement in fit.

We have found in Box 3.5 that the linear term was significant, and it is consequently inevitable that the quadratic and cubic regressions, which contain the linear term, will also be significant, and this is reflected in the high values of the upper of the two F ratios in each ANOVA table. However, we have found by testing for significance of increase in fit that the quadratic equation is the most efficient for modelling the data.

Note, though, that the quadratic curve can only have any validity within the range of the x values: even more than the linear, it will soon extrapolate to give predictions of copper $> 100\%$!

cannot be regarded as different in type. Two approaches are available: the major axis and the reduced major axis (RMA).

These types of regression are often used for analysis of isometry and allometry: see Box 3.8.

Major axis

The major axis line through the bivariate data scatter minimises the sum of the direct distances (perpendicular to the line) from the line to each point (Fig. 3.8). The major axis is equivalent to the first PC: this is fully discussed in Section 8.5.

Box 3.8 Isometry and allometry

Linear relationships between two variables can be described in terms of isometry and allometry. Although originally intended to describe shape change with growth in palaeontology, the concepts are usefully applicable in the general case. An isometric relationship between two variables is one in which the ratio between them is constant: the regression line is rectilinear and passes through the origin. If the data are measurements of fossils from a population, this means no change in shape with growth. The alternatives, curved lines or non-zero intercepts, are anisometric. In palaeontology, these show change in shape with growth.

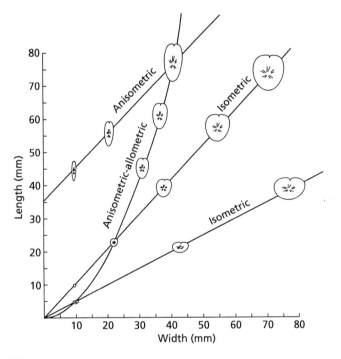

Fig. B3.8.1

A commonly occurring anisometric relationship is conveyed by the allometric equation:

$$y = ax^b$$

This can readily be investigated by plotting and regressing $\log(y)$ vs. $\log(x)$, as

$$\log(y) = \log(a) + b \log(x)$$

so the gradient of the regression yields b and the intercept yields $\log(a)$.

Relationships of the form $y = ax^b$ are not unusual in nature, so this

continued on p. 176

Box 3.8 *Continued*

transformation is often worth investigating. In palaeontology, the classic allometric relationship is between bone length and bone width of large vertebrates. The cross-sectional area of the bone, related to width^2, determines its strength, and the strength of the bone evolves in relation to the weight it supports. As this weight is proportional to volume and volume is proportional to measurements of length^3, the bone length vs. width relationship typically contains values of b significantly different from 1.

Reduced major axis

In this method, the criterion by which the regression line is calculated is the minimisation of the sum of the areas of triangles between the data points and the line (see Fig. 3.8). Although this sounds complicated, it can be shown that the gradient b_1 is simply calculated by:

$$b_1 = s_x/s_y$$

As with other regression lines, the line passes through the joint means of x and y, so that the intercept b_0 needed to complete the regression equation is:

$$b_0 = y - b_1x$$

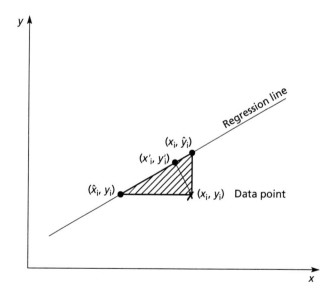

Fig. 3.8 Regression criteria: in classical regression, the sum of the errors $(\hat{y}_i - y_i)^2$ is minimised; in the major axis method the sum of distances from (x_i, y_i) to (x'_i, y'_i) is minimised; in the reduced major axis method, the sum of the areas of the triangles is minimised.

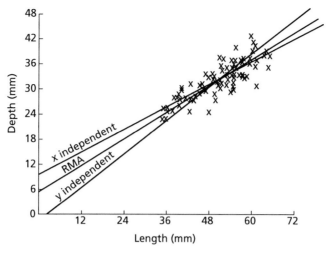

Fig. 3.9 Comparison of RMA regression with classical regression (x on y and y on x) applied to length and depth measurements of the echinoid *Micraster*. The three lines intersect at the joint mean.

The RMA line is shown compared with other regression lines in Fig. 3.9.

With elliptical-shaped data scatters having a vertical or horizontal long axis, RMA will obviously yield a b_1 coefficient related to the axial ratio of the ellipse, regardless of the fact that there is no real gradient. This regression result would be deceptive, so a test of significance is required. A rough test is:

$$s_e = b_1 \sqrt{\frac{1 - r^2}{n}}$$

s_e is the standard error of b_1: if the range $b_1 + 1.96s_e$ to $b_1 - 1.96s_e$ encloses zero (i.e. zero gradient) then the regression is not significant at the 5% level.

3.3.3 Weighted least squares methods

This technique requires that there are independent data on the error in the x and/or y directions, either from replicated measurements or from knowledge of the accuracy of a measurement device. It is primarily useful in achieving a much greater precision in the result, and consequently it is extensively used in geochronology, where the date given can depend sensitively on the regression line and a measure of the error in the result is demanded.

Clearly, a regression line should be constrained to go nearer to the points having the least error (Fig. 3.10a), and a modification of the classical linear method is available to calculate this. However, this is hardly used at all in geology, as error in the y variable can usually be assumed to be consistent across the range of x, and because the additional effort of repli-

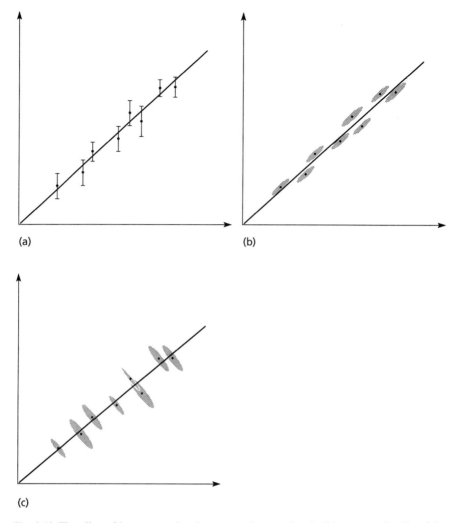

(a)

(b)

(c)

Fig. 3.10 The effect of known correlated errors on the error involved in a regression line: (a) Error in y only, shown by standard deviation error bars. (b) Error in x and y, shown by standard deviation ellipses, are positively correlated. The regression line has high error: it does not intersect many of the error ellipses. (c) Error in x and y, shown by standard deviation ellipses, are negatively correlated. The regression line has low error: it intersects all of the error ellipses.

cating measurements at relatively few values of x is usually better spent on recording y at additional values of x (which will also give information on error).

In geochronology, the ratio of two measured isotopes (e.g. Rb and Sr) is the required result, but each variable has the same error status, so classical regression cannot be used. The situation is further complicated by the need to consider not just the error in each variable but the covariance between the errors. If errors are correlated parallel to the direction of the regression line, the error in the regression will be greater than if the errors are correlated perpendicular to the line (Fig. 3.10b,c). Specialised methods are required for this type of problem.

FURTHER READING

Aitchison J. (1984) The statistical analysis of geochemical compositions. *Mathematical Geology*, **16**(6), 531–564.
Aitchison J. (1986) *The Statistical Analysis of Compositional Data*. Wiley, New York.

these are the key references for geologists who need to get rigorous results from compositional data.

Cardott B.J. and Lambert, M.W. (1985) Thermal maturation by vitrinite reflectance of Woodford shale, Anadarko basin, Oklahoma. *Bulletin of the American Association of Petroleum Geologists*, **69**, 1982–1998.
Ghosh A.K. (1965) A statistical approach to the exploration of copper in the Singhbhum shear zone, Bihar, India. *Economic Geology*, **60**, 1422–1430.
Till R. (1971) Are there geochemical criteria for differentiating reef and non-reef carbonates? *Bulletin of the American Association of Petroleum Geologists*, **55**(3), 523–530.

some examples of diverse applications of bivariate methods.

4   Non-parametric Statistics

4.1   INTRODUCTION

4.1.1   Why use non-parametric statistics?

The parameters which non-parametric methods specifically *don't* use are the parameters of the normal distribution: the mean and standard deviation. Non-parametric methods must be used in the following two circumstances.

1   The measurement scale of the data is ordinal rather than interval or ratio (see Section 1.3). With ordinal scales, the units along the scale are not constant, so the difference between 9 and 10 isn't necessarily the same absolute magnitude as the difference between 1 and 2. Examples are subjective measures of grain roundness and stratigraphic position as a measure of time. Only the relative position of data points is meaningful, and arithmetic operations are not applicable: calculation of mean and standard deviation is therefore redundant.

2   The measurements are on interval or ratio scales but investigation of the frequency distribution shows a marked departure from the normal distribution. The central limits theorem (Section 2.4.5.2) allows us to use parametric methods on quite markedly non-normal distributions, but only when sample sizes are large. Non-parametric methods are valid regardless of sample size and distribution, but with small sample sizes (<20), parametric statistics should only be used if tests such as normal scores or Kolmogorov–Smirnov (KS) (Section 2.5.6) lead to acceptance of the null hypothesis of normal distribution. Remember that various transformations should be applied to the data in an attempt to normalise the data before parametric methods are abandoned.

As the mean has no relevance in non-parametric calculations, tests are based on the median (see Section 2.1.4). Unlike the mean, the median is found without calculations and is based only on relative positions. Non-parametric univariate statistical tests often have a null hypothesis of equality of population medians, and there are non-parametric alternatives to all of the standard parametric tests.

Non-parametric methods are based on ranks of the data values: each value is replaced by a number giving its place in the sequence from highest to lowest value. The calculations are based only on the ranks. In the case of ordinal scale data, the ranks contain all the meaningful information, but with interval and ratio scales there is loss of information.

It is valid to apply non-parametric methods to normally distributed data, and this is often tempting as we can avoid testing for normality and calculations are often easier. However, the loss of information means that the result is conservative: this means that rejection of a null hypothesis is less

Table 4.1 Comparison of parametric and non-parametric methods

	Method	
Data	Parametric	Non-parametric
Normal, large n	Most precise and reliable result possible	If H_0 rejected: result would be the same with parametric test If H_0 accepted: result not reliable
Non-normal, small n	Result completely unreliable: often leads to false rejection of H_0	Best result possible given quality of data

likely, so parametric methods should be used for a precise result. The general applicability of non-parametric methods allows them to be referred to as robust; a conservative result is the inevitable consequence. Table 4.1 provides a summary.

If a data set contains many variables with a variety of shapes of frequency distributions, it is often efficient to apply non-parametric methods to all variables to avoid the time-consuming process of assessing normality. If null hypotheses are rejected, this may be an acceptable result (although the level of significance at rejection may be underestimated). Variables giving failure to reject the null hypothesis can be assessed for suitability of non-parametric methods. Note that transformations make no difference at all to the results of non-parametric tests.

4.1.2   Use of ranks

Ranks are simply the position of a data value in the ordered sequence of highest to lowest data values (or lowest to highest). For a data value x_i, the rank is symbolised by $R(x_i)$. For example:

$x_1 = 4.3$ $R(x_1) = 5$
$x_2 = 9.3$ $R(x_2) = 8$
$x_3 = 0.3$ $R(x_3) = 1$
$x_4 = 2.9$ $R(x_4) = 3$
$x_5 = 3.2$ $R(x_5) = 4$
$x_6 = 7.7$ $R(x_6) = 7$
$x_7 = 5.0$ $R(x_7) = 6$
$x_8 = 0.4$ $R(x_8) = 2$

These have been ranked from lowest to highest, the lowest value having a rank of 1. If we had chosen to rank from highest to lowest, the ranks would obviously be different but non-parametric tests would yield the same result. Notice that, if x_2 had been 9300, its rank would still have been 8: this illustrates the irrelevance of the absolute scale, and the loss of information if we have interval or ratio data.

If there are two or more identical values, ranks are tied, and the rank allocated is the average of the ranks for the pair or group. For example:

x_i	56	42	61	61	42	55	35	42	39	65
$R(x_i)$	7	4	8.5	8.5	4	6	1	4	2	10

Some test statistics require an adjustment to be made if tied ranks occur, but this is not usually crucial.

4.2 UNIVARIATE METHODS

4.2.1 Sign test for hypothetical median

This is the non-parametric analogue of the one-sample t-test. It has the null hypothesis that the data were drawn from a population with a specified hypothetical median. The test might be used to assess whether or not a sample of geochemical measurements could have been drawn from a population with a median compatible with economic viability. The median is defined as the point exactly half-way through the rank order: where the sample size n is odd, it is the value of the point with rank $(n + 1)/2$; where n is even, it is between the values of the points with ranks $n/2$ and $(n + 2)/2$.

Non-parametric tests are derived directly from probability theory, and this is particularly clear in this case. If a sample is drawn from a population with a specified median, each item of data has a prior probability of 0.5 of being larger (or smaller) than that median. This is an identical situation to coin-tossing experiments. The probability of having, say, six out of 20 values greater than the population median is the same as obtaining six heads in 20 tosses of a coin. This is found by use of the binomial probability distribution (see Section 2.3.7).

The only relevant attribute of the ranks of data is whether or not the data are larger or smaller than the hypothetical median. This can be symbolised by the signs + and −. The further the sample median is from the hypothetical population median, the greater the imbalance between the number of + and −, and the less likely it is that the sample came from that population.

Suppose we have the following measurements of mercury in ppm:

56 42 61 61 42 55 35 42 39 65 44 51
32 82 41

and the hypothetical median is 40 ppm. We find that we have 3 − and 12 + values. If we require a two-tailed test (H_0: source population median = 40) with a 5% level of significance, then we can establish a confidence interval in terms of numbers of + and −. If we regard a larger (+) value as a success, we find from tables of the binomial distribution where $p = 0.5$:

probability of 7 successes from 15 trials = 0.19638
      "   " 8 "   " " " = 0.19638

"	"	6	"	"	"	"	= 0.15274
"	"	9	"	"	"	"	= 0.15274
"	"	5	"	"	"	"	= 0.09164
"	"	10	"	"	"	"	= 0.09164
"	"	4	"	"	"	"	= 0.04166
"	"	11	"	"	"	"	= 0.04166

(from Appendix 2.1). The sum of these (0.96484) exceeds 0.95, so there is a less than 5% chance that any other number of successes could have arisen under the null conditions. We can be 95% confident that any other number of + values (including our count of 12) represents a population with a different median.

If we require a one-tailed test, for example where we have an alternative hypothesis that the population median exceeds 40 ppm (this may be an economic threshold), the confidence limit is found as follows:

probability of	0	successes from 15 trials					= 0.00003
"	"	1	"	"	"	"	= 0.00046
"	"	2	"	"	"	"	= 0.00320
"	"	3	"	"	"	"	= 0.01389
"	"	4	"	"	"	"	= 0.04166
"	"	5	"	"	"	"	= 0.09164
"	"	6	"	"	"	"	= 0.15274
"	"	7	"	"	"	"	= 0.19638
"	"	8	"	"	"	"	= 0.19638
"	"	9	"	"	"	"	= 0.15274
"	"	10	"	"	"	"	= 0.09164
"	"	11	"	"	"	"	= 0.04166

The sum of these (0.98242) exceeds 0.95, so the probability of a greater number of successes is less than 0.05 and we reject the null hypothesis: the median of the source population probably exceeded 40 ppm.

The above demonstrations show a rather tedious procedure, but in practice the summing is avoided by the use of tables of cumulative binomials. Furthermore, there is a useful approximation for larger sample sizes.

Sign test for hypothetical median

H_0: population median = specified hypothetical median
H_1: population median ≠ specified hypothetical median

We reject H_0 at $\alpha = 0.05$ if the hypothetical median is outside the range of values corresponding to $t_{0.05, n-1}\sqrt{n}/2$ ranks either side of the sample median. If $n < 20$, tables of cumulative binomials should be used.

Worked example: see Box 4.1.

Box 4.1 Worked example: confidence interval of median

Appendix 3.3 gives data on the depths in a borehole at which specimens of two species of ostracods (which we will call A and B) were found, together with their average size at those depths. A monograph specifies that the median size of species A is 1.1 mm: are the data compatible with this specification?

The 95% confidence interval for the median is $t_{0.05, n-1} \sqrt{n}/2$ ranks either side of the median. The data, in rank order, are:

```
0.7  0.8  0.8  0.9  0.9  0.9  1.0  1.1  1.2  1.2  1.2  1.2  1.3  1.3
1.3  1.3  1.4  1.5  1.5  1.5  1.5  1.6  1.6  1.6  1.6  1.6  1.7  1.7
1.8  1.8  1.8  1.9
```

There are 33 values, so the median is the value with the $(33 + 1)/2 = 17$th rank; this is 1.4 mm.

$$t_{0.05, 32} = 1.693$$

so we require $1.693\sqrt{33}/2 = 4.863$ or, conservatively, five ranks either side of the median. The 95% confidence interval is therefore between the 12th and the 22nd ranks:

```
0.7  0.8  0.8  0.9  0.9  0.9  1.0  1.1  1.2  1.2  1.2  1.2  1.3  1.3
1.3  1.3  1.4  1.5  1.5  1.5  1.5  1.6  1.6  1.6  1.6  1.6  1.7  1.7
1.8  1.8  1.8  1.9
```

(median in bold, 95% confidence interval italicised).

The specified value 1.1 mm is outside the confidence interval, so we are 95% sure that the data were not drawn from a population with a median size of 1.1 mm. We may have a different species, or the specified median may be incorrect, or the data may have been taken with a bias in favour of larger specimens.

Note here that the properties of the data are satisfactory for parametric methods, so, unless a question was specifically phrased in terms of medians, we would normally prefer to use confidence intervals of the mean based on t (see Box 2.20).

4.2.2 Mann–Whitney test

This is the non-parametric analogue of the two-sample t-test, and is probably the most widely used non-parametric test. It tests the null hypothesis of equality of medians of populations from which two samples were drawn. We might, for example, use it to assess whether or not there is any difference in the occurrence through time of two species of microfossil. Having data on depths at which each species is found, we can use depth as an ordinal measure of time and test the hypothesis that the median depths are equal.

The calculation uses ranks of the data values, the ranking being done on the combined pair of samples. In this case the derivation from probability theory is complicated, but the equation is easy to calculate:

Mann–Whitney statistic for equivalence of medians

H_0: median of population x = median of population y
H_1: median of population $x \neq$ median of population y

$$T = \sum_{i=1}^{n} R(x_i) - \frac{n(n + 1)}{2}$$

where $R(x_i)$ are the ranks of sample x and n is the sample size of x. The sample size of y is m, which is required when comparing the calculated T with critical values from tables: see Appendix 2.10. (The ranks of sample y are unnecessary information, being dictated by n, m and the $R(x_i)$ values.)
    Worked example: see Box 4.2.

4.2.3   Kruskal–Wallis test

This test is a simple extension of the Mann–Whitney test for the multisample situation: it is therefore analogous to one-way analysis of variance. The null hypothesis is that all samples were drawn from populations with identical medians. This is rejected if at least one of the medians is different. The calculation requires summing the (pooled) ranks of each sample.

Kruskal–Wallis statistic for equivalence of medians

H_0: the medians of all source populations are equal
H_1: the median of at least one source population is different

$$H = \frac{12}{N(N + 1)} \sum_{j=1}^{k} \frac{\left(\sum_{i=1}^{n_j} R(x_{ij}) \right)^2}{n_j} - 3(N + 1)$$

where there are k samples of sizes $n_1, n_2, \ldots, n_k$, making a total sample size of N. $R(x_{ij})$ is the rank of the ith data point in the jth sample.
    The calculated H can be compared with critical values from tables: H is distributed very similarly to χ^2 with $k - 1$ degrees of freedom (Appendix 2.6).
    Worked example: see Box 4.3.

4.3   NON-PARAMETRIC CORRELATION

The parametric correlation coefficient has the full title of Pearson's product-moment correlation coefficient. The most popular non-parametric alternative

Box 4.2 Worked example: Mann–Whitney test

Using the data in Appendix 3.3 on depths in a borehole of specimens of two species of ostracod, is there any significant difference in the occurrence in time between the two species? We can use depth as an ordinal measure of time, and use a null hypothesis of equality of median depths.

The data and their ranks are as follows.

Species A		Species B	
Depth	Rank	Depth	Rank
242	5	202	1
253	7	203	2
271	9.5	208	3
292	13	233	4
305	15	251	6
332	21	258	8
335	22	271	9.5
337	23	282	11
338	24	283	12
350	25.5	301	14
357	28	308	16
364	29	314	17
365	30	327	18
371	31	329	19
372	32	330	20
385	34.5	350	25.5
401	39	356	27
402	40	378	33
410	41	385	34
412	43	386	36
418	44	387	37
423	46	399	38
427	47	411	42
429	49	422	45
432	50	428	48
446	51.5	446	51.5
451	53		$\Sigma = 578$
454	54		
460	55		
470	56		
474	57		
481	58		
497	59		

Using:

$$T = \sum_{i=1}^{n} R(x_i) - \frac{n(n + 1)}{2}$$

we have $\Sigma R(x_i) = 578$ for species B, so

$T = 578 - 26(26 + 1)/2 = 578 - 351 = 227$

Box 4.2 *Continued*

For $n = 26$ and $m = 33$ we find from statistical tables (Appendix 2.10) that this value is outside the range defined by the upper and lower critical values of T, so we reject the null hypothesis and conclude that the two samples came from populations with different medians, i.e. the two species are differently distributed through time.

Box 4.3 Worked example: Kruskal–Wallis test

We wish to investigate the distribution of silver in the MASSON data set (Appendix 3.4) on soil geochemistry on Masson Hill, Derbyshire. Data were collected along a series of seven traverses. Is there any difference between the traverses in terms of amounts of silver? This might indicate the location of underlying hydrothermal veins.

Using a parametric method we can obtain using statistical software:

Analysis of variance (ANOVA) on Ag

Source	d.f.	SS	MS	F
Trav.	6	51.42	8.57	1.33
Error	42	271.52	6.46	
Total	48	322.94		

Traverse	N	Mean	St. dev.	95% Confidence intervals
				--- + --------- + --------- + --------- + ---
1	9	1.504	0.164	-------- * -------
2	6	4.895	6.842	--------- * ----------
3	9	2.417	2.109	------- * --------
4	6	1.693	0.283	--------- * ----------
5	8	1.997	0.393	-------- * --------
6	5	1.936	0.154	----------- * ---------
7	6	1.777	0.130	---------- * --------
				--- + --------- + --------- + --------- + ---
			Ag	0.0 2.0 4.0 6.0

The calculated F of 1.33 does not exceed the critical value at $\alpha = 0.05$ ($F_{0.05, 6, 42} = 2.33$, Appendix 2.5), so we fail to reject the null hypothesis and conclude that there is no evidence for differences between the means of the traverses.

continued on p. 188

Box 4.3 *Continued*

However, using the Kruskal–Wallis test on the same data, we obtain:

Traverse	N	Median
1	9	1.470
2	6	2.395
3	9	1.700
4	6	1.660
5	8	1.910
6	5	1.980
7	6	1.780

$H = 19.10$

Critical values of H can be approximated by χ^2 with degrees of freedom $\nu = k - 1$ (k = number of samples or traverses).

$\chi^2_{0.05,\,6} = 12.59$ (Appendix 2.6)

so the calculated H exceeds the critical value, and we reject the null hypothesis of equality of medians.

The parametric approach based on means and standard deviations has given a different result from the non-parametric alternative. Which is the best result?

The dot plot of the data is shown in Fig. B4.3.1.

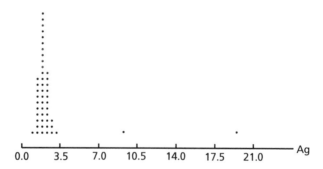

Fig. B4.3.1

This shows two extreme outliers. The parametric ANOVA has been adversely affected by the distribution: the within-sample variance has been inflated so that differences between sample means do not appear significant. The Kruskal–Wallis test, however, is not sensitive to the magnitude of outliers and the conclusion that the amounts of silver vary significantly between traverses is the better one.

Box 4.4 Worked example: Spearman's rank correlation coefficient

Using the data in Appendix 3.3 on occurrences and sizes of ostracods, is there any size change through time of species A?

Again, we use depth as an ordinal measure of time. We obtain ranks of depth and size separately, and then use:

$$r' = 1 - \frac{6 \sum_{i=1}^{n} (R(x_i) - R(y_i))^2}{n(n^2 - 1)}$$

with the null hypothesis that the sample was drawn from a population with zero correlation between time (depth) and size.

Species A

x (depth)	$R(x)$	y (size)	$R(y)$	$(R(x) - R(y))^2$
242	1	1.3	15.5	210.25
253	2	0.9	5	9
271	3	0.7	1	4
292	4	0.8	2.5	2.25
305	5	0.8	2.5	6.25
332	6	1.2	11.5	30.25
335	7	0.9	5	4
337	8	1.1	11.5	0.25
338	9	1.6	25	256
350	10	1.6	25	225
357	11	1.0	7	16
364	12	1.2	11.5	0.25
365	13	1.3	15.5	6.25
371	14	1.4	18	16
372	15	1.1	8.5	42.25
385	16	0.9	5	121
401	17	1.3	15.5	2.25
402	18	1.5	20.5	6.25
410	19	1.8	31	144
412	20	1.6	25	25
418	21	1.2	11.5	90.25
423	22	1.5	20.5	2.25
427	23	1.5	20.5	6.25
429	24	1.7	28.5	20.25
432	25	1.9	33	64
446	26	1.5	20.5	30.25
451	27	1.6	25	4
454	28	1.2	11.5	272.25
460	29	1.6	25	16
470	30	1.7	28.5	2.25
474	31	1.8	31	0
481	32	1.8	31	1
497	33	1.3	15.5	306.25
				$\Sigma = 1941.5$

continued on p. 190

> **Box 4.4** *Continued*
>
> $r' = 1 - (6 \times 1941.5)/(33(33^2 - 1)) = 1 - (11649/35904) = 0.676$
>
> From tables of critical values (Appendix 2.11) we obtain
>
> $r'_{0.05,\,33} = 0.29$
>
> so we reject the null hypothesis and conclude that the data were drawn from a population with a correlation between time and size. The ostracods become significantly larger further down the borehole.

has the uncomfortably similar name of Spearman's rank correlation coefficient. It was noted in Section 3.2 that it is not often in geology that the parametric test of correlation can be applied, as it requires both variables to be normally distributed. Parametric correlation is particularly dangerous in that it readily yields spurious high correlations where there are outliers. Consequently, rank correlation is a useful and popular tool for general applications.

Starting from bivariate x, y data pairs, the x variable and the y variable are ranked separately, giving $R(x_i)$ and $R(y_i)$ data. If, for each i, the ranks of x equal the ranks of y, the rank correlation is regarded as perfect. The statistic is based on the sum of the difference between the corresponding ranks of x and y. Values of the coefficient are scaled between 0 (no correlation) and 1 or -1 (perfect correlation): this is analogous to the parametric version.

Spearman's rank correlation coefficient

$H_0: \rho' = 0$
$H_1: \rho' \neq 0$

$$r' = 1 - \frac{6 \sum\limits_{i=1}^{n} (R(x_i) - R(y_i))^2}{n(n^2 - 1)}$$

The calculated r' is compared with critical values from tables (Appendix 2.11).

Worked example: see Box 4.4.

As with other non-parametric methods, values of x and y can vary extensively without affecting the result: this is illustrated in Fig. 4.1. The analyst must remember that a high r' does not imply a good linear relationship: rather than linearity, we say that y has a high monotonicity with x. It is quite feasible for a high Pearson's to be associated with a low Spearman's and, to a lesser extent, vice versa (Fig. 4.2). Despite this, in our experience

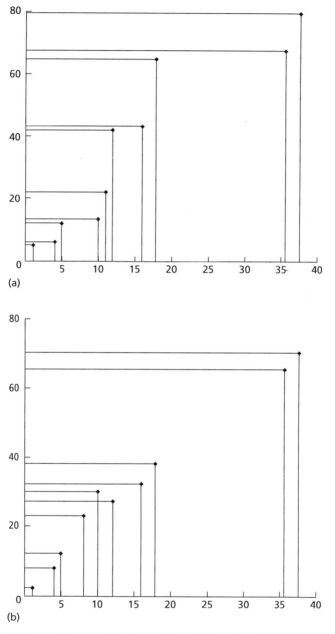

Fig. 4.1 Rank correlation coefficients. In (a) the rank correlation is 1: the points are in the same rank order along the x and y scales; this is shown by no crossing of the construction lines. In (b) the rank correlation is less than 1: there is one pair of points with reversed x and y ranks.

of geological data, it is unusual for Pearson's to provide a statistical test result markedly superior to Spearman's, even with normally distributed data. There seem to be few real cases where Pearson's leads to rejection of H_0 when Spearman's doesn't: the conservatism of the non-parametric

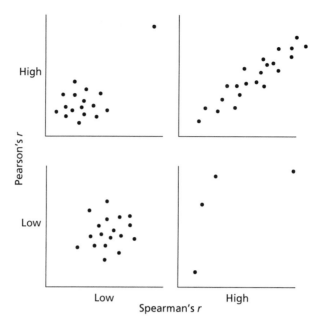

Fig. 4.2 Some hypothetical data scatters, showing how the relative magnitude of Pearson's and Spearman's correlation coefficients may differ.

approach (noted in Section 4.1.1) is outweighed by robustness and is not a serious handicap here.

FURTHER READING

Rock N. (1988) *Numerical Geology*. Lecture Notes in Earth Science no. 18, Springer Verlag, Berlin.

particularly useful as a source for non-parametric methods.

Al-Turki K.I. and Stone M. (1978) Petrographic and chemical distinction between the mega-crystic members of the Carnmenellis granite, Cornwall. *Proceedings of the Ussher Society*, **4**, 182–189.

an example of geological use of non-parametric methods.

5   Directional Data and Circular Statistics

5.1   INTRODUCTION

5.1.1   Definitions and data types

In geology, more than any other subject, the directions or orientations of objects or structures are frequently of major interest. The crucial characteristic of such data is that they are measured in terms of angles, often bearings from north. The commonest type of directional data are dip-and-strike data, conveying (by the use of various conventions) the direction in three dimensions of the line of maximum dip on a surface. This chapter, however, deals principally with the simpler case of data such as palaeo-current measurements for which only one directional angle needs to be specified. Such data can be ordinated on a circle: hence the term circular statistics. Dip-and-strike data need to be ordinated on a sphere or, in practice, a spherical projection, and, although some of the techniques of circular statistics can be extended to spherical data, the more complex spherical case is best analysed using eigenvector methods: see Section 8.5.3.

Circular data are of two types: directional data (*sensu stricto*) and oriented data. Directional data are used to convey linear phenomena in which one end of the 'line' is distinguishable from the other: this is likely to be the case where the distinction is due to flow in the specified direction. Examples are illustrated in Fig. 5.1(a). Oriented data are appropriate for phenomena without such directional distinction. This being the case, it is arbitrary whether the orientation is recorded in one direction or its opposite, 180° around. This effective equality of measurements 180° apart is the reason why oriented data need to be distinguished in analysis. Examples of oriented phenomena are illustrated in Fig. 5.1(b). There may be ambiguity as to whether geological features should be treated as oriented or directional: a general guideline is that objects with approximate twofold rotational symmetry must be regarded as oriented: their 'front' is similar to their 'back'. However, even for objects without this symmetry, it is possible that the 'front' and 'back' are effectively identical with respect to the geological process. The blind use of directional methods on oriented phenomena is likely to result in failure to detect significant trends. The use of oriented data analytical procedures on directional phenomena merely results in ambiguity as to which of two opposite directions is the true mean. Geological common sense may be able to resolve this!

Before analysing circular directions, it is often essential to remove tectonic dip. The apparent bearing of, say, groove marks on turbidites folded along north–south axes will rotate more towards north–south the greater the dip. The plunge of folds will also corrupt the sedimentological

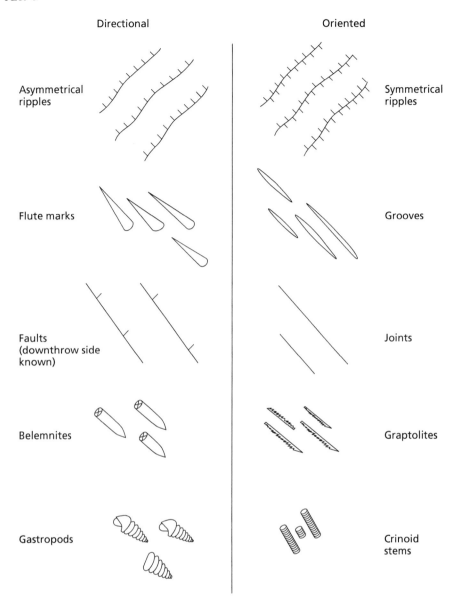

Directional | Oriented

Asymmetrical ripples | Symmetrical ripples

Flute marks | Grooves

Faults (downthrow side known) | Joints

Belemnites | Graptolites

Gastropods | Crinoid stems

Fig. 5.1 Examples of directional and oriented phenomena. In directional features we can distinguish one end from the other, or left from right.

information. Mathematical correction for dip and plunge is quite complex, but it may be possible, in simple cases, to measure angles on tilted bedding planes in such a way as to make correction easy: see Box 5.1.

5.1.2   **Graphical presentation**

The simplest graphical representation of directional data is as points marked on the periphery of a circle, perhaps with lines to the centre. As with simple

Box 5.1 Tectonic dip correction

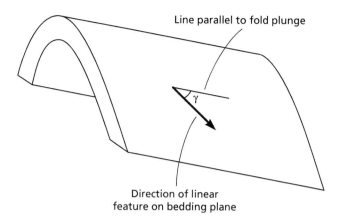

Line parallel to fold plunge

γ

Direction of linear
feature on bedding plane

Fig. B5.1.1

If we measure the direction of a linear feature on a bedding plane, we need
to know the plunge amount and the plunge direction of the fold system in
order to restore the original direction. If we rotate, in a vertical plane,
the plunge back to horizontal and then rotate the bedding plane back
to horizontal about the hinge, the angle between the linear feature and
the original plunge direction is retained, and the plunge retains its direction
in plane view. Consequently, the original bearing of the linear feature
equals the plunge direction plus the angle in the plane of the bedding
between the linear feature and the plunge.

If we have:

θ = direction of plunge of feature (measured in horizontal plane)
α = amount of plunge of feature (measured in vertical plane)
ϕ = direction of plunge of fold (measured in horizontal plane)
β = amount of plunge of fold (measured in vertical plane)

then the unit vector (x_1, y_1, z_1) for the linear feature is found by:

$x_1 = \cos \alpha \sin \theta$
$y_1 = \cos \alpha \cos \theta$
$z_1 = \sin \alpha$

and the unit vector (x_p, y_p, z_p) for the fold plunge is found by:

$x_p = \cos \beta \sin \phi$
$y_p = \cos \beta \cos\phi$
$z_p = \sin \beta$

If the angle between the linear feature and the fold plunge in the plane
of the bedding is γ, we can set up a right-angled triangle with the hypotenuse
of unit length and:

$\cos \gamma = a$

continued on p. 196

Box 5.1 *Continued*

where *a* is the length of the projection, at right angles, of the one vector on to the other.

Multiplying the vectors gives the length of the projection:

$$\cos \gamma = \cos \alpha \, \sin \theta \, \cos \beta \, \sin \phi + \cos \alpha \, \cos \theta \, \cos \beta \, \cos \phi + \sin \alpha \, \sin \beta$$

So the original bearing is given by

$$\phi + \cos^{-1}(\cos \alpha \, \sin \theta \, \cos \beta \, \sin \phi + \cos \alpha \, \cos \theta \, \cos \beta \, \cos \phi + \sin \alpha \, \sin \beta)$$

This is only valid for measurements of features lying in the plane of bedding, and requires the assumption that only simple rotations are involved in folding: if there is significant shear, angles in the plane of bedding are not preserved. Even modest amounts of deformation will render approximate any estimate of original bearing.

---

univariate data, the distribution of points can be clarified by dividing the scale into classes and plotting the frequency in each class. In the case of ordinary univariate data, this results in a histogram comprising rectangles with height proportional to frequency; with circular data, the analogous graph is made of segments of circles with radii proportional to frequency: the rose diagram (Fig. 5.2a). However, unlike the histogram, the rose

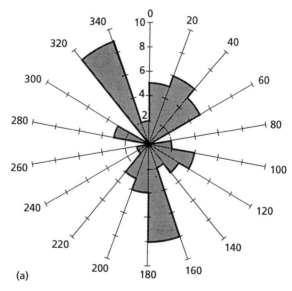

(a)

Fig. 5.2 Rose and kite diagrams for directional data, using the data from the Bearraraig sandstone (Appendix 3.6). (a) Conventional rose diagram, with radius proportional to frequency in each class. This visually overemphasises high frequencies. (b) Rose diagram with radius proportional to the square root of frequency. The visual impression is accurate. (c) Kite diagram, showing unattractive spiky appearance, but good representation of the data.

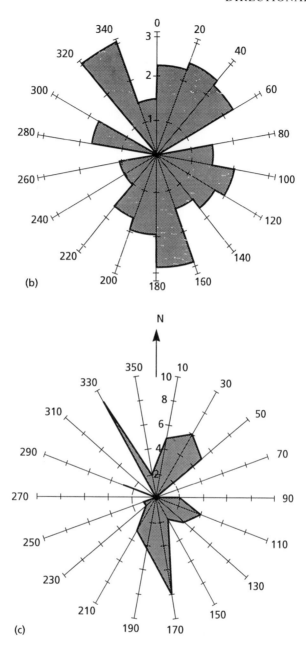

(b)

(c)

Fig. 5.2 *Continued*

diagram in this form is not acceptable as an unbiased representation of data. This is because the visual impression of a segment is proportional to its area, and the area is proportional to radius squared (whereas, in a histogram, the rectangles do have areas directly proportional to height and frequency). The rose diagram, therefore, visually overemphasises high frequencies and

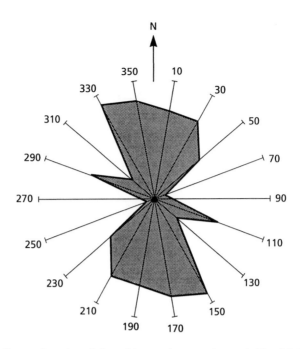

Fig. 5.3 Kite diagram for oriented data: this uses the same data as in Fig. 5.2, but the construction is undertaken as if the data were oriented. Opposite frequencies are added, so the graph repeats through 180°.

underemphasises low frequencies. This can lead to false impressions of preferred directions in effectively random data. This should be avoided, either by plotting radius proportional to the square root of frequency, or by means of a kite diagram: see Fig. 5.2(b,c).

For oriented data, opposite directions 180° apart are equivalent, so the graphical portrayal should either be restricted to half of the complete circle, or have rotational symmetry so that opposite classes have the same frequency (Fig. 5.3).

5.2   STATISTICS ON DIRECTIONAL DATA

It might initially be asked: why do we need special methods for circular data? The reason becomes clear if we imagine we have two directional measurements: 1° and 359°. These are both more or less due north, but a simple arithmetic mean of the two gives us $(1 + 359)/2 = 180°$, which is due south! We need a method which treats 1 and 359 as similar numbers, and 0 and 360 as identical numbers. The answer is to use trigonometrical functions: sine, cosine and tangent functions all repeat every 360°, so that $\sin(0°) = \sin(360°)$ and the same for cos and tan. First, though, the methods to be used can be justified graphically.

Geometrical concept of summing directions

The calculation of the mean of a set of ordinary numbers involves first obtaining the sum. Suppose the numbers are: 70, 160, 80, 70. The sum can be portrayed by adding these values as lengths end to end along an axis: see Fig. 5.4(a). With directional data, we can use a similar end-to-end construction except that each measurement defines a direction, not a length; the length is constant: see Fig. 5.4(b). Here the analogue of the sum is the line connecting the starting and finishing points: this is called the resultant. Its direction is the mean direction. Already, then, we have a concept of a mean direction, but we also need a measure of variance. Suppose a second data set is: 130°, 330°, 20°, 170°. The graphical construction is shown in Fig. 5.4(c). These data result in a mean direction (conveyed by the direction of the resultant) similar to that of the first data set, but the difference here is the length of the resultant (and the mean resultant): the greater dispersion of the second data set gives a shorter resultant. This forms the basis for a measure of variance.

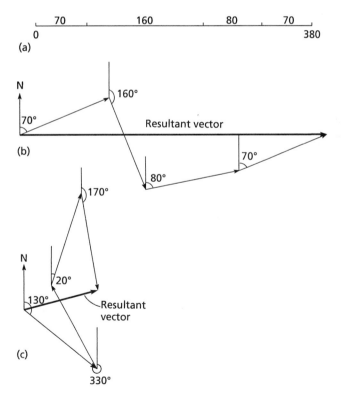

Fig. 5.4 (a) Addition of four numbers on an ordinary data scale. (b) The concept of addition of the same numbers as directional data. Each angle is represented by a unit vector. The result is called the resultant vector. (c) Addition of directional data with higher variance. The resultant vector is shorter.

Calculation of means and variances of directional data involves treating the data as a set of unit vectors (see Appendix 1): the bearings are converted into x, y components. Summing the x components and the y components is the same as drawing the vectors end to end: the result gives the x, y components of the resultant vector.

5.2.2 **Calculation of mean direction and measures of dispersion**

The mean direction may be the most important measure required in a study, particularly in sedimentology: it can be used to infer shapes of sand bodies and palaeogeographical features. Dispersion is probably much more important in directional data than in ordinary univariate data: it is used, for example, to diagnose braided alluvial systems (low dispersion) and meandering systems (high dispersion).

If the data are symbolised by:

$$\theta_1, \theta_2, \theta_3, \ldots, \theta_n$$

each unit vector has components:

$$x_i = \sin \theta_i$$
$$y_i = \cos \theta_i$$

(see Fig. 5.5) so that the resultant vector has components:

$$x_r = \Sigma \sin \theta_i$$
$$y_r = \Sigma \cos \theta_i$$

The mean direction $\bar{\theta}$ is then the direction of the hypotenuse.

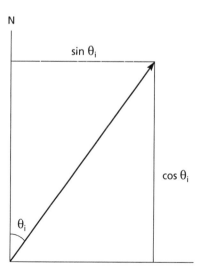

Fig. 5.5 Components of a unit vector defined by the bearing θ.

Mean direction

$$\bar{\theta} = \tan^{-1}(x_r/y_r)$$

Warning

The $\tan^{-1}$ function, also known as arctan or atan, is likely to yield misleading results from a computer or calculator: a value will be produced in the range $-90°$ to $90°$ while the actual result must be in the range $0°$–$360°$. This is, of course, due to the repetition of the tangent function every $180°$. The correct value can be found by adding either $0°$, $180°$ or $360°$ to the returned value, according to the signs of the x and y components (Table 5.1). Note also that, by default, most computers work in radians, not degrees (1 radian $= 57.2956°$).

Table 5.1 Adjustment to inverse tangents according to signs of x and y

x	y	Add
+	+	$0°$
+	−	$180°$
−	−	$180°$
−	+	$360°$

The length of the resultant R is clearly:

$$R = \sqrt{(x_r^2 + y_r^2)}$$

As was noted above, this length contains information about the dispersion of the data. However, it is dependent on the sample size n, so instead we use the mean resultant length:

$$\bar{R} = R/n$$

This has the unfortunate property that higher $\bar{R}$ indicates less variance, so the circular variance is often preferred:

$$\sigma_0 = 1 - \bar{R}$$

which has a range 0 to 1, where zero indicates that all directions are identical. For an example of the calculation of $\bar{R}$ and $\bar{\theta}$, see Box 5.2(a).

5.2.3 Frequency distributions

With ordinary univariate data, it was clear that a variable having random values must have a frequency distribution showing a preferred value, with lesser frequencies at higher and lower values, simply because it is inconceivable that there could be a uniform probability of values extended to infinity

Box 5.2 Worked example: Rayleigh's test

(a) Appendix 3.5 contains 41 palaeocurrent measurements (from ripple forests) from a sandstone. Is there any evidence of a preferred trend?

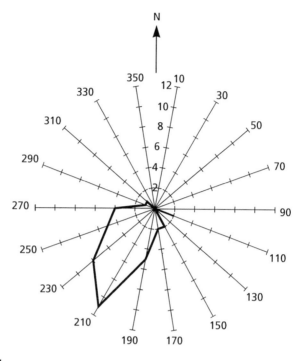

Fig. B5.2.1

H_0: $\kappa = 0$
H_1: $\kappa \neq 0$

Using simple spreadsheet operations, we find:

$\Sigma \sin \theta = -19.8604$
$\Sigma \cos \theta = -25.5326$
$$R = \sqrt{(\Sigma \sin \theta^2 + \Sigma \cos \theta^2)} = \sqrt{(19.8604^2 + 25.5326^2)} = 32.347$$
$$\bar{R} = R/n = 32.347/41 = 0.789$$

For $n = 41$ and $\alpha = 0.05$, the critical value of $\bar{R}$ from tables (Appendix 2.12) is

$\bar{R} = 0.27$

As the calculated value exceeds the critical value, we reject the null hypothesis. There appears to be a preferred trend, so the currents were reasonably consistent in direction. The direction of this preferred trend, $\bar{\theta}$, is found by:

$$\bar{\theta} = \tan^{-1}(\Sigma \sin \theta / \Sigma \cos \theta) = \tan^{-1}(-19.8604/-25.5326) = 217.877°$$

Box 5.2 *Continued*

(Calculators and computers are likely to give an answer of 37.877 to this inverse tangent; the negative signs of $\Sigma\sin\theta$ and $\Sigma\cos\theta$ indicate that the correct answer is in the 180–270 quadrant.)

(b) Appendix 3.6 contains 58 palaeocurrent measurements (from trough cross bedding) from the Middle Jurassic Bearraraig sandstone of Skye, Scotland (see Fig. 5.2). Is there any evidence of a preferred trend in these data?

H_0: $\kappa = 0$
H_1: $\kappa \neq 0$

Using the same procedures as above:

$\Sigma\sin\theta = 6.246$
$\Sigma\cos\theta = 4.253$
$$R = \sqrt{(\Sigma\sin\theta^2 + \Sigma\cos\theta^2)} = \sqrt{(6.246^2 + 4.253^2)} = 7.5589$$
$$\bar{R} = R/n = 7.5589/58 = 0.13$$

For $n = 58$ and $\alpha = 0.05$, the critical value of $\bar{R}$ from tables (Appendix 2.12) is

$$\bar{R} = 0.23$$

As the calculated value does not exceed the critical value, we fail to reject the null hypothesis. There appears not to be a preferred trend, but we must remember that the existence of more than one modal direction can result in failure to reject H_0 in Rayleigh's test.

in one or both directions. Consequently, our null expectation of a univariate frequency distribution is the normal distribution, or some transformation of it. However, with circular data, this is no longer the case: it is perfectly conceivable to have a uniform probability distribution over the entire possible range of values, as this is only 0 to 360°: we do not need to be concerned about infinity with circular data! So, the null random frequency distribution for directional data is simply a uniform distribution, without the kind of modal 'peak' value that we expect in the univariate situation.

However, it remains the case that many geological situations result in directional data with a dispersion of values around a preferred direction, perhaps indicating a prevailing current direction. We can therefore use a normal distribution, or something like it, to assist in statistical calculations. If we apply the normal distribution to directional data, it is inevitable that the tails of the distribution will wrap around the circular scale, so that values corresponding to the extreme tails of the distribution will begin to approach the mean from the opposite direction! A dispersion greater than 180° from the mean direction clearly has little meaning. Although a 'wrapped normal' distribution is sometimes used, an alternative, the von Mises distribution, is more popular as it can be used rather similarly to a normal distribution in univariate statistics. The von Mises has two parameters: the mean direction $\bar{\theta}$ and the concentration parameter κ. The latter can be estimated from the mean resultant length and has values which increase as the dispersal decreases: $\kappa = 0$ is a uniform distribution.

5.2.4

Tests of significance of mean direction

Test of mean resultant length

Any directional data set has a calculable mean, and we may be interested in interpreting this as indicating a prevailing current direction, a palaeoslope or flow in a magma. However, the mean will not be useful if the data are random, which might mean that there is no prevailing current, no palaeoslope or no flow. Furthermore, the mean will be misleading if the data have more than one mode. The mean resultant length $\bar{R}$ is greater with a stronger single preferred direction. Statistical tables are available which give critical values of $\bar{R}$, to test the null hypothesis that there is no single preferred direction: this is Rayleigh's test (Appendix 2.12).

> *Rayleigh's test for significance of mean direction*
>
> $$\bar{R} = \frac{1}{n}\sqrt{\left(\sum_{i=1}^{n} \sin\theta_i\right)^2 + \left(\sum_{i=1}^{n} \cos\theta_i\right)^2}$$
>
> Worked example: see Box 5.2.

Tests of uniformity

Randomness in a directional data set is likely to result from a lack of flow in the geological medium, whether it is water flow indicated by belemnite orientations, or magma flow indicated by phenocrysts. Randomness results in a frequency distribution not significantly different from uniform. It is clear that, if random, the expected frequency in each class is equal and is simply n/k (k = number of classes), so the actual observed frequencies can be readily compared with these using a χ^2 test (see Section 2.5.6) with a null hypothesis of randomness.

Worked example: see Box 5.3.

Bimodal and polymodal distributions

Note that, although we will fail to reject the null hypotheses with both Rayleigh's and the χ^2 test if we have random data, and we will reject null hypotheses with both tests if we have a strongly preferred mean direction,

Box 5.3 χ^2 Test for uniformity

The data on the Bearraraig sandstone (Box 5.2b) have been shown not to have a preferred directional trend, but is the distribution uniform?

H_0: data drawn from uniformly distributed population
H_1: data drawn from non-uniformly distributed population
$$\chi^2 = \Sigma(O_j - E_j)^2/E_j$$

As the 'expected' model is uniform, E_j is the same for all classes, so we can simplify to:

$$\chi^2 = (1/E_j)\Sigma(O_j - E_j)^2$$

Using 12 classes of 30° width (so $E_j = 58/12 = 4.833$), we have:

j	Interval	O_j	E_j	$(O_j - E_j)^2$
1	0–30	9	4.833	17.361
2	30–60	7	"	4.694
3	60–90	1	"	14.694
4	90–120	5	"	0.028
5	120–150	5	"	0.028
6	150–180	8	"	10.028
7	180–210	6	"	1.361
8	210–240	2	"	8.028
9	240–270	1	"	14.694
10	270–300	3	"	3.361
11	300–330	2	"	8.028
12	330–360	9	"	17.361
		$\Sigma = 58$		$\Sigma = 99.666$

$$\chi^2 = 99.666/4.833 = 20.62$$

The critical value of χ^2 at $k - 1 = 11$ d.f. and $\alpha = 0.05$ is 19.68 (Appendix 2.6).

The calculated value exceeds the critical value, so we reject the null hypothesis of uniformity. This and the result in Box 5.2(b) indicate that there were two or more preferred current directions. These have been related to tidal effects in a strait.

these two tests will not always yield the same results. In the case of bi- or polymodal data, Rayleigh's test may not lead to a rejection of the null hypothesis, and certainly will not if the modes are opposite, as is likely in, for example, tidal regimes. Referring back to the geometrical construction at the beginning of this section, it is clear why this should be: opposite directions cancel out in terms of adding to the length of the resultant.

However, bi- or polymodality is likely to differ significantly from the uniform model in a χ^2 test: see Boxes 5.2(b) and 5.3.

5.2.5 Testing for hypothetical mean direction

One sample

We can use a crude analogue of the *t*-test for assessing whether or not data could have been drawn from a specified mean direction. For example, suppose that we know of the existence of a strait between two land areas in a palaeogeographic reconstruction; we may wish to know if the measured ripple foresets are compatible with the idea of currents flowing through the strait. This can be done by calculating the standard error (s_e) of the mean, which can be used in a similar way to the standard deviation:

Confidence interval of mean direction

At α = 0.05, the true population mean will be in the range $\bar{\theta} + 1.96s_e$ to $\bar{\theta} - 1.96s_e$, where

$$s_e = 1/\sqrt{n\bar{R}\kappa}$$

where n is the sample size, κ is the concentration parameter (Appendix 2.13) and $\bar{R}$ the mean resultant length. (The figure 1.96 comes from the Z or normal distribution at α = 0.05.) We reject H_0 (sample comes from population with specified hypothetical mean) if the specified hypothetical mean lies outside the interval.

Warnings

1 The units of s_e are radians, not degrees. To convert, multiply by 57.2956.
2 The data should very approximately fit the von Mises distribution.
3 Rayleigh's test should be applied first to check that there is a preferred direction. If there isn't, the standard error will be large and the hypothetical mean will be likely to fall within the confidence interval, but the result will be meaningless.
4 Remember that the probability attached to failure to reject H_0 is unknown, so we cannot be confident about the match of the data to the specified mean. This is appropriate because, even with data that passes Rayleigh's test, s_e is often large and produces a confidence interval that spans a large proportion of the circle!
    Worked example: see Box 5.4.

Two samples

Flute and groove casts from two turbidite layers may be compared and there may be a suggestion that the directions differ, perhaps indicating different

Box 5.4 Worked example: confidence interval of mean direction

A sedimentologist suspects that the ripples measured from a sandstone (see Box 5.2a) were formed by longshore currents parallel to an inferred palaeocoastline trending 200°. Do the data support this idea?

We can use the hypotheses:

H_0: sample drawn from population with mean direction = 200°
H_1: sample drawn from population with mean direction ≠ 200°

Using

$$s_e = 1/\sqrt{(nR\kappa)}$$

we have (from Box 5.2a) $n = 0.41$, $\bar{R} = 0.789$, $\bar{\theta} = 217.877°$ and, from tables (Appendix 2.13):

$\kappa = 2.75$

so

$$s_e = 1/\sqrt{(41 \times 0.789 \times 2.75)} = 0.106 \text{ radians} = 6.075°$$

The 95% confidence interval is between $\bar{\theta} - 1.96s_e$ and $\bar{\theta} + 1.96s_e$.

$1.96 \times 6.075 = 11.907$

So the interval is

205.97° to 229.78°

This does not include the hypothetical mean of 200°, so we reject the null hypothesis. We are 95% confident that the sample was not drawn from a population with a mean of 200°. It is likely that either the currents were not parallel with the coastline, or the inferred coastal trend is wrong.

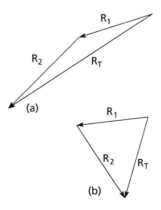

Fig. 5.6 Comparison of resultant lengths (R_1 and R_2) of two samples with the pooled total resultant length (R_T). In (a), the mean directions of the two samples are similar, so R_T is close to $R_1 + R_2$; in (b) they are different, so R_T is much shorter.

provenance directions. We can test this by using an analogue of the two-sample t test, with a null hypothesis that two samples came from populations with the same mean direction. We can compare the resultant lengths for the two samples, R_1 and R_2, with the resultant length for samples 1 and 2 combined, R_T. If the sample means are close, the sum of lengths $R_1 + R_2$ will be close to R_T (see Fig. 5.6a), but, if the two means are very different, the combined resultant vector will be much shorter (Fig. 5.6b). The test, then, is based on the value $R_1 + R_2 - R_T$.

Test for equality of mean directions

H_0: samples come from populations with same mean direction
H_1: samples come from populations with different directions

$$F = \left(1 + \frac{3}{8\kappa}\right)\frac{(n-2)(R_1 + R_2 - R_T)}{(n - R_1 - R_2)}$$

κ is found from Appendix 2.13 using $\bar{R}_T$. The calculated F is compared with critical values from the standard F tables (Appendix 2.5), with degrees of freedom $= 1, n - 2$. If the calculated F exceeds the critical F, we reject H_0.

Warnings

1 Both samples must come from a von Mises distribution.
2 The above equation is strictly only valid for $10 > \kappa > 2$. $\kappa > 10$ is very unlikely, and if $\kappa < 2$ it is likely that there is no preferred direction or that the two mean directions are obviously different. In other cases, special tables are required: refer to Mardia (1972).
  Worked example: see Box 5.5.

5.2.6 **Some quick non-parametric tests**

Non-parametric directional methods have the advantage that they do not require the assumption of a von Mises distribution and they are often easy and quick to apply manually. A large number are available; the following are among the simplest.

Hodges–Ajne test for preferred trend

For this test it is best to imagine (or actually plot) the data on the perimeter of a circle. If there is a preferred trend, there will tend to be a concentration of points on one side of the circle. For this test, we rotate a diameter around the circle until we obtain the maximum difference between the number of points on one side and the other. The Hodges–Ajne statistic M_{min} is the

Box 5.5 Worked example: testing equality of two mean directions

The data set introduced in Box 5.2 represents a sandstone formation A, overlain by a second sandstone B from which further data were taken (Appendix 3.5). Is there evidence for a change in the current direction?

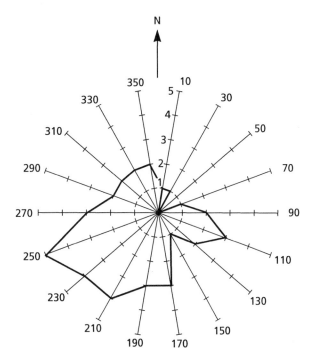

Fig. B5.5.1

$H_0: \bar{\theta}_A = \bar{\theta}_B$
$H_1: \bar{\theta}_A \neq \bar{\theta}_B$

Test statistic:

$$F = \left(1 + \frac{3}{8\kappa}\right)\frac{(n - 2)(R_A + R_B - R_T)}{n - R_A - R_B}$$

From Box 5.2 and further similar spreadsheet calculations on data set B and the combined data set T we have:

$R_A = 32.347$ $R_B = 12.882$ $R_T = 45.14$ $n = 82$ $\kappa = 1.326$
$F = 1.2828 \times 80 \times 0.089/36.771 = 0.2484$

With d.f. $= 1, 80$ and $\alpha = 0.05$ the critical value of F is 3.95 (but we don't actually need to look this up as all critical F values exceed 1). The calculated value does not exceed the critical value, so we cannot reject H_0; we conclude that the two samples could have been drawn from populations with the same mean direction.

Box 5.6 Worked examples: non-parametric tests

(a) Hodges–Ajne test for preferred direction

Using the data sets introduced in Box 5.2, we can use the Hodges–Ajne statistic to test the null hypothesis of no preferred direction.

For the ripple data, we can find a 180° range between, say, 110° and 290° which includes all but one of the points: the statistic is therefore 1. At $n = 41$, we find from tables that the critical value at $\alpha = 0.05$ is 10, so we can easily reject the null hypothesis and conclude that there is a preferred direction.

For the Beararaig sandstone data, it is difficult to find the best diameter, but we can soon establish that the minimum number in one half must be greater than the critical value for $n = 58$ and $\alpha = 0.05$ of 14, so we fail to reject the null hypothesis that there is no preferred direction.

(b) Runs test for equality of directions

Using the same data and with the same objective as in Box 5.5, we can try the non-parametric alternative.

The pooled ranked data are as follows, with sample B data in italics:

15	35	72	93	96	105	106	111	118	*118*	123	135	141
148	155	*163*	*171*	172	177	*179*	*180*	182	189	*190*	191	192
196	198	201	201	205	208	210	212	214	*215*	216	*216*	217
217	217	218	219	221	*222*	222	222	223	*223*	227	228	233
234	*237*	238	242	*242*	244	245	248	*251*	251	255	*255*	257
260	262	269	*271*	275	276	*278*	280	288	290	*300*	305	*313*
320	*333*	*341*	*349*									

The number of runs, U, is 38. (Note that, as the data are circular, we must start the count at a genuine beginning of a run, not at 0/360°. Also, where values are tied between different groups, the order selected is arbitrary and does not bias the result towards a larger or smaller number of runs.)

The test statistic is

$$Z = \frac{U - \bar{U}}{\sigma_{\bar{U}}}$$

The expected number of runs, with $n_1 = n_2 = 41$, is

$$\bar{U} = 1 + \frac{2n_1 n_2}{n_1 + n_2} = 42$$

and the expected variance is

$$\sigma_{\bar{U}}^2 = \frac{2n_1 n_2 (2n_1 n_2 - n_1 - n_2)}{(n_1 + n_2)^2 (n_1 + n_2 - 1)} = \frac{3362 \times 3280}{6724 \times 81} = 20.247$$

Box 5.6 *Continued*

So

$\sigma_{\bar{U}} = 4.5$

and

$Z = (38 - 42)/4.5 = -0.888$

In this context, the test is one-tailed: we only reject the null hypo-thesis of equality if there is an anomalously low number of runs, indi-cating that the samples are not well mixed. From tables of the normal distribution (Appendix 2.2), we find that, at $\alpha = 0.05$, the critical value of Z is -1.65; the calculated value is not outside this, so we fail to reject H_0 and conclude that the two samples could have the same preferred direction.

---

minimum number of points on one side. Critical values can be found in Mardia (1972), but the test is not very sensitive.

Worked example: Box 5.6(a).

Runs test for equality of mean directions

In this adaptation of the normal runs test (see Section 6.3.4.2), we combine the data from two samples and organise the data in order. A 'run' in this context constitutes a series of consecutive items from the same sample. If two samples come from populations with identical means, then we expect that the samples would be intermingled and the number of runs would be large, but, if the two directions were widely separated, we may get only two runs. However, we need the dispersions of the two samples to be broadly similar because we could also obtain an anomalously small number of runs, even if the mean directions were identical, if all the observations from one of the samples were tightly clustered together. The number of runs is tested using the same method as described in Section 6.3.4.2, except that the test is one-tailed: we are not concerned with anomalously large numbers of runs.

Worked example: Box 5.6(b).

5.3 ## STATISTICS ON ORIENTED DATA

Calculations on oriented data proceed in an identical manner to directional data, except that all the angles are doubled in advance, and all angular results, such as mean directions and standard errors, are consequently halved. The effect of doubling the angle is to make directions 180° apart (identical orientations) become 360° apart, and therefore identical in the trigonometrical calculations. All other tests can be applied validly using the doubled data, and can be imagined as applying to a special circular orientation scale with 180° forming the circle.

FURTHER READING

Mardia V. (1972) *Statistics of Directional Data*. Academic Press, London.

the source book for theory and methods.

Jones J.A. (1968) Statistical analysis of orientation data. *Journal of Sedimentary Petrology*, **38**, 61–67.
Maill A.D. (1974) Palaeocurrent analysis of alluvial sediments: a discussion of directional variance and vector magnitude. *Journal of Sedimentary Petrology*, **44**(4), 1174–1185.

useful descriptions and discussions for geologists.

Maill A.D. and Gibling M.R. (1978) The Silur–Devonian clastic wedge of Somerset Island. *Sedimentary Geology*, **21**, 85–127.
Michelson P.C. and Dott R.H. (1973) Orientation analysis of trough cross-stratification in Upper Cambrian sandstones in West Wisconsin. *Journal of Sedimentary Petrology*, **43**(3), 784–794.

case studies.

6 Data Through Time

One of the ironies of geological data analysis is the status of time: although fundamentally important to geological thought, it is rare for time to be measured with sufficient accuracy to allow straightforward analyses. The methods described in this chapter, then, include a variety ways of coping with rather restricted information.

6.1 MARKOV CHAINS

6.1.1 Concepts

The poorest quality of data giving information on changes through time contains no numerical values for time or any other variable. The data may consist merely of a series of descriptions of successive states of a system. This situation is familiar to geologists in the form of a sedimentological log of a section or a borehole core: the data are simply a succession of rock descriptions or lithotypes. This is a classic situation for the application of Markov chain analysis: this method tests for patterns in the record of transitions between one state (= rock type) and others. It is used to investigate cyclicity or rhythmicity, caused perhaps by eustatic sea-level changes or deltaic evulsion, but, of course, there will be no means of ascertaining whether or not such cycles were regular through time.

The lack of numerical information for analysis is overcome by the use of simple probability theory. Consider two sequences of letters, which may be lithology codes:

Sequence 1: ADCCBCADBBBABDDDCBAACDAA
Sequence 2: ABCDABCDABCDABCDABCDABCD

The first contains no clear pattern, the second is perfectly rhythmic. We can begin to see how probability can distinguish the two. Take the letter B: this occurs six times in each sequence of 24 letters, so the probability of encountering it at random is 6/24 = 0.25. In the first sequence B is preceded by A one time out of its six occurrences (1/6 = 0.1666), by B twice (2/6 = 0.333), by C twice (2/6 = 0.333) and by D once (1/6 = 0.1666). These figures are not greatly different from the random expectation of 0.25. However, in the second sequence, B is preceded by A in all of its six occurrences, which is markedly different from random.

The complete information on the succession of states is conveyed by the transition frequency matrix. This records the number of observed transitions between each possible pair of states. For the two sequences, the transition frequency matrices are:

Sequence 1:

To

		A	B	C	D
From	A	2	1	1	2
	B	2	2	1	1
	C	1	2	1	1
	D	1	1	2	2

Sequence 2:

To

		A	B	C	D
From	A	0	6	0	0
	B	0	0	6	0
	C	0	0	0	6
	D	6	0	0	0

The pattern in the second matrix is clear, but it can be appreciated that the first may be close to random. However, rigour is required in investigating the null hypothesis of randomness. The full procedure will be described after a consideration of the problems of dealing with real geological data.

6.1.2   Coding geological sequences

We need to convert a sedimentological log into a form which can be conveyed as a sequence of states. These may be coded as letters, so that the result may look like the simplistic sequences used above. The problems are: how do we classify the lithological units on a log into discrete states? and what is the criterion for recording a transition between two states?

A sedimentologist should not have difficulty in allocating the observed lithologies to one or other of a set of predefined categories: the problem is what degree of detail to use. This could be anything from a simple sandstone, shale, limestone, etc. classification to a scheme which distinguishes fine/medium/coarse grain sizes, packstones/wackestones/grainstones, sedimentary structures and fossil content. It is desirable to use potentially important detail in the classification, but a large number of lithological categories results in a large transition frequency matrix, and, unless the sequence is especially long, the matrix will be sparsely occupied by observations. This effectively means that there is a small sample size and, as in other statistical procedures, the result will be correspondingly unreliable. The objective should be an average of five transitions per entry in the matrix: in other words, the number of observed transitions should exceed

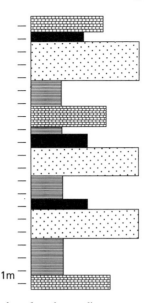

1m

Fig. 6.1 Idealised sedimentary log of sandstone, limestone, coal and shale. This can be coded for Markov analysis by recording the lithological transitions every metre or where lithology changes.

$5 \times$ (number of lithological categories)^2. This usually means that the classification is of the sandstone/shale/limestone type!

There is a choice of three criteria for defining a transition (Fig. 6.1).

1 Where the lithology changes. This is the easiest approach but has the consequence that changes from a state to itself cannot be recorded. This causes difficulties in statistical testing.

2 At bedding planes. This method is rarely applied due to difficulties in defining a bedding plane for these purposes: if we had a flaggy sandstone, we might record enormous numbers of sandstone-to-sandstone transitions.

3 Using regular specified intervals through the sequence. In this approach, thin beds may be missed and thick beds will give many self-transitions. This may be regarded as inappropriate if certain thin beds are crucial for geological interpretation, or if the thickness of thick beds is mundane information.

6.1.3   Markov analysis

6.1.3.1   The transition frequency matrix

If there are m lithologies, the transition frequency matrix is an $m \times m$ matrix with a row and a column for each lithology. The matrix is conventionally constructed and read 'from' row 'to' column, so that a transition where limestone overlies shale would be recorded in the shale row and limestone column. The transitions are counted manually, by maintaining a

Box 6.1 Worked example: Markov transition probability matrix

The sequence of limestone, sandstone, shale and coal in Fig. 6.1, taking observations every metre at the ticks on the left of the log, is:

L Sh Sh Sh S S C Sh Sh S S C Sh L Sh Sh S S S C L

Counting transitions, we obtain the observed transition frequency matrix:

		To				
		Sh	S	C	L	Total
From	Sh	4	3	0	1	8
	S	0	4	3	0	7
	C	2	0	0	1	3
	L	2	0	0	0	2
	Total	8	7	3	2	20

Dividing elements by row totals gives the observed transition probability matrix:

		To			
		Sh	S	C	L
From	Sh	0.5	0.375	0	0.125
	S	0	0.571	0.428	0
	C	0.666	0	0	0.333
	L	1	0	0	0

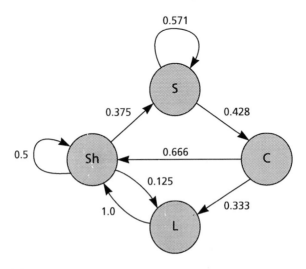

Fig. B6.1.1

Box 6.1 *Continued*

The information in this matrix allows us to construct a flow diagram which may elucidate the nature of the pattern (Fig. B6.1.1). This shows that there are two cycles (excluding self-transitions) that are more likely to occur:

Sh → S → C → Sh

and

L → Sh → S → C → L

Both suggest a deltaic sequence of coarsening up, with decreasing marine influence, culminating in a delta-top coal, followed by marine transgression.

---

tally while moving through the stratigraphic succession. As a check, the row and column totals should be found. The sum of the row totals should equal the sum of the column totals, and both should equal one less than the number of beds. If the first lithology in the succession is the same as the last, the row totals will be identical to the corresponding column totals; otherwise, there will be a discrepancy of one in two of the lithologies.

Worked example: see Box 6.1.

6.1.3.2   The transition probability matrix

The properties of the transition frequency matrix can be expressed in a more generally comparable form by conversion to probabilities. If there are 10 transitions from shale to something (10 will be the row total), and there are three transitions from shale to limestone, the conditional probability of the next bed being limestone if we start from shale is $3/10 = 0.3$; using the notation introduced in Section 2.3.4:

$Pr(limestone \mid shale) = 0.3$

The transition probability matrix is simply calculated by dividing each element in the transition frequency matrix by its row total.

The properties of this matrix can be shown by using a type of tree diagram, where we lay out the lithologies and show by mean of annotated arrows the likely 'flow' of lithological succession. If there is any significant pattern in the transitions, this diagram is used to facilitate geological interpretation: the most likely cycle or rhythm becomes apparent. The probability that a series of, say, four transitions will match a specified cycle is readily calculated by multiplying the four probabilities.

Worked example: see Box 6.1.

6.1.3.3   **Testing the transition frequency matrix**

There is no point in looking for patterns in the transition frequency and probability matrices if the data are random. The test of the null hypothesis of randomness involves comparing the observed transition frequency matrix with the matrix that would have been expected if there were no pattern. When working out what is expected, we must include consideration of the observed relative frequency of each lithology.

First, the expected random probability matrix is calculated. If the succession were random, the 'next' lithology would not be influenced by the underlying lithology: the probabilities would be unconditional. So the probability of finding a limestone after a shale would be the same as the probability of finding a limestone by blindly sticking a pin into the succession. As the probability of a transition 'to' something is the same regardless of the 'from' state, it should be clear that a column in the expected random probability matrix will have the same probability repeated in each row. This probability will be the proportion of all transitions which are 'to' that lithology, which is the column total divided by the total number of transitions.

The expected random probability matrix is then converted to the expected random transition frequency matrix by converting back to counts. If the random expectation of the transition shale to limestone is 0.2 and there are nine beds of shale observed in the succession, the expected frequency of the transition is $9 \times 0.2 = 1.8$. In general, the elements in the expected random transition frequency matrix are found by multiplying the elements in the expected random probability matrix by the corresponding row totals from the observed transition frequency matrix: this is much clearer in the context of an example! – see Box 6.2.

At this stage, we have the observed transition frequency matrix and a directly comparable matrix conveying the expected random model. These can be compared statistically by means of the χ^2 test.

χ^2 *Test for randomness in transition frequency matrix*

With $m \times m$ classes (m: umber of states) corresponding to the elements in the matrices, we can use O_j as the observed transition frequency in the jth class and E_j as the corresponding frequency from the expected random transition frequency matrix.

To test:

H_0: random succession of states

we use

$$\chi^2 = \sum_{j=1}^{m^2} (O_j - E_j)^2 / E_j$$

with $(m - 1)^2$ degrees of freedom.

Box 6.2 Worked example: test for significance of expected transition frequency matrix

The sequence presented in Box 6.1 appears to have a cyclic pattern, but is the succession significantly different from random?

Dividing each column total of the observed transition frequency matrix by the total number of transitions, we calculate the fixed probability vector:

Sh 0.4
S   0.35
C   0.15
L   0.1

This shows the probability of going 'to' each lithology if that probability is independent of the 'from' state.

The expected random transition probability matrix is then determined by these probabilities:

To

		Sh	S	C	L
From	Sh	0.4	0.35	0.15	0.1
	S	0.4	0.35	0.15	0.1
	C	0.4	0.35	0.15	0.1
	L	0.4	0.35	0.15	0.1

These probabilities are converted into expected counts by multiplying by row totals from the observed transition frequency matrix (Box 6.1) to give the expected random transition frequency matrix:

To

		Sh	S	C	L
From	Sh	3.2	2.8	1.2	0.8
	S	2.8	2.45	1.05	0.7
	C	1.2	1.05	0.45	0.3
	L	0.8	0.7	0.3	0.2

We now have an observed transition frequency matrix and an expected random transition frequency matrix in the same form. The observed and the expected counts can then be compared using χ^2.

H_0: the data come from a population of transitions that are random; the probability of encountering a lithology is not dependent on the underlying lithology

continued on p. 220

Box 6.2 *Continued*

H_1: the data come from a population of transitions that are non-random

Using:

$$\chi^2 = \Sigma(O_j - E_j)^2/E_j$$

We have:

Class	O_j	E_j	$(O_j - E_j)^2/E_j$
Sh–Sh	4	3.2	0.2
Sh–S	3	2.8	0.014
Sh–C	0	1.2	1.2
Sh–L	1	0.8	0.05
S–Sh	0	2.8	2.8
S–S	4	2.45	0.981
S–C	3	1.05	3.621
S–L	0	0.7	0.7
C–Sh	2	1.2	0.533
C–S	0	1.05	1.05
C–C	0	0.45	0.45
C–L	1	0.3	1.63
L–Sh	2	0.8	1.8
L–S	0	0.7	0.7
L–C	0	0.3	0.3
L–L	1	0.2	3.2
			$\Sigma = 19.229$

Using d.f.

$$v = ((\text{no. of lithologies}) - 1)^2$$

we have

$$v = (4 - 1)^2 = 9$$

The critical value is

$$\chi^2_{0.05,9} = 16.92 \quad (\text{Appendix 2.6})$$

The calculated value exceeds the critical value, so we reject the null hypothesis and conclude that there is a significant Markov property: the occurrence of lithologies is, to an extent, dependent on preceding lithology.

However, note that:

1 We have not met the guideline that the expected values should all be greater than 5. This undermines the validity of the result. A larger number of transitions is required, but was not presented here to avoid the illustrative material becoming too unwieldy!

2 A rejection of the null hypothesis of randomness isn't the same thing as accepting a cyclic scheme such as those shown in Box 6.1.

Worked example: see Box 6.2.

Warnings

1 There should be at least five transitions expected in most of the classes, so the number of transitions should be at least $5m^2$.

2 If the criterion for transition used is a change in lithology (see Section 6.1.1), there must be zeros in the leading diagonal elements of the observed matrix, but there will inevitably not be zeros in the leading diagonal of the expected random matrix, so the expected matrix is not a proper null model. There is a rather tedious iterative method available for compensating for this; use of specialised software is advised.

3 It should be remembered that the test leads to rejection of H_0 if there is any type of non-randomness, not just cyclicity or rhythmicity. If the criterion for transition used is regular intervals or bedding planes (see Section 6.1.1), a likely (and rather uninteresting) type of non-randomness is great thicknesses of the same lithology!

The above test investigates the structure of the whole matrix. Some authors have introduced tests for individual elements in the matrix (specific transitions) based on binomial probabilities: we can assess the probability of the observed transition count (O_j) having arisen from a 'population' with the expected transition probability. This would clearly involve more calculation to assess the whole matrix, and is particularly not recommended because of the interrelatedness of the probabilities in the matrix and the problems of choosing an appropriate level of significance when m^2 repetitions of the test are being done. If there are, say, 25 elements in the matrix, we expect at least one element to be significant at $\alpha = 0.05$ when the sample is drawn from a random sequence, so a rejection of H_0 would be erroneous.

6.2   SERIES OF EVENTS

6.2.1   Objectives and applications

The simplest type of time-based data is merely a list of times or dates. In geology, this form is used for analysis of historical events such as earthquakes and volcanic eruptions, and for more distant geological phenomena such as meteorite impacts and mass extinctions. The data are regarded as points in time: the event must be very short compared with the length of the time interval being considered. It would be difficult to regard an orogeny as a point event on any time scale, but a meteorite impact is certainly abrupt enough! Many 'events' are just peaks on a continuously varying curve: a good example is 'earthquake' – this is only an 'event', rather than just a peak on a continuous seismic record, when it is defined in terms of exceeding a specified threshold value.

A peculiarity of series of events analysis is the unusual importance of prediction. The end of a time record of events is usually the present, and there is great human interest in extrapolating into the future a record of earthquakes, eruptions, meteorite impacts or mass extinctions! Indeed, human interest may determine the thresholding of seismic activity: if it is life-threatening, it is an event.

Events may be randomly distributed through time, in which case prediction on the basis of the past pattern of events is useless. Non-randomness can be in many forms: regularity, trends or patterns, and each can help in prediction. Note that, in the context of this book, we are only considering prediction on the basis of the sequence of events, and not from the special geological monitoring that may be used in areas of volcanic or seismic hazard.

A generally important point to bear in mind when preparing data for analysis is the definition of the start and end of the time interval for which data exist. Events themselves should not be used to demarcate the terminations; this would introduce an immediate bias in the record, increasing the event density and changing the pattern. The time interval should be defined independently: in the case of historical data, the beginning and end of available records may be suitable.

6.2.2 Testing for randomness

Randomness in event data means that the occurrence of an event does not affect the probability of occurrence of other events. This seems inherently unlikely in geological situations: an earthquake or eruption may change probabilities of other events by releasing stress or causing instability. However, the natural processes acting in such situations may be so complex that the result is indistingishable from randomness.

Imagine 10 random events in 100 years: if we divide the time-scale up into regular intervals (say, decades) and count the number of events in each interval, it is likely that there will be one or two decades with zero events, many with one event and a few with two events. Numbers of events more than two would rapidly become increasingly unlikely. The frequency distribution of random data in this form fits the Poisson model – see Section 2.3.7.3. The Poisson distribution has one parameter (γ) describing the average density of points. The test for goodness of fit to the Poisson model can be done using the χ^2 test. The number of intervals expected to have j events, according to the Poisson model, is determined from the total number of events (n) and number of intervals T by:

$$Te^{(-n/T)}(n/T)^j/j!$$

A series of events can differ from the Poisson model by being more clustered in time, in which case there will tend to be anomalously high numbers of intervals with zero and many events, or by being more regular,

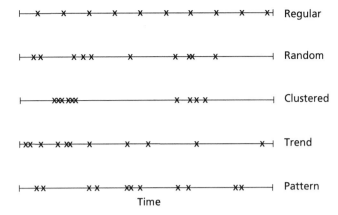

Fig. 6.2 Examples of types of possible distribution of events in time. A random sequence is shown together with four different types of non-randomness.

in which case the distribution will be more peaked, with all intervals tending to have the same number of events: see Fig. 6.2.

Testing for Poisson distribution

H_0: events randomly distributed through time
H_1: events clustered or regular

With n events in T regular intervals, the observed number (O_j) of intervals with j events is compared with the expected number (E_j) predicted on the basis of the Poisson distribution, where

$E_j = Te^{(-n/T)}(n/T)^j/j!$

by using the χ^2 test:

$\chi^2 = \Sigma(O_j - E_j)^2/E_j$

The number of classes (and hence number of summations) used will depend on the data. Most classes will correspond to one value only of j, but at the ends of the distribution the j numbers should be combined together in classes under the general guideline that the expected frequency E_j in each class should be five or more.

Note that, for $j = 0$, we use $j! = 1$.

Degrees of freedom $v =$ (number of classes used in χ^2) $- 2$

Worked example: see Box 6.3.

Warning

The test is only sensitive to the frequency distribution of interval counts; once the number in intervals is assessed, the intervals are effectively dissociated from each other and from the time-scale. Consequently, there is no sensitivity to non-randomness in the form of trends of increasing or decreasing event frequency through time.

Box 6.3 Worked example: test for Poisson distribution in series of events

10 m

Fig. B6.3.1

In a sequence of 45 m of Devonian carbonates, thin tuff horizons were observed at the positions shown, in Fig. B6.3.1. Measured from the base of the sequence, the positions were (m):

```
0.5   2.3   3.2   4.2   4.9   7.0   11.4  12.7  14.6
16.0  21.5  22.5  25.8  30.3  31.9  36.2  42.8
```

Do the tuffs occur at random through the sequence? If so, the number of tuffs observed in each of regular intervals should fit a Poisson distribution. We must remember that this type of data will be the result of sedimentation rates as well as the timing of events, so any result will not be immediately attributable to the characteristics of the volcanic activity.

If we use 3 m intervals, we can make the following count:

Interval	Count (j)
0–3	2
3–6	3
6–9	1
9–12	1
12–15	2
15–18	1

Box 6.3 *Continued*

Interval	Count (*j*)
18–21	0
21–24	2
24–27	1
27–30	0
30–33	2
33–36	0
36–39	1
39–42	0
42–45	1

So the frequency distribution of counts is:

j	O_j
0	4
1	6
2	4
3	1
4	0

To calculate the number expected according to the Poisson distribution, we use

$$E_j = Te^{-n/T}(n/T)^j/j!$$

For these data we have

Number of intervals $T = 15$
Number of events $n \quad = 17$

so

$$Te^{-n/T} = 15 \times e^{-17/15} = 4.829$$

and we can calculate:

j	$(n/T)^j$	*j*!	$(n/T)^j/j!$	E_j
0	1.0	1	1.0	4.829
1	1.133	1	1.133	5.470
2	1.284	2	0.642	3.101
3	1.456	6	0.243	1.173
4	1.649	24	0.069	0.333
5	1.870	120	0.013	0.063

continued on p. 226

Box 6.3 *Continued*

For the χ^2 test:

H_0: data drawn from population with Poisson distribution (tuffs are randomly distributed)
H_1: data not drawn from population with Poisson distribution
$\chi^2 = \Sigma(O_j - E_j)^2/E_j$

j	O_j	E_j	$(O_j - E_j)^2/E_j$
0	4	4.829	0.142
1	6	5.470	0.051
2–∞	5	4.701	0.019
Total	15	15.000	$\chi^2 = 0.212$

Note that we have had to combine classes in order to (nearly!) fulfil the criterion for χ^2 that the number expected in each class should be at least 5.

From tables (Appendix 2.6), the critical value of χ^2 for $\alpha = 0.05$ and d.f. = (no. of classes) − 1 = 2 is 5.99.

Although the test is not very reliable for such a small number of classes, it is clear that χ^2 is very small and we fail to reject H_0. The data fit the Poisson distribution well and so may be random.

6.2.3   **Testing for trends**

Regardless of any other pattern, trends of decreasing or increasing frequency of events can be of overriding importance in prediction and in developing an understanding of the processes involved.

Changes in the frequency of events can be quantified by changes in the length of the interval between events. A graph of the event number (in order from first to last) against the interval between events will show any trends that exist, and correlation can be used to quantify this. The 'event number' scale is best regarded as ordinal, so we should use a non-parametric correlation coefficient to test for significance. With a null hypothesis of no trend of increasing or decreasing event frequency, we can apply the test using Spearman's rank correlation coefficient (Section 4.3).

Spearman's rank correlation coefficient for trend in interval lengths

H_0: no trend of change in interval lengths
H_1: interval lengths become shorter/longer

$$r' = 1 - \frac{6\sum_{i=1}^{n}(i - R(h_i))^2}{n(n^2 - 1)}$$

where h_i = length of ith interval, and n = number of intervals = (number of events) − 1.

Compare calculated values with critical value in tables (Appendix 2.11). Worked example: see Box 6.4.

Box 6.4 Worked example: test for trends in series of events

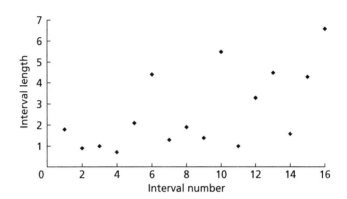

Fig. B6.4.1

Using the tuff band data from Box 6.3, is there any trend in increasing or decreasing frequency? If so, there should be a correlation between the event number and the interval between events. The intervals are:

Event:　　0.5　　2.3　　3.2　　4.2　　4.9　　7.0　　11.4　　12.7　　14.6
Interval:　　　　1.8　　0.9　　1.0　　0.7　　2.1　　4.4　　1.3　　1.9　　1.4
Event:　　16.0　21.5　22.5　25.8　30.3　31.9　36.2　42.8
Interval:　　　5.5　　1.0　　3.3　　4.5　　1.6　　4.3　　6.6

We need to assess the correlation between the event number and interval length using Spearman's rank correlation coefficient:

$$r' = 1 - \frac{6\sum_{i=1}^{n}(i - R(h_i))^2}{n(n^2 - 1)}$$

H_0: no trend in interval length
H_1: trend in interval length

continued on p. 228

Box 6.4 *Continued*

The interval numbers and their ranks with the squared differences are:

Number i	Interval h_i	Interval rank $R(h_i)$	$(i - R(h_i))^2$
1	1.8	8	49
2	0.9	2	0
3	1.0	3.5	0.25
4	0.7	1	9
5	2.1	10	25
6	4.4	13	49
7	1.3	5	4
8	1.9	9	1
9	1.4	6	9
10	5.5	15	25
11	1.0	3.5	56.25
12	3.3	11	1
13	4.5	14	1
14	1.6	7	49
15	4.3	12	9
16	6.6	16	0
			$\Sigma = 287.5$

$r' = 1 - (6 \times 287.5)/16 \times (16^2 - 1) = 1 - (1725/4080) = 0.577$

From tables (Appendix 2.11), we find that the critical value of r' at $\alpha = 0.05$ and $n = 16$ is 0.427.

The calculated value exceeds the critical value, so we reject H_0 and conclude that there is a trend. Because r' is positive, this means that interval lengths increase, so events become significantly more spread out later in the sequence.

6.2.4 Testing for uniformity

Uniformly distributed events are likely to result from geological processes in which the occurrence of an event reduces the probability of another in the near future, but with increasing probability thereafter. This is likely where events release stress which accumulates in the intervening time.

The Kolmogorov–Smirnov (KS) test (Section 2.5.6.2) can be used for a simple test with a null hypothesis of uniformity. The null model is expressed on a cumulative frequency plot (with time as the horizontal axis) as a straight line: see Fig. 6.3. The cumulative plot for the data will inevitably differ from this, and the maximum vertical discrepancy is the basis of the KS statistic. Critical values of KS are based on the magnitude of discrepancy

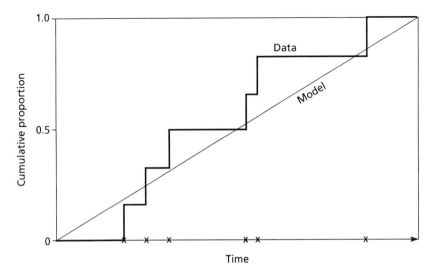

Fig. 6.3 Comparison of series of events with a uniform model. Both are expresssed on a cumulative scale for comparison.

that would be expected with random data: if this is exceeded, we reject the null hypothesis.

A useful property of this test is its sensitivity to trends as well as clusters. The KS calculation involves comparing, for each event, the proportion of events elapsed with the proportion of time elapsed. For the ith event in a series of n events in a time record of length T, where t_i is the time elapsed since the start of the record, we compare i/n with t_i/T. If the events were perfectly uniform, t_i/T would, for every event, equal $(i - k)/n$, where k could be any constant between 0 and 1. For the KS test, we calculate

$$(i/n) - (t_i/T)$$

and

$$((i - 1)/n) - (t_i/T)$$

for each of the events: these form the corners of the steps on the cumulative plot (Fig. 6.3). If we have, for example, perfect uniformity (regularity) in the form of 10 events in a 100-year record, the events will occur at 5, 15,

Table 6.1

Event	Year	i/n	$(i - 1)/n$	t_i/T
1	5	0.1	0	0.05
2	15	0.2	0.1	0.15
3	25	0.3	0.2	0.25
⋮	⋮	⋮	⋮	⋮
10	95	1.0	0.9	0.95

25, ... , 95 years after the start. For this situation, the differences will be minimum for every event (Table 6.1). So the less the uniformity, the greater some of the values will be. The largest absolute value forms the KS statistic.

Kolmogorov–Smirnov test for uniformity

H_0: events uniform or random
H_1: events clustered or with trend of changing frequency
$KS = \sqrt{n} \max |((i - 1)/n) - (t_i/T), (i/n) - (t_i/T)|$

KS is compared directly with critical values from tables (Appendix 2.7).
        Worked example: see Box 6.5.

Box 6.5 Worked example: test for uniformity in series of events

Using the same data as in Box 6.3, are the events distributed uniformly or are there concentrations at some positions?
        Using Kolmogorov–Smirnov (KS), we compare the actual cumulative position with those expected from the uniform model. With n events numbered i at position t_i observed in a period of length T, the expected positions are between i/n and $(i - 1)/n$

i	t_i/T	i/n	$(i - 1)/n$	$t_i/T - i/n$	$t_i/T - (i - 1)/n$
1	0.011	0.059	0	−0.048	0.011
2	0.051	0.118	0.059	−0.067	−0.008
3	0.071	0.176	0.118	−0.105	−0.047
4	0.093	0.235	0.176	−0.142	−0.083
5	0.109	0.294	0.235	−0.185	−0.126
6	0.156	0.353	0.294	−0.197	−0.138
7	0.253	0.412	0.353	−0.159	−0.100
8	0.282	0.471	0.412	−0.189	−0.130
9	0.324	0.529	0.471	−0.205	−0.147
10	0.356	0.588	0.529	−0.232	−0.173
11	0.478	0.647	0.588	−0.169	−0.110
12	0.500	0.706	0.647	−0.206	−0.147
13	0.573	0.765	0.706	−0.192	−0.133
14	0.673	0.823	0.765	−0.150	−0.092
15	0.709	0.882	0.823	−0.173	−0.114
16	0.804	0.941	0.882	−0.137	−0.078
17	0.951	1.000	0.941	−0.049	0.010

The KS statistic is the maximum of the absolute values of the numbers in the two right-hand columns multiplied by $\sqrt{n}$, so

$KS = 0.232 \times 4.126 = 0.957$

From tables (Appendix 2.7), we find that the critical value of KS for $\alpha = 0.05$ and $n = 17$ is 0.318.

Box 6.5 *Continued*

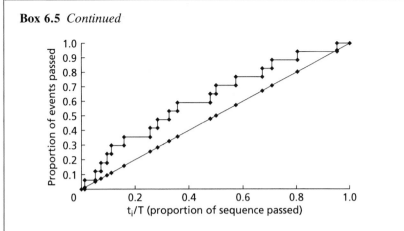

Fig. B6.5.1

The calculated value exceeds the critical value, so we reject the null hypothesis and conclude that the events are not uniform. This result is compatible with that from Box 6.4; the lack of uniformity is due to a higher density of events at the beginning of the series.

Note that regularity, the extreme case of uniformity, will yield extremely low values of KS in this test, but cannot easily be statistically distinguished from uniformity. Tests with a null hypothesis of randomness lead to rejection if data are either uniform or regular, and tests with a null hypothesis of uniformity lead to failure to reject if data are uniform or regular. In the case of events which we may wish to define as regular, we need to permit some random variation or error around the perfect periodic pattern, but larger and larger amounts of such variance create a continuum of possibilities from regular through uniform to random. There is no objective way of placing a boundary (in terms of amount of acceptable error) between regular and uniform.

6.2.5   **Testing for patterns**

There are types of non-randomness which would remain undetected by the methods so far described. These include short-term cycles of changing event frequency and patterns of alternately long and short intervals. This is common in geology, where uniformity due to stress accumulation and release is combined with clustering of events in phases of activity.

Patterns may be looked for in the succession of interval lengths (h). The relationship between successive interval lengths can be investigated by correlation coefficients: if there were no pattern in successive intervals, there would be zero correlation between h_i and h_{i+1}. As there is no reason to expect a normal distribution, we should use a non-parametric method. A significant negative correlation indicates alternate long and short intervals; a positive correlation indicates that successive interval lengths are similar (so interval lengths must change only gradually).

Spearman's rank correlation coefficient for pattern in interval lengths

H_0: no relationship between successive interval lengths
H_1: correlation between successive interval lengths

$$r' = 1 - \frac{6 \sum_{i=1}^{n} (R(h_i) - R(h_{i+1}))^2}{n(n^2 - 1)}$$

where n = (number of events) − 2. (Refer to Section 4.4.)
   Compare calculated values with critical value in tables (Appendix 2.11).
   Worked example: see Box 6.6.

Box 6.6 Worked example: test for pattern using serial correlation

Using the data on tuff bands introduced in Box 6.3, is there any pattern in the occurrence of the tuffs?
   We can test for relationship between successive interval lengths by assessing the non-parametric correlation between intervals i and $i + 1$.

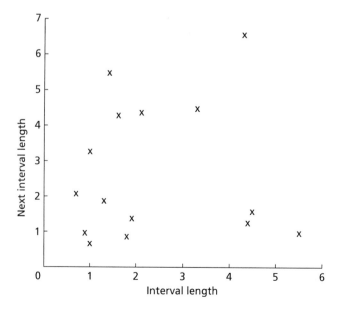

Fig. B6.6.1

Box 6.6 *Continued*

Interval h_i	Rank $R(h_i)$	Next int. h_{i+1}	Rank $R(h_{i+1})$	$(R(h_i) - R(h_{i+1}))^2$
1.8	8	0.9	2	36
0.9	2	1.0	3.5	2.25
1.0	3.5	0.7	1	6.25
0.7	1	2.1	9	64
2.1	10	4.4	12	4
4.4	13	1.3	5	64
1.3	5	1.9	8	9
1.9	9	1.4	6	9
1.4	6	5.5	14	64
5.5	15	1.0	3.5	132.25
1.0	3.5	3.3	10	42.25
3.3	11	4.5	13	4
4.5	14	1.6	7	49
1.6	7	4.3	11	16
4.3	12	6.6	15	9
				$\Sigma = 511.0$

Using Spearman's rank correlation coefficient:

$$r' = 1 - \frac{6 \sum_{i=1}^{n} (R(h_i) - R(h_{i+1}))^2}{n(n^2 - 1)}$$

$$r' = 1 - (6 \times 511)/15 \times (15^2 - 1) = 1 - (3066/3360) = 0.0875$$

From tables (Appendix 2.11), we find that the critical value of r' at $\alpha = 0.05$ and $n = 15$ is 0.443.

The calculated value does not exceed the critical value, so we fail to reject H_0 and conclude that there is no serial correlation: consecutive intervals seem to be independent.

Note that this is similar to the autocorrelation at lag $= 1$: see Section 6.3.3. Whereas autocorrelation at lag $= 1$ is usually mundane due to inherent smoothness in data, here it is significant: consistency of successive interval lengths may be of great interest.

An alternative approach which pays less attention to the absolute lengths of intervals is the runs test. Successive interval lengths are coded simply as increasing (up, or U) or decreasing (D). The series of events is then represented as a series of Us and Ds, which themselves are analysed for patterns. This process is described in further detail in Section 6.3.4.2.

6.3   **TIME SERIES ANALYSIS**

Time series techniques deal with data which record the change of a variable through time. In some analyses, such data can be regarded as simple bivariate data where one variable happens to be time. However, time is a special variable in that: (a) the closeness in time of two observations has particular importance, unlike similarity with respect to normal types of variable; (b) data have the property of being inherently ordered with respect to time; and (c) many natural processes engender cyclicity through time. The property cited in (a) is analogous to closeness of distance, and we will find that many time series techniques and concepts will be applicable to data on traverses and spatial data (Chapter 7).

The obvious importance of the time dimension in much of geology might suggest that time series techniques should find widespread use in the subject, but in fact the imperfect calibration of time in the rock record restricts the method to special cases. In general, applications are to three types of data.

1 Data on very long time-scales (10^6–10^9 years), where radiometric dating provides sufficient resolution. Such data inevitably tend to convey major, global changes, such as magnetic reversals, eustatic sea-level changes and rates of extinction in the fossil record.

2 Geophysical data, principally seismic, where time is readily and accurately measured on very short time-scales. Many of the techniques explained in this section are used in seismic data processing, but this is a complex and specialised field that will not be specifically dealt with in this book.

3 Data on intermediate time-scales (1–10^6 years). On these scales, time can only be known with sufficient resolution and accuracy in the case of historical data (e.g. volcanic and earthquake activity). However, geological interpretations of sedimentological or other sequences may allow a speculative time-scale to be used: see Box 6.7. This may sometimes be done with confidence, as with varved lake sediments, where the process producing the annual layers is well known and criteria for their recognition are established. Increasingly often, however, geologists are approximating time by stratigraphic position, particularly in attempts to detect Milankovich climatic cycles. This involves the assumption that the sedimentation rate is constant and that the sequence is continuous, or that perturbations in these respects are not substantive (see Box 6.8). These assumptions may be justified retrospectively by positive results, but all such analyses should be treated with caution.

6.3.1   **Preparation of data**

Many simple procedures are available for enhancing time series data. This may be done preparatory to further analysis, or it may be an end in itself.

Box 6.7 Time interpretations of geological sequence

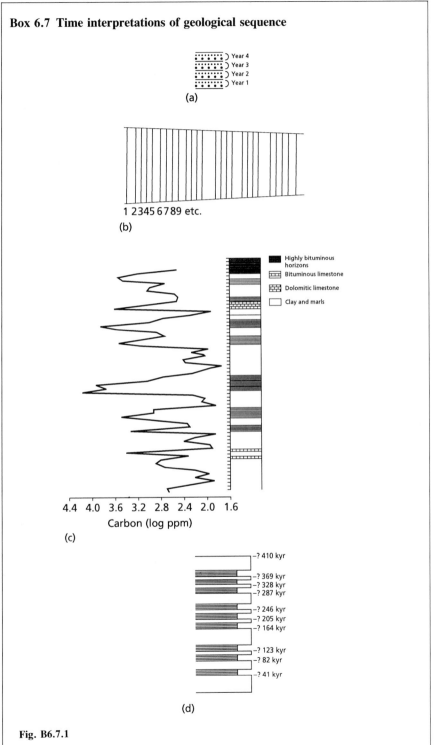

Fig. B6.7.1

continued on p. 236

Box 6.7 *Continued*

1 Varves. Using recent analogues, 1 varve = 1 year. The time series consists of a sequence of annual measurements of varve thickness.
2 Growth lines on fossils. Recent marine organisms are known which secrete shell material in rhythms corresponding to years, lunar months, days and tidal cycles. The time series consists of a sequence of growth line spacings, each probably corresponding to a consistent interval of time, but what length of interval it is may not be known a priori.
3 Stratigraphic log conveying aspect of sediment composition (e.g. electric log, geochemical log). If sedimentation rate is assumed to be constant, time is proportional to sedimentary thickness, so a distance measurement (e.g. depth) can substitute for time. (After Dunn, 1974.)
4 Sequence of bed thicknesses. If controlled by Milankovitch or other climatic cyclicity, beds may be interpreted as due to peaks on relatively high-frequency climatic fluctuations exceeding a threshold in the sedimentological process. If, for example, each bed corresponds to a peak on a 41 kiloyear (ky) cycle, the time series consists of a regular sequence of bed thicknesses recording intervals of 41 ky, and can be analysed to reveal lower frequency cycles.

Box 6.8 Corruptions of the time record

Corruptions of time record

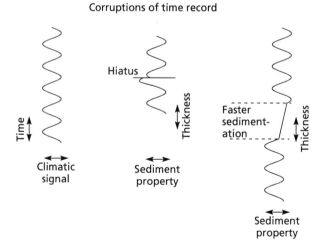

Fig. B6.8.1

Figure B6.8.1 shows corruption of the sediment composition record by changing sedimentation rate. This primarily affects interpetation of type **3** in Box 6.7.

Box. 6.8 *Continued*

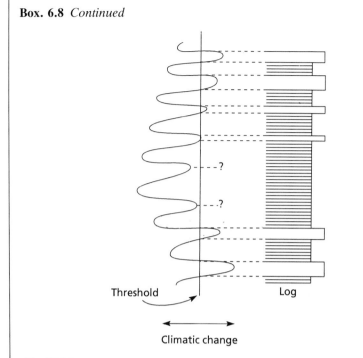

Fig. B6.8.2

Figure B6.8.2 shows corruption of bed thickness data (type **4** in Box 6.7) due to failure to observe low peaks on cycles: there would be less of a problem if the low peaks were recorded as beds of zero thickness, but this, of course, cannot be done objectively! Similar problems are likely to occur in types **1** and **2** of Box 6.7.

The selection described here give an indication of some of the possibilities. These methods tend to be custom-built to serve a specific purpose; consequently they are not often to be found in software packages. Fortunately, it is not usually difficult, even for the novice programmer, to write programs to achieve the desired result.

Interpolation

Variables changing through time can usually be regarded as doing so continuously, so values could in principle be recorded at infinitely short intervals. (An exception to this are inherently discrete data such as varves, in which the shortest time interval resolvable is 1 year.) In practice, of course, a time series is represented by a finite number of observations. Most time

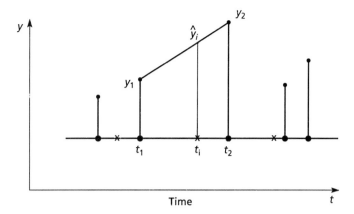

Fig. 6.4 Simple linear interpolation procedure. Dots indicate times of observations; crosses show the regularly spaced times at which data are needed. Value $\hat{y}_i$ at time t_i is here being estimated by interpolation.

series analytical techniques demand that these should be equally spaced through time. This is an obvious strategy for data collection, especially where electronic instrumentation is used, but occasionally the practical constraints of data collection result in irregularly spaced data. It may, then, be necessary to obtain a regularly spaced estimated sequence by appropriate interpolation between irregularly spaced data. A simple approach is linear interpolation (Fig. 6.4). This yields an estimated value ($\hat{y}_i$) for the variable at time t_i by:

$$\hat{y}_i = y_1 + \frac{t_i - t_1}{t_2 - t_1}(y_2 - y_1)$$

where t_1 and t_2 are the times at adjacent data points and y_1 and y_2 are the values at these points.

This yields reasonable estimates provided that the time trend is smooth relative to the spacing between data points. There are an abundance of more sophisticated methods available which use polynomial regression and splines: such techniques are more commonly used in estimation of point values from spatial data and so are appraised in Section 7.4. However, regardless of the sophistication of the method, reliable estimates cannot be made if the irregularity of the time series is on a fine scale relative to the spacing between estimated and observed data points. This is related to the important concept of autocorrelation: see Section 6.3.3.

Smoothing

Raw data tend to be composed of two components: the 'signal' generated by the geological process and the 'noise' caused by random interference. Much of statistics aims to separate these two. Smoothing is a crude but effective way to reduce noise in time series data, especially for the purpose of graphical clarity. The principle is that noise tends to occur at high frequencies and hence has inconsistent effects on adjacent observations; consequently, it can be reduced by 'averaging out' over a short series of observations. Care must be taken that the averaged series isn't so long as to smooth out the signal as well! In practice, then, the observed value at each point is replaced by an estimate of the noise-free value. This estimate is calculated by arithmetic averaging of the observed value at that point and values at adjacent points. It is clear that the observed value at the point itself should have the greatest influence in the estimate, and that neighbouring points should have more influence than distant points: this is done by appropriate weighting of the values. The problem is to decide: (a) the number of observations to be involved in the weighted averaging; and (b) the values of the weightings. The number of observations must be odd (in order to be symmetrical about the central value) but is otherwise dictated by the user's appreciation of the data and the likely underlying signal: a short series (3–7) will be needed to ensure that a relatively high frequency signal will remain, but otherwise longer series will smooth more effectively. The choice of weights is not a serious problem: ideal weights based on estimation by polynomial regression have been calculated by mathematicians. For example, Table 6.2 gives weights derived from quadratic polynomials for the central point and two points either side. The estimate y_i' at time t_i is calculated by:

$$y_i' = (-3y_{i-2} + 12y_{i-1} + 17y_i + 12y_{i+1} - 3y_{i+2})/35$$

Note that 35 is the sum of the weights: division by the sum of the weights is essential to produce a total weight of 1 and hence to prevent systematic overestimation.

For longer weighted series, the quadratic weights are as in Table 6.3.
A result of a nine-term smoothing is shown in Fig. 6.5.

Table 6.2 Quadratic polynomial smoothing: 5 terms

t_{i-2}	t_{i-1}	t_i	t_{i+1}	t_{i+2}
−3	12	17	12	−3

Table 6.3 Quadratic polynomial smoothing: 5–9 terms

No. of terms	t_i	t_{i+1} t_{i-1}	t_{i+2} t_{i-2}	t_{i+3} t_{i-3}	t_{i+4} t_{i-4}
5	17	12	−3		
7	7	6	3	−2	
9	59	54	39	14	−21

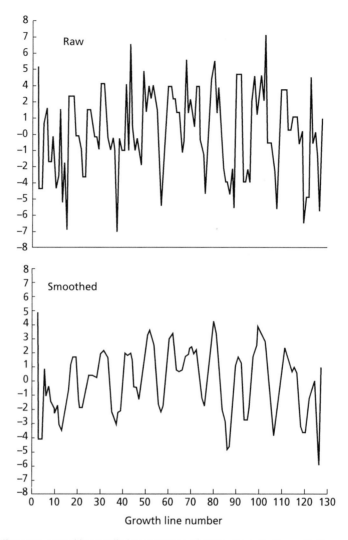

Fig. 6.5 Nine-term smoothing applied to sequence of measurements of growth line spacing from a Silurian nautiloid (Appendix 3.8). Smoothing removes the blurring effect of random noise.

Note that, for a smoothing series of $2n - 1$ terms, the first and last n points of the original time series data cannot be operated on in the same manner. These may be smoothed using fewer terms or left unaltered.

Windows and filters

The smoothing procedure described in the previous section is an example of an algorithm where a pattern (in that case, the series of weights) is applied to each point of a time series in succession in order to transform the data into a form that is in some way enhanced. The range of points in the time

Table 6.4 Terms for simple difference filter

t_{i-1}	t_i
-1	1

Table 6.5 Terms for polynomial differential filter

t_{i-2}	t_{i-1}	t_i	t_{i+1}	t_{i+2}
-2	-1	0	1	2

series over which such algorithms operate in each iteration (= the number of terms in smoothing) is often described as a window, and the specific pattern and algorithm comprise a filter (the smoothing procedure 'filters out' noise). These can be custom-built to enhance any aspect of a time series: if a specific pattern in the input data is of interest, a filter can be designed which marks each occurrence of the pattern with a conspicuous spike in the output. This procedure is used primarily with seismic data and examples may be sought in geophysics textbooks.

A more generally applicable use of filters and windows is in the enhancement of gradients in time series data. This can be used for measurement of rates of change of the variable, or for the division of the time series into internally consistent zones delimited by phases of rapid change. The latter approach has clear applications with stratigraphic data.

A simple approach to this problem is to produce a series of differences between successive observations. In terms of a filter, the weights are as in Table 6.4, and the calculation is simply

$$y'_i = y_i - y_{i-1}$$

where y'_i forms the new time series of differences.

This produces a spiky result (Fig. 6.6a). A more rigorous and satisfactory result is obtained by using a filter based on the differential of the polynomial functions used in smoothing filters. For a five-term window, the weights for the filter are as in Table 6.5, so the equation is simply

$$y'_i = 2y_{i+2} + y_{i+1} - y_{i-1} - 2y_{i-2}$$

A result of the application of this is shown in Fig. 6.6(b).

For further enhancement of steep local gradients for the purpose of zonation, a 'split moving window' technique is available. Here, a window divided into two equal halves is moved along the data array. At each position, the mean and variance of the data are calculated in each half separately. In positions corresponding to transitions between zones, the

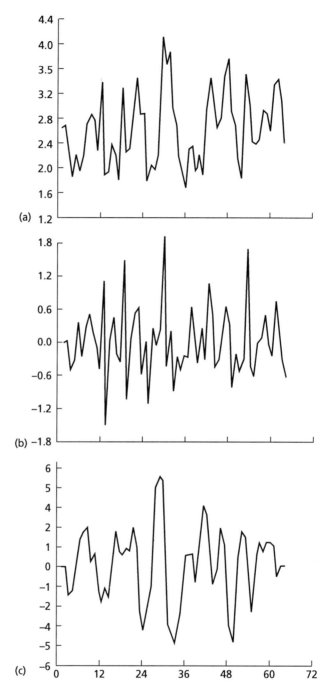

Fig. 6.6 Filtering to enhance rates of change: a sequence of data (a) consisting of measurements of organic carbon from a section in the Upper Jurassic Kimmeridge Clay (Appendix 3.7) is transformed to show rates of change of organic carbon using a simple two-term difference filter (b) and a five-term filter (c).

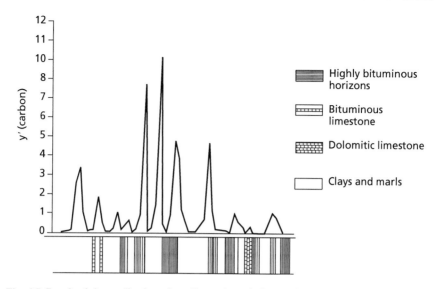

Fig. 6.7 Result of the application of a split moving window to the sequence introduced in Fig. 6.6. The sedimentary log is shown for comparison. Peaks on the curve mark points of rapid change; these may be used to break the sequence up into beds or zones.

difference between two means is large relative to the two variances. This can be quantified by:

$$y_i' = \frac{\bar{y}_1 - \bar{y}_2}{s_1^2 + s_2^2}$$

where $\bar{y}_1$, s_1^2 and $\bar{y}_2$, s_2^2 are the means and variances of the variable in the two halves of the window.

The consequent plot of y' (Fig. 6.7) shows sharp spikes at zonal boundaries, but the interpretation of the geological significance of these may remain subjective.

6.3.2 Trends through time

The term 'trend' applied to time series data is sometimes used to mean just the 'signal' as opposed to the 'noise'. Here we restrict its use to long-term linear trends rather than cycles. These are not discernible using the types of analysis covered in Section 6.3.3 and such trends actually need to be removed if such analyses are to be applied.

Linear trends are readily detected using simple bivariate linear regression, with time as the independent variable, plus the analysis of variance (ANOVA) F-test for significance (refer to Section 3.3). If a linear trend does exist, it can be removed by using the residuals from the regression rather than the raw data. Thus, if the regression equation with time (t) as the independent variable is:

$$y = a + bt$$

the residuals y' will be calculable by:

$$y_i' = y_i - bt_i - a$$

Worked example: see Box 6.9.

Box 6.9 Worked example: removal of trend from time series

Is there a trend in data on organic carbon from the Kimmeridge Clay (Appendix 3.7)? If so, it should be removed prior to further analysis.
Raw data are shown in Fig. B6.9.1.

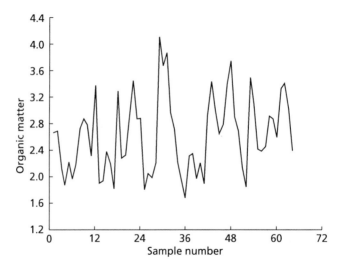

Fig. B6.9.1

Regression equation from statistics package:

$$y = 2.36 + 0.00805x$$

F test from ANOVA:

Calculated $F = 4.24$

From tables (Appendix 2.5):

Critical $F = 4.00$

Result: trend is significant.
Transform data by:

$$y' = y - 0.00805x - 2.36$$

Box 6.9 *Continued*

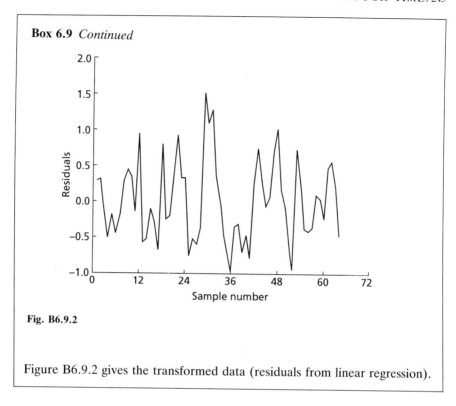

Fig. B6.9.2

Figure B6.9.2 gives the transformed data (residuals from linear regression).

Although the use of the residuals in further analysis is not necessary if the regression gradient is not significantly different from zero, and it is often not necessary to subtract the constant (*a*), it is nevertheless good practice to use this as a routine procedure.

Polynomial regression is not often useful in detecting trends in time series data as curved trends through time tend to be better described in terms of cycles.

6.3.3   **Detection of cycles**

There is an inherent tendency for many types of earth science time series data to be cyclic because of the pervasive effect of astronomical cycles. Additionally, geological systems such as magma chambers, hydrothermal systems and stress–release structures may also settle into cyclic behaviour. In seismic processing, it is important to analyse the cycles resulting from seismic waves passing a sensor.

Cyclic astronomical influences on the earth are diverse in terms of time-scale, influence and scientific status. On relatively short time-scales, cycles are well known in ordinary experience: daily, tidal, lunar and yearly cycles have obvious effects. These are, however, only of interest to the geologist in cases where exceptional detail is preserved in the record. Classic examples are varves and growth bands on fossils. Longer astronomical cycles, such as the precession, obliquity and orbital Milankovich cycles (10^4–10^6 years) are known processes but the degree of their influence in the geological record is under debate. Even larger cycles (*c.* 30 megayears) have been claimed to occur in a variety of global phenomena, but here both the evidence and the suggested causative process (cyclic phases of meteorite impacts) are controversial. Statistical treatment of these types of data is essential. Although a variety of specialised statistics have been used, the standard techniques for analysis of cyclicity are autocorrelation and Fourier analysis. These, however, require good-quality data, with no distortion in the time dimension. If the time record is uncertain, as with stratigraphic data, time is an ordinal measurement, and better results may be achieved using the pattern detection techniques described in Section 6.3.4.

6.3.3.1 Autocorrelation

Autocorrelation, as the name implies, involves correlating a sequence of data with itself. If this is done on the raw data, the result is obviously a correlation coefficient of 1.0. The process only becomes interesting when the correlation is calculated between duplicates of the time series which are displaced relative to each other. In other words, the two sets of numbers to be correlated are arrived at by pairing each value y_i with $y_{i+\tau}$, where i gives the time or position in the time series and τ is an integer value of displacement known as the lag. The correlation coefficient between the time series and a displaced copy of itself is known as the autocorrelation coefficient, r_τ. It can be calculated at successive lags – we can imagine this as sliding the time series past itself – and the resulting series of r_τ values reveals useful information on the structure of the data. These are plotted on an r_τ vs. τ graph called an autocorrelogram (Figs 6.8 & 6.9). This procedure is included on many statistical software packages. There are two important features of the autocorrelogram.

1 It is inevitable that the autocorrelation coefficient at zero lag is 1.0 and that this will decline as the lag begins to increase. However, the rate of decline of r_τ will depend on the smoothness of the data. Totally random data will show an arbitrarily small r_τ at lag $\tau = 1$, and this will fluctuate randomly at higher lag. However, if there is a degree of smoothness, values of data points are not totally different from those immediately before and after and this will be reflected in a positive autocorrelation coefficient at low values of lag. Eventually, with increasing lag, the r_τ values decline into insignificance, but the data can be described as being autocorrelated over time intervals up to that point. The concept of the autocorrelated variable is important in spatial analysis, where it is expressed in relation to distance rather than time, because the degree of smoothness of a variable determines the ability to estimate unknown values (see Section 7.4).

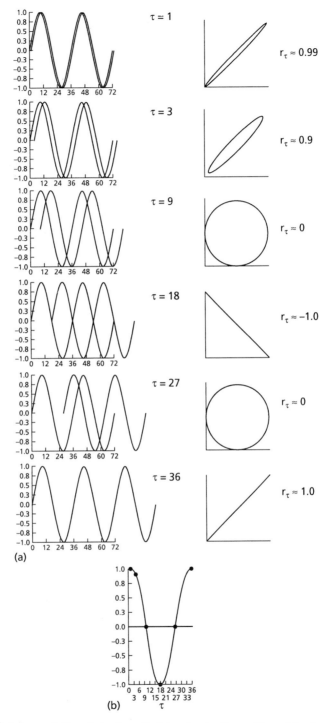

Fig. 6.8 (a) Autocorrelation of simple cyclic data, showing the position of the data and its displaced duplicate at various increasing lags, together with the corresponding shape of the bivariate scatter and the correlation coefficient. (b) Autocorrelogram resulting from the data shown in (a). The lag at the peak corresponds to the wavelength in the data. Dots show the positions illustrated in (a).

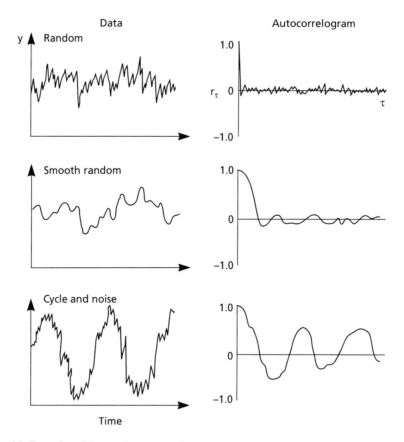

Fig. 6.9 Examples of data and corresponding autocorrelograms.

2 With increasing lag, once the initial autocorrelation due to smoothness has subsided, any further peaks on the autocorrelogram with sufficiently high positive r_τ can indicate a degree of cyclicity in the data. This is because, after displacement equal to the lag at such a peak, the data has been slid past itself through a whole wavelength of a cycle in the data so that highs and lows once again match. The lag at the peak, then, is equal to the wavelength or period of the cyclicity. Unless the cyclicity is unnaturally perfect, the resulting r_τ will be less than 1. A test for significance of the autocorrelation coefficient is needed to discriminate real cyclicity from random effects.

Test for significance of autocorrelation coefficient

$$Z_\tau = r_\tau \sqrt{(n - \tau + 3)}$$

where τ is the lag, r_τ is the autocorrelation coefficient at that lag and n is the number of observations.

H_0: r_τ is attributable to randomness

Z_τ is a standard Z statistic which will be significant at the 5% level if it exceeds 1.96 (Appendix 2.2).

Worked example: see Box 6.10.

Warnings

1 The observations must be regularly spaced through time.

2 Any linear trend in the data should be removed in advance (see Section 6.3.2). Linear trends will cause a gradual decline in peaks on the auto-correlogram with increasing lag.

3 In order for there to be sufficient comparisons in the calculation of the coefficient, the rules of thumb are: (a) there should be at least 50 observations in the time series; and (b) the lag should not exceed $n/4$.

4 Significantly high r_τ values at small lags may not reflect cyclicity but just smoothness in the data.

5 Although significantly negative Z_τ values are possible, these are not important as they correspond to negative autocorrelations, themselves due to peak–trough correspondences in the data: these will inevitably occur in association with high positive (peak–peak, trough–trough) autocorrelations and offer no additional information.

A related technique is cross-correlation: this differs in that two separate time series are compared. This has been claimed to be useful for stratigraphic correlation, but this cannot be recommended given the likelihood of differing sedimentation rates. Numerical stratigraphic correlation is better achieved using multivariate similarity (Section 8.6.3), but there are nevertheless some specialised applications for cross-correlation in matching well-log data which have been arbitrarily displaced by errors in depth measurement.

6.3.3.2   **Fourier analysis**

Fourier methods attempt to decompose a time series into a suite of wave-forms. Fourier, an eighteenth–nineteenth-century mathematician, originally established that any sequence of data could be regarded as the sum of a series of sinusoidal functions (sine and cosine waves) with specified amplitude, frequency and phase. Fourier analysis calculates the amplitude and phase for each frequency of wave, and the data can then be represented in this new form – the data are said to be transformed into the frequency domain. Normally, the object is to find any waves with a significantly high amplitude: the period of such a wave should then correspond to the period of cyclicity in the data.

The cosine wave forms the basis of Fourier analysis. The generalised equation for cosine waves is:

$$y = A\cos(k\theta - \phi)$$

Box 6.10 Worked example: autocorrelation

Is there any cyclicity in the sequence of data recording organic carbon from the Kimmeridge Clay? (See Fig. 6.7.)

Using the data with removed trend (see Box 6.9), the autocorrelogram can be obtained using a standard statistics package:

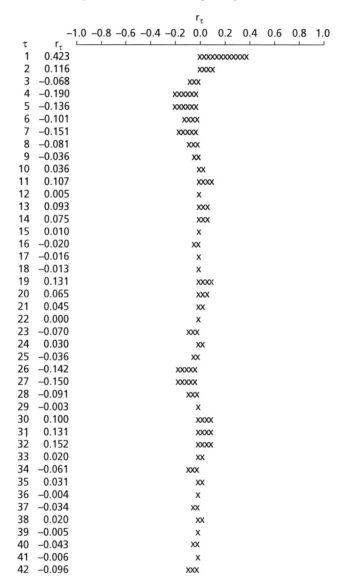

Fig. B6.10.1

To test for cyclicity of period = 11, corresponding to lag = 11 above:

H_0: autocorrelation due to randomness

Box 6.10 *Continued*

Test statistic:

$Z_\tau = r_\tau \sqrt{(n - \tau + 3)}$
r_τ = 0.107
τ = 11
n = 64
$Z_\tau = 0.107 \times \sqrt{(64 - 11 + 3)} = 0.8007$

At $\alpha = 0.05$:

Critical $Z_\tau = 1.96$ (Appendix 2.2)

The calculated Z does not exceed critical Z: fail to reject null hypothesis. The observed autocorrelation could be due to random effects.

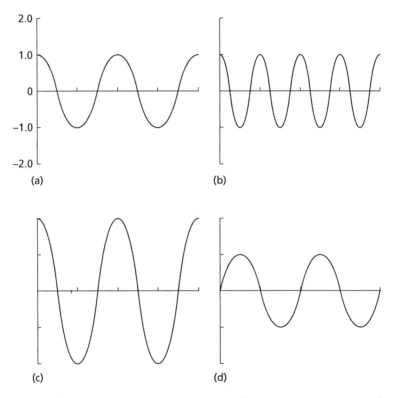

Fig. 6.10 Effect on basic cosine wave of changing amplitude, frequency and phase. (a) A, 1; k, 2; ϕ, 0. (b) A, 1; k, 4; ϕ, 0. (c) A, 2; k, 2; ϕ, 0. (d) A, 1; k, 2; ϕ, $\pi/2$.

θ in trigonometry is the angle; in the present context it represents time but needs to be expressed in radians (2π radians = 360°). If the length of the time series is T and time is t, $u = 2\pi(t/T)$. A is the amplitude of the wave: half of the peak-to-trough distance in the y direction. k is the frequency: the number of times the wave is repeated. ϕ is the phase, or displacement of the wave in the θ direction, in the same units as θ. (Refer to Fig. 6.10.)

A time series, then, can be represented by the sum of a suite of cosine waves of different frequencies:

$$y = \sum_{k=0}^{\infty} A_k \cos(k\theta + \phi_k)$$

in which the unknowns are the amplitude and phase for waves of each frequency. This equation can be shown to be equivalent to:

$$y = \sum_{k=0}^{\infty} (\alpha_k \cos(k\theta) + \beta_k \sin(k\theta))$$

where $\alpha_k = A_k \cos \phi_k$ and $\beta_k = A_k \sin \phi_k$.

The solution of the problem yields values for α_k and β_k for each frequency: geometrically, each pair of these can be regarded as the coordinates of a point on a circle with radius $= A$ and angular position corresponding to ϕ. The amplitude and the phase can be calculated by:

$$A_k = \sqrt{(\alpha_k^2 + \beta_k^2)}$$
$$\phi_k = \tan^{-1}(\beta_k/\alpha_k)$$

In practice, the most useful measure of the strength of a frequency is not the amplitude but a value known as the power, which is equivalent to the variance at that frequency:

$$s_k^2 = (\alpha^2 + \beta^2)/2$$

Computation of α_k and β_k for each value of k is quite complex and requires access to appropriate software. Regrettably, Fourier procedures are not as often included in standard statistics packages as the geologist would like. A popular and efficient version is the fast Fourier transform (FFT), for which FORTRAN program listing is available in Kanesewich (1981). The FFT offers rapid execution at the expense of the following constraints.

1 The data must be regularly spaced through time and the number of observations must be 2^n where n is an integer.

2 There should be no trend in the data (see Section 6.3.2).

3 Only integer frequencies are calculated.

4 As a consequence of 3, the method assumes that any cyclicity in the data is present in numbers of whole cycles; the beginning and end of the time series should be at the same point in the cycle.

In most applications, it is the power or amplitude more than the phase which is of interest, and the results are conventionally shown as a power spectrum. This is simply a graph showing the power at each frequency. Examples of power spectra resulting from some idealised situations are given in Fig. 6.11. Where only integer frequencies are calculated, the result is a discrete power spectrum, which may be represented by discrete vertical lines or by a straight-line graph. Fourier methods that produce a continuous

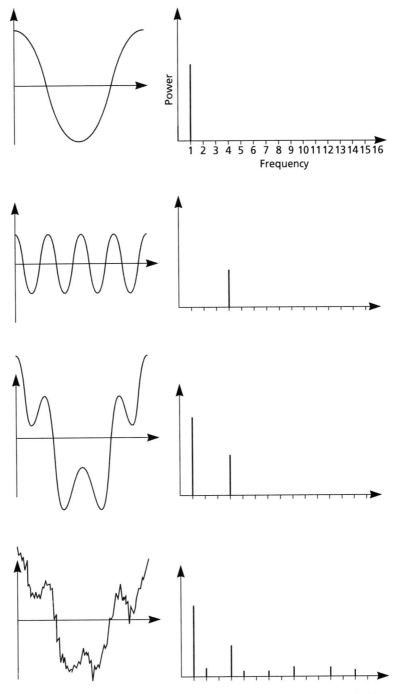

Fig. 6.11 Fourier power spectra (right) resulting from various idealised data sets (left).

power spectrum, representable by a smooth curve, show power as proportional to the area under the graph between limiting frequencies. Special smoothing algorithms can be used to convert a discrete power spectrum resulting from the FFT to a continuous spectrum. These can pin-point the positions of peaks at non-integer frequencies.

Cyclicity in a time series will be represented by a significant spike on the power spectrum. The wavelength or period of the cyclicity is readily calculated: if T is the length of the time series in real time units (i.e. seconds, years), and k is the frequency at which a specific spike occurs, the period of cyclicity is simply T/k. It should be noted that the maximum frequency resolvable (the Nyquist frequency) is half the number of observations in the time series, so the minimum wavelength is double the interval between observations. This is because a minimum of two observations is needed to define the waveform (one at a peak, one at the adjacent trough). However, higher frequencies, if sampled at regular intervals, can cryptically simulate waveforms at lower frequencies – an effect known as aliasing. This cannot readily be detected.

Once the signal in a time series has been identified, knowledge of the amplitudes and phases of the significant frequencies can be used to reconstruct the signal, free of noise. Fourier processing, then, can be used as a filter to remove noise from time series data.

Power spectra from real data will typically be noisy, with many minor spikes due to random effects. In order to discriminate signal from noise, two tests for significance may be attempted: the g statistic and the white noise test.

The g statistic

The g statistic uses the interpretation of power as variance; it is based on the ratio of the maximum variance at a frequency (s_{max}^2) to the total variance of the series (s^2), which is equivalent to the sum of the powers at all frequencies.

Test for significance of peak power

$$\hat{g} = \frac{s_{max}^2}{2s^2}$$

H_0: power at frequency attributable to randomness
H_1: cyclicity exists at that frequency

Critical values of g are estimated by:

$$g = 1 - e^{\frac{\ln p - \ln m}{m - 1}}$$

where p = level of significance and m = (number of observations)/2.
  Worked example: see Box 6.11.

Warning

The outstanding problem with this test is that it tests one frequency only. If a real signal has two or more superimposed wavelengths (which is usually the case), the subsidiary peaks will detract from the apparent significance of the main peak. This means that the test is conservative, i.e. it requires a particularly strong signal to give rejection of H_0.

Box 6.11 Worked example: fast Fourier transform

Are there significant cycles in the series of growth lines from a Silurian nautiloid? (Appendix 3.8.)

Using the fast Fourier transform (algorithm from Kanesewich, 1981) applied to the data after removal of the linear trend, the power spectrum given in Fig. B6.11.1 was obtained.

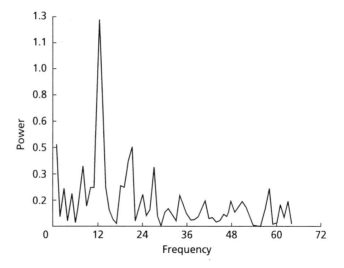

Fig. B6.11.1

For highest peak:

$$\text{Power} = 1.28 \text{ at } k = 12$$
$$\text{Total variance} = 9.9$$
$$n = 128$$
$$m = n/2 = 64$$

To test the significance of the peak:

H_0: data due to randomness

$$\hat{g} = \frac{s_{max}^2}{2s^2}$$

continued on p. 256

Box 6.11 *Continued*

$\hat{g} = 1.28/(9.9 \times 2) = 0.07$

Critical value of g:

$$g = 1 - e^{\frac{\ln p \ln m}{m-1}}$$

$$g = 1 - e^{\frac{\ln 0.05 - \ln 64}{63}}$$

$$= 1 - e^{-0.113} = 1 - 0.892 = 0.107$$

Calculated $\hat{g}$ does not exceed critical g: fail to reject null hypothesis.
Conclusion: the analysed peak could have resulted from random data.

White noise test

'White noise' is a term used to describe data in which power is uniformly spread across the spectrum. White noise can clearly be used as a null hypothesis of randomness. The test is based on the KS statistic (Section 2.5.5.2): the null situation is represented by a straight inclined line from 0 to 1 on a graph of cumulative power vs. frequency. The null hypothesis is rejected if the cumulative curve for the data departs significantly from the null line. The cumulative property of this test allows the detection of significant concentrations of power at parts of the spectrum, rather than testing one individual peak.

White noise test

H_0: white noise
H_1: not white noise

The cumulative proportion of power γ_k at frequency k is:

$$\hat{\gamma}_k = \frac{\sum_{i=1}^{k} s_i^2}{\sum_{i=1}^{n/2} s_i^2}$$

The expected average cumulative power if the data were drawn from 'white noise' is:

$$\gamma_k = \frac{2k}{n}$$

At $\alpha = 0.05$, the confidence limits are:

$$\frac{2k}{n} - \frac{1.36}{\sqrt{\frac{n}{2} - 1}}$$

and

$$\frac{2k}{n} + \frac{1.36}{\sqrt{\frac{n}{2} - 1}}$$

(see Jenkins and Watts, 1968).

If any γ_k falls outside the confidence interval, we are 95% confident that the data were not the result of a 'white noise' process.

Warning

Although this detects non-randomness, it should be noted that the H_1 cited above does not equate with 'cyclicity'. With most types of geological data, the H_0 of white noise is rejected due to a tendency for powers to be greater at the low-frequency end of the spectrum. This need not be the result of cyclicity; it is more likely to be reflecting a degree of smoothness in the data. Any diffusive geological process (e.g. resulting from fluid migration) will smooth off high-frequency information, so the rejection of the hypothesis of white noise can be rather mundane! However, this does not apply to data on measurements of successive bed thicknesses or growth lines, in which 'smoothness' need not be a trivial attribute.

Worked example: see Box 6.12.

6.3.3.3 The Walsh spectrum

Stratigraphic sequences of beds of varying lithology have been analysed for cyclicity by coding lithotypes at regular intervals with integers (say, 0 = shale; 1 = limestone) and representing the sequence of changing codes as a 'square wave'. It is not appropriate to fit Fourier's cosine waves to such data (unless heavily smoothed) because artefacts occur on the power spectrum where cosine waves are attempting to accommodate the corners of the square waveforms! Instead, a variety of the Fourier technique, the Walsh power spectrum, is used. This models the data in terms of square waves but is otherwise directly analogous to the normal Fourier method: the concepts of amplitude, frequency and phase still apply. A Walsh algorithm is available in Kanesewich (1981).

6.3.3.4 Outline analysis

Outlines of geological objects such as sand grains and fossils have proved difficult to quantify in an accurate and informative way. However, if the outline is conveyed in polar coordinates (radii to the circumference of the object at regular angular increments through 360°), it can be treated as a time series, with the useful property that the beginning and end 'wrap

Box 6.12 Worked example: white noise test

Are the data on nautiloid growth lines introduced in Box 6.11 explainable by random 'white noise'?

The cumulative power γ_k at all frequencies k were calculated from the Fourier transform output using a simple program, and are compared to the upper and lower limits of the 95% confidence interval given by:

$$2k/n - 1.36/\sqrt{((n/2) + 1)} = 2k/128 - 0.171$$

and

$$2k/n + 1.36/\sqrt{((n/2) + 1)} = 2k/128 + 0.171$$

as follows:

k	Lower limit	γ_k	Upper limit	k	Lower limit	γ_k	Upper limit
1.00	−0.1557	0.0511	0.1870	33.00	0.3443	0.7431	0.6870
2.00	−0.1401	0.0571	0.2026	34.00	0.3599	0.7627	0.7026
3.00	−0.1245	0.0812	0.2182	35.00	0.3755	0.7765	0.7182
4.00	−0.1088	0.0845	0.2338	36.00	0.3912	0.7837	0.7338
5.00	−0.0932	0.1055	0.2495	37.00	0.4068	0.7875	0.7495
6.00	−0.0776	0.1076	0.2651	38.00	0.4224	0.7917	0.7651
7.00	−0.0620	0.1257	0.2807	39.00	0.4380	0.7973	0.7807
8.00	−0.0463	0.1634	0.2963	40.00	0.4537	0.8084	0.7963
9.00	−0.0307	0.1759	0.3120	41.00	0.4693	0.8246	0.8120
10.00	−0.0151	0.2005	0.3276	42.00	0.4849	0.8293	0.8276
11.00	0.0005	0.2247	0.3432	43.00	0.5005	0.8348	0.8432
12.00	0.0162	0.3536	0.3588	44.00	0.5162	0.8375	0.8588
13.00	0.0318	0.4350	0.3745	45.00	0.5318	0.8410	0.8745
14.00	0.0474	0.4618	0.3901	46.00	0.5474	0.8486	0.8901
15.00	0.0630	0.4721	0.4057	47.00	0.5630	0.8543	0.9057
16.00	0.0787	0.4767	0.4213	48.00	0.5787	0.8703	0.9213
17.00	0.0943	0.4785	0.4370	49.00	0.5943	0.8788	0.9370
18.00	0.1099	0.5042	0.4526	50.00	0.6099	0.8910	0.9526
19.00	0.1255	0.5288	0.4682	51.00	0.6255	0.9066	0.9682
20.00	0.1412	0.5686	0.4838	52.00	0.6412	0.9186	0.9838
21.00	0.1568	0.6181	0.4995	53.00	0.6568	0.9248	0.9995
22.00	0.1724	0.6210	0.5151	54.00	0.6724	0.9255	1.0151
23.00	0.1880	0.6321	0.5307	55.00	0.6880	0.9255	1.0307
24.00	0.2037	0.6523	0.5463	56.00	0.7037	0.9260	1.0463
25.00	0.2193	0.6585	0.5620	57.00	0.7193	0.9361	1.0620
26.00	0.2349	0.6689	0.5776	58.00	0.7349	0.9602	1.0776
27.00	0.2505	0.7057	0.5932	59.00	0.7505	0.9615	1.0932
28.00	0.2662	0.7124	0.6088	60.00	0.7662	0.9636	1.1088
29.00	0.2818	0.7127	0.6245	61.00	0.7818	0.9776	1.1245
30.00	0.2974	0.7214	0.6401	62.00	0.7974	0.9828	1.1401
31.00	0.3130	0.7327	0.6557	63.00	0.8130	0.9984	1.1557
32.00	0.3287	0.7398	0.6713	64.00	0.8287	1.0000	1.1713

The cumulative power is outside the upper limit between frequencies of 13 and 43, so we are 95% confident that the data do not fit the 'white noise'

Box 6.12 *Continued*

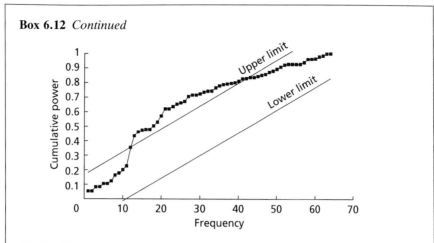

Fig. B6.12.1

model. This could be due to a specific peak at frequencies 12 and 13, representing cyclicity, but this result could also simply mean that there is less power at high and low frequencies, meaning in this context that there tend not to be very narrow or very broad growth lines. This is as likely to be due to a simple physiological constraint as it is to cyclicity in the environment.

around' and are equivalent. This allows processing by FFT, and the resulting power spectra can be used to resolve important features (see Box 6.13). Such techniques are becoming important in automated data retrieval through video image processing.

6.3.4 General pattern detection: stratigraphic position as ordinal data

It was noted at the beginning of Section 6.3 that accurate time measurements are not usually available for analysis of stratigraphic and analogous sequences, and that approximation of time by rock thickness is unlikely to be reliable (Box 6.8). In these circumstances, failure to detect significant cycles may be due to distortion of regular time cycles by changing sedimentation rate. In these circumstances, the time variable has to be regarded as ordinal (see Chapter 1). Data do not then need to be regularly spaced, but, as a consequence, techniques dealing with such data will not be able to distinguish regular cyclicity from other repetitive patterns or rhythms.

6.3.4.1 Graphical treatment

A useful simple graph is the serial correlation plot. In this, each observation (*y*) is paired with the immediately succeeding observation and plotted as a

Box 6.13 Shape analysis using Fourier power spectra

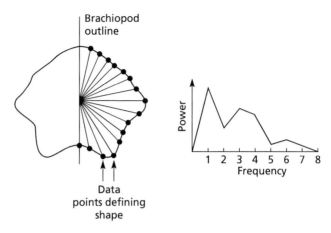

Fig. B6.13.1

1 Palaeontology (Fig. B6.13.1a). The outline of the brachiopod can be expressed in polar coordinates. Only half needs to be analysed in this case due to bilateral symmetry. The Fourier transform, applied to these data, produces a power spectrum that characterises the shape: for example, the peak at frequency = 3 relates to the uniquely lobed shape.

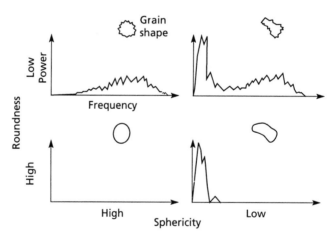

Fig. B6.13.2

2 Sedimentology (Fig. B6.13.2). Grain outlines can also be expressed as polar coordinates and analysed by Fourier transform. This provides a useful means of quantifying roundness and sphericity: lack of roundness appears as power in high frequencies; lack of sphericity corresponds to power in low frequencies.

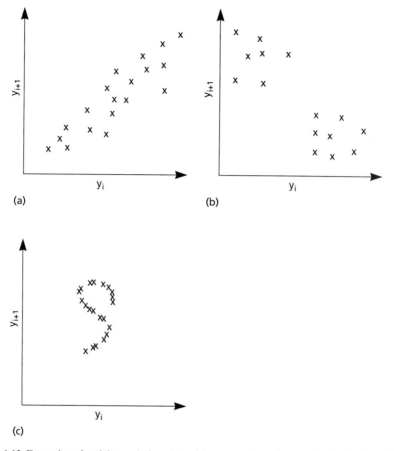

Fig. 6.12 Examples of serial correlation plots: (a) consecutive values tend to be similar; (b) alternating high–low pattern; (c) simple chaotic system.

point on a bivariate scatter of y_i vs. y_{i+1} (Fig. 6.12). This may show correlations, which will correspond to the autocorrelation property at lag $= 1$. Positive correlations may just indicate a degree of smoothness in the data; significant negative correlations will result from a pattern of alternately large and small values. More cryptic behaviour may also become evident: this procedure has been used to reveal 'strange attractors' in chaotic systems.

6.3.4.2   **Runs tests**

Runs tests are based on simple probability theory (Section 2.3) applied to successions of observations of a two-state situation. In the present context, two alternative schemes for defining the two states are useful:

State 1 – the value is larger than the previous value $(+)$
State 2 – the value is smaller than the previous value $(-)$
(consecutive equal values are ignored)

or:

State 1 – the value is larger than the median (+)
State 2 – the value is smaller than the median (−)

Runs tests are based on the number of runs in a succession of observations of the same state. For example: + + + + − − − − has two runs; + − + + − + + + − − has six runs. Runs tests look for significant departures from the number of runs that would be expected in a random sequence. This may result from anomalously high numbers of runs (e.g. + − + − + − + − + − = 10 runs), indicating a rapidly fluctuating system, or from anomalously low numbers of runs (e.g. + + + + + − − − − − = two runs), indicating relatively long-term trends. The strength of this approach is that these types of pattern are likely to survive minor corruption of the record, of the sort shown in Box 6.8. There are, however, inevitably some patterns which yield numbers of runs insignificantly different from random: the technique is not sensitive to these.

The observed number of runs is compared with the expected number of runs in a Z statistic. The equation for the expected number is derived from probability theory.

Z statistic for runs test

$$Z = \frac{U - \bar{U}}{\sigma_{\bar{U}}}$$

where U is the observed number of runs, $\bar{U}$ is the expected number of runs and $\sigma_{\bar{U}}$ is the expected standard error.

H_0: data result from randomness

If Z exceeds 1.96, there is a significantly large number of runs (using $\alpha = 0.05$); if Z is smaller than −1.96, there is a significantly small number of runs.

The expected number of runs is calculated by:

$$\bar{U} = 1 + \frac{2n_1 n_2}{n_1 + n_2}$$

and the expected variance by:

$$\sigma_{\bar{U}}^2 = \frac{2n_1 n_2 (2n_1 n_2 - n_1 - n_2)}{(n_1 + n_2)^2 (n_1 + n_2 - 1)}$$

where n_1 and n_2 are the numbers of observations of each state.
Worked example: Box 6.14.

This statistic is appropriate for any sequence of measurements: successive bed thicknesses, electric log readings from a borehole, compositional changes during crystal growth, etc. However, although an anomalously high number

Box 6.14 Worked example: runs tests

Is there a pattern in the thicknesses of successive limestone beds in the Lower Jurassic Blue Lias?

Identification of runs:

Consecutive bed thicknesses (cm): 17 14 13 17 9 15 36 13 26 10

$$-\quad-\quad+\quad-\quad+\quad+\quad-\quad+\quad-\quad+$$

Change: 25 13 25 8 28 15 19 34 22

$$-\quad+\quad-\quad+\quad-\quad+\quad+\quad-$$

Number of runs = 15

H_0: data are random

Test statistic:

$$z = \frac{U - \bar{U}}{\sigma_{\bar{U}}}$$

where:

$$\bar{U} = 1 + \frac{2n_1 n_2}{n_1 + n_2}$$

$$\sigma_{\bar{U}} = \sqrt{\frac{2n_1 n_2 (2n_1 n_2 - n_1 - n_2)}{(n_1 + n_2)^2 (n_1 + n_2 - 1)}}$$

Calculation:

$$U = 15$$
$$n_1 = 9$$
$$n_2 = 9$$
$$\bar{U} = \frac{2 \times 9 \times 9}{9 + 9} + 1 = (162/18) + 1 = 10$$
$$\sigma_{\bar{U}}^2 = \frac{2 \times 9 \times 9 \times (2 \times 9 \times 9 - 9 - 9)}{(9 + 9)^2 \times (9 + 9 - 1)}$$
$$= (162 \times 144)/(324 \times 17) = 4.235$$
$$\sigma_{\bar{U}} = \sqrt{4.235} = 2.058$$
$$Z = \frac{15 - 10}{2.058} = 2.429$$

Critical $Z = 1.96, -1.96$

The calculated Z exceeds critical Z so we reject the null hypothesis. There is a significantly large number of runs, corresponding to a tendency for alternating thick/thin limestone beds.

of runs will always point towards a genuine pattern, an anomalously low number of runs needs more careful interpretation. With geochemical and petrophysical measurements it is likely that a low number of runs reflects

---

Box 6.15 Worked example: Noether's test

Are there a significantly high number of triplets in the sequence of Blue Lias limestone bed thicknesses? (Same data as Box 6.14.)

H_0: data due to randomness

Data: 17   14   13   17   9   15   36   13   26   10   25   13   25   8   28   15   19   34

Triplet: | | | | | | |

Monotonic: √   x   x   x   x   √

From cumulative binomial tables: for two successes in six trials at $p = 0.333$, probability $\gg 0.05$; we fail to reject the null hypothesis. The sequence of triplets may have arisen from random processes.

---

nothing more than a degree of inherent smoothness (or autocorrelation) in the data, as noted with the white noise test in Section 6.3.3, perhaps resulting from a diffusive diagenetic process, or, in the case of closely sampled stratigraphic sequences, bioturbation. This, however, would not apply to an analysis of successive bed thicknesses.

6.3.4.3   **Noether's test**

This test is similar to runs tests except that it is designed to pick up short sequences, and thus fills a gap in the range of sensitivity of runs tests. The sequence of observations is divided into triplets of three observations each. The number of triplets that are monotonic is counted. Monotonicity means that, within the triplet, the second observation is bigger than the first and the third is bigger than the second, or that the second and third are successively smaller. Tied values are ignored. If we rank each triplet, the six alternative possibilities are: 1, 2, 3; 1, 3, 2; 2, 1, 3; 2, 3, 1; 3, 1, 2; 3, 2, 1, of which only two (1, 2, 3 and 3, 2, 1) are monotonic. Thus, in a random sequence, 1/3 of the triplets will be monotonic. To assess whether or not the observed number of monotonic triplets is significantly greater than this we use the following.

Noether's test

The probability of obtaining x' monotonic triplets from n triplets, with $p = 0.333$, is given by cumulative binomials – see tables in Appendix 2.1.

H_0: sequence is random

If the probability found from cumulative binomials is smaller than 0.05, then we reject the null hypothesis and conclude that a pattern exists.
     Worked example: Box 6.15.

Warnings

1 This is subject to the same reservations regarding naturally smooth data as were noted with respect to runs tests.
2 The result is very sensitive to incompleteness of the sequence: a perfect sequence of monotonic triplets would become insignificant if one data value in the middle were to be removed — half the sequence would become out of phase and not monotonic.

FURTHER READING

Chatfield C. (1989) *The Analysis of Time Series: an Introduction.* Chapman and Hall, London.
Jenkins G.M. and Watts D.G. (1968) *Spectral Analysis and its Applications.* Holden-Day, London.
Kanasewich E.R. (1981) *Time Sequence Analysis in Geophysics.* University of Alberta Press.

source books on theory and methods.

Doveton J.H. (1971) An application of Markov chain analysis to the Ayrshire coal measures succession. *Scottish Journal of Geology*, **7**, 11–27.

probably the best Markov chain case study.

Schwarzacher W. (1964) An application of statistical time series analysis to a limestone–shale sequence. *Journal of Geology*, **72**, 195–213.

autocorrelation.

Cohn B.P. and Robinson J.E. (1976) A forecast model for Great Lakes water levels. *Journal of Geology*, **84**, 455–465.

a good example of spectral analysis and filtering.

Dunn C.E. (1974) Identification of sedimentary cycles through Fourier analysis of geochemical data. *Chemical Geology*, **13**, 217–232.

one of the earliest attempts at resolving Milankovich cycles in the geological record.

Fischer A.G. and Bottjer D.J. (eds) (1991) Orbital forcing special issue. *Journal of Sedimentary Petrology*, **61**, 7 (various papers).

for background on Milankovich cycles.

Schwarcz H.P. and Shane K.C. (1969) Measurement of particle shape by Fourier analysis. *Sedimentology*, **13**, 213–231.

outline analysis.

7 Geographically Distributed Data

DISTRIBUTION OF POINTS

Types of geological point data

All geologists are familiar with geological data presented in a spatial form as maps, cross-sections or block diagrams, in which the geology is portrayed as filling a two- or three-dimensional space. Such data are dealt with in Sections 7.2–7.4; this section deals with the special (and less common) situations where geological observations or phenomena can be represented as points in two dimensions, usually on a map. Examples of natural data of this type are: volcanoes, earthquake epicentres, fossils on a bedding plane and mineral grains on a slab of granite. The ability to represent geological features as points depends on the scale used: an oilfield may be representable as a point on a global scale, but probably not on a map of the basin.

Point distributions are, however, more often artificial: sample site locations in a geochemical survey, for example. The desirable properties of a sampling scheme have been outlined in Chapter 1; in some cases sampling schemes will have been designed with these criteria in mind, but in other cases an arbitrary or pragmatically determined scheme may need to be tested to ascertain whether or not the criteria are met.

The raw data required for analysis of point distributions are simply the coordinates, often grid references, of each point. The purpose of analysis is to identify and distinguish homogeneity vs. heterogeneity, isotropy vs. anisotropy, and randomness. As with many other types of data analytical jobs, the results of point distribution analysis may contribute to an understanding of causative geological processes: we can ask why the earthquake epicentres are clustered, or why the brachiopods are uniformly distributed.

Randomness

Randomness in point distributions means that the process which influenced the position of a point has operated independently of the positions of other points and has involved equal probability of occurrence in each equal subdivision of the total area. A consequence of such a random process is that there is always a small probability that the resulting distribution will appear highly non-random: for example, all of eight random points will fall in the same half of an area one time in 128. Statistical inference is clearly required to resolve this.

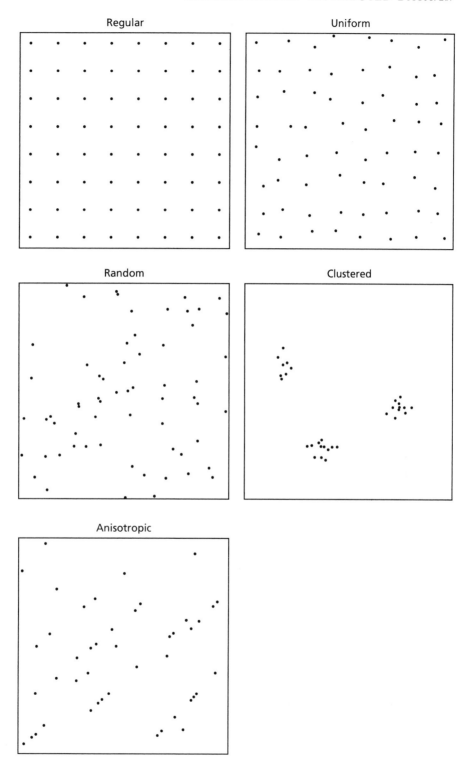

Fig. 7.1 Types of point distribution.

Non-randomness

Non-randomness can take many forms (Fig. 7.1). The degree of homogeneity can be greater or less than that of a random distribution: although probability is homogeneous in random distributions, the result has a degree of heterogeneity. Distributions tending towards more homogeneity are described as uniform (or, in extreme cases, regular); with more heterogeneity they become clustered. Random distributions are isotropic, having no directional relationships between points – no fabric. Other distributions may be anisotropic by virtue of points forming lineations or tending to be closer in some directions than others.

7.1.2   Testing for uniformity

Uniform distributions have points distributed with fairly constant density over the area: they therefore show homogeneity. Random distributions only necessarily have uniformity with respect to *probability* of occurrence of points. Paradoxically, random distributions are often appraised subjectively as being highly heterogeneous, and highly uniform distributions may appear random. Uniformity is only obvious when geometrically regular: a special case which can be identified using a technique described in Section 7.1.4. Uniform distributions contrast with clustered distributions, in which the density of points varies significantly around the area.

Uniformity is usually a desirable property of a sampling scheme: it assures that all parts of the area are represented and that there is no bias due to over-representation of some areas. Uniformity of geological point phenomena indicates effective homogeneity of the geological environment, whether it be the sea-bed, magma or a continental plate. The crucial difference from randomness is that uniformity suggests lack of independence of points from each other: the establishment of the geological point phenomenon in a certain position decreases the probability of other points developing in that vicinity. This can arise, for example, by competition between organisms on the sea-bed, by diffusive effects in crystallisation and by local release of stress or fluid pressure.

The recommended test for uniformity is highly intuitive and straightforward. The area is divided into a number of equal, usually square or rectangular, subareas, and the number of points in each subarea is counted. The test involves χ^2: we are familiar with χ^2 for comparing data distributions with some models (Section 2.5.6.1); here the model is uniformity, which we represent by dividing the points equally between the subareas. Hence we can calculate χ^2 on the basis of the observed number of points in a subarea (O_j) as compared with what we expect according to the model:

χ^2 test for uniformity of point distribution

H_0: point distribution uniform
H_1: point distribution heterogeneous

$$\chi^2 = \frac{1}{E}\sum_{j=1}^{k}(O_j - E)^2$$

where k is the number of subareas and E = (total no. of points)/k.
    Worked example: see Box 7.1.

Box 7.1 Worked example: χ^2 test for uniformity of point distribution

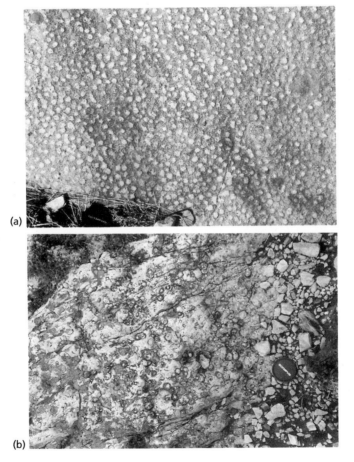

Fig. B7.1.1

The photographs in Fig. B7.1.1 show two bedding planes intersected by vertical burrows from the Skolithos quartzite (pipe rock) of the Cambrian of northwest Scotland. In Fig. B7.1.1(a), the burrows are simple cylindrical

continued on p. 270

Box 7.1 *Continued*

structures; in Fig. B7.1.1(b) they have funnel-shaped apertures where they meet the bedding plane. For this and subsequent analyses, the positions of burrows have been interpreted as points. In Fig. B7.1.1(b) only the large funnel-type burrows are used. (It should be noted that there are practical problems at this first stage in objective specification of the point locations, especially in Fig. B7.1.1(b) and that the burrows are not really true point phenomena!) Are the point distributions uniform?

For the bedding plane in Fig. B7.1.1(a)

Fifteen squares were used, dividing part of the area into 5×3 subareas. The frequencies in these are:

21 18 19 16 14
20 16 18 17 16
16 18 16 18 13

H_0: uniform point distribution
H_1: non-uniform point distribution

There are 256 points in 15 subareas, so the null model for the data has $256/15 = 17.066$ points in each subarea.

Using $\chi^2 = \Sigma(O_j - E_j)^2/E_j$, the calculation is as follows:

O_j	E_j	$(O_j - E_j)^2/E_j$
21	17.066	0.906
18	17.066	0.051
19	17.066	0.219
16	17.066	0.066
14	17.066	0.551
20	17.066	0.504
16	17.066	0.066
18	17.066	0.051
17	17.066	0.000
16	17.066	0.066
16	17.066	0.066
18	17.066	0.051
16	17.066	0.066
18	17.066	0.051
13	17.066	0.969
		$\chi^2 = 3.687$

d.f. $v = k - 1 = 14$
$\alpha = 0.05$
Critical $\chi^2 = 23.68$

Fail to reject H_0: data are compatible with uniform distribution.

Box 7.1 *Continued*

For the bedding plane in Fig. B7.1.1(b)

Using nine subareas in 3×3 array, the frequencies are:

```
3  5  3
6  7  9
7  5  7
```

Using the H_0 and χ^2 test as above, there are 52 points in nine areas, so we expect $52/9 = 5.777$ points in each subarea according to the null hypothesis.

O_j	E_j	$(O_j - E_j)^2/E_j$
5	5.777	0.105
3	5.777	1.335
3	5.777	1.335
6	5.777	0.009
7	5.777	0.259
9	5.777	1.797
7	5.777	0.259
5	5.777	0.105
7	5.777	0.259
		$\chi^2 = 5.463$

$$v = k - 1 = 8$$
$$\alpha = 0.05$$
Critical $\chi^2 = 15.51$

Fail to reject H_0: data are compatible with points being uniform.
    For a summary of results, see Box 7.3.

7.1.3

Testing for randomness

Randomness in geological point phenomena, like uniformity, implies effective homogeneity of the geological medium, but additionally implies independence of points from each other: the geological process determining position of points is blind to the positions of other points. Truly random natural point phenomena are probably rare in geology, but randomness in geochemical survey sample site distribution is often advocated.

The distribution of random points on a plane has much in common with random points along a line, and hence with point events through time (see Section 6.2): the characteristics of the distribution can be fitted to a Poisson model. Whereas, with events through time, the time line is divided up for analysis into equal short intervals, with point distributions the area must be divided into equal small quadrat subareas. The number of these is normally more than would be used for the uniformity test (Section 7.1.2): it is important that the number of subareas is large. As is usual for this type of test, the number of subareas with zero points is counted, and then one point, two points and so on. The resulting profile can then be compared with a Poisson model, for which expected frequencies are calculable. Significant departure from the model implies non-randomness; this can be in the direction of greater homogeneity (uniformity) or heterogeneity (clustering).

χ^2 test for randomness in point distributions

H_0: point distribution random
H_1: point distribution tending towards uniformity or clustering

$$\chi^2 = \sum_{j=1}^{k} \frac{(O_j - E_j)^2}{E_j}$$

where O_j is the number of subareas containing j points, and E_j is the number expected according to the Poisson model.

There is normally one class for each number of points per subarea, but the number of classes k is determined so that each class has an expected frequency of five or more.

The expected number of subareas containing j points is calculated from the following, based on the Poisson equation:

$$E_j = Te^{-n/T} \frac{(n/T)^j}{j!}$$

where n is the total number of points and T is the number of subareas. For $j = 0$, $j!$ is taken to be 1.

Worked example: see Box 7.2.

7.1.4

Testing for clustering and regularity

There is a spectrum of possible point distributions ranging from clustered to random to uniform to regular. The hypothetical end member at the clustered end has all points coincident; at the regular extreme there is perfect equilateral spacing. Clustered natural point phenomena demonstrate heterogeneity of the geological medium: the heterogeneity is in the distribution of probability as well as of points. It would be no surprise to find grains of a certain mineral clustered in xenoliths, fossil sessile organisms to be clustered on local hard substrates or hydrothermal ore deposits clustered around intrusions; the probability of a point forming is not constant over the whole area in these circumstances. Conversely, regular distributions are rare and imply strong constraints on point distribution.

Box 7.2 Worked example: testing for randomness

The test will use the same data as introduced in Box 7.1.

For the bedding plane in Fig. 7.1.1(a)

H_0: points distributed randomly
H_1: points non-random

Using χ^2 to test for Poisson distribution, where the expected number E_j of subareas with r points is

$Te^{(-n/T)} (n/t)^r/r!$

Using grid squares in a 10×6 array, we have:

$$T = 60$$
$$n = 256$$
$$Te^{(-n/T)} = 0.84171$$

r: no. in subarea	O_j: no. of subareas	E_j	$(O_j - E_j)^2/E_j$
0	0 ⎱	0.842 ⎱	
1	0 ⎰	3.591 ⎰	4.433
2	1	7.665	5.792
3	10	10.896	0.074
4	29	11.623	25.982
5	13	9.918	0.957
6	6	7.053	0.157
7	1	4.299	2.532
8	0 ⎱	2.293 ⎱	
9–∞	0 ⎰	1.824 ⎰	4.117
	60	60.00	$\chi^2 = 44.04$

Note: as usual for the χ^2 test, the number expected in each class should be more than five, so some classes have been amalgamated in this and the following examples to get closer to this rule of thumb. The final class includes the tail of the Poisson distribution, and the expected frequency is calculated simply by subtracting the sum of others from the total, in this case 60.

$$\text{d.f. } \nu = k - 2 = 6$$
$$\alpha = 0.05$$
$$\text{Critical } \chi^2 = 12.59$$

Calculated value exceeds critical value, so we reject H_0; points are not randomly distributed.

continued on p. 274

Box 7.2 *Continued*

For the bedding plane in Fig. B7.1.1(b)

Using χ^2 to test the same hypothesis as above, using a 6 × 6 array of grid squares:

$$T = 36$$
$$n = 52$$
$$Te^{(-n/T)} = 8.4916$$

r	O_j	E_j	$(O_j - E_j)^2/E_j$
0	9	8.492	0.030
1	13	12.266	0.044
2	7	8.859	0.390
3	4	4.265	
4	2	1.540	
5	1	0.445	0.059
6–∞	0	0.134	
	36	36.0	$\chi^2 = 0.5237$

$$v = k - 2 = 2$$
$$\alpha = 0.05$$
$$\text{Critical } \chi^2 = 5.99$$

Fail to reject H_0; data compatible with points being randomly distributed. For summary of results, see Box 7.3.

The methods introduced in Sections 7.1.2 and 7.1.3 may yield results suggesting clustering or regularity, but the 'nearest-neighbour' analysis described in this section produces a clearer measure for these purposes. Here, the measure is based on distance from a point to its nearest neighbouring point. This results in low mean values where points are clustered together, and the method contrasts this with regularity, where points are evenly spaced and thus show maximum mean nearest-neighbour distances. Clearly, random distributions lie between these extremes.

The nearest-neighbour test for clustering and regularity

H_0: point distribution is random
H_1: point distribution tends towards regularity or clustering

$$Z = \frac{\bar{d} - \bar{\delta}}{s_e}$$

where $\bar{d}$ is the observed mean nearest-neighbour distance, $\bar{\delta}$ is the expected mean nearest-neighbour distance and s_e is the standard error of the mean nearest-neighbour distance.

We calculate these by:

$$\bar{d} = \frac{1}{n}\sum_{i=1}^{n} d_i$$

where d_i = distance from point i to the closest other point – its nearest neighbour – and n = number of points;

$$\bar{\delta} = \frac{1}{2}\sqrt{A/n}$$

where A is the area of the region in which point distribution data are available (usually the map area); and

$$s_e = \frac{0.26136}{\sqrt{n_2/A}}$$

The Z statistic, as usual, has critical values of 1.96 and -1.96 at $a = 0.05$. If $Z < 1.96$, we reject H_0 and accept an alternative hypothesis of clustering; if $Z > 1.96$, we reject H_0 and accept an alternative hypothesis of uniformity or regularity.

Worked example: see Box 7.3.

Warning

There is a bias in this statistic due to the existence of the edges of the area being investigated: $\bar{d}$ will tend to be overestimated if close neighbours are ignored because they are off the edge of the map! This is best overcome by establishing a buffer zone: the edge of the area for which the test is to apply is withdrawn from the limit of the area for which data are available. Nearest-neighbour measurements are then permitted to straddle the new edge.

Box 7.3 Worked example: nearest-neighbour test

The test will use the data introduced in Box 7.1.

For the bedding plane in Fig. B7.1.1(a)

H_0: points distributed randomly
H_1: points non-random

$$Z = \frac{\bar{d} - \bar{\sigma}}{s_e}$$

$$\bar{\delta} = \frac{1}{2}\sqrt{\frac{A}{n}}$$

continued on p. 276

Box 7.3 *Continued*

$$s_e = \frac{0.26136}{\sqrt{\dfrac{n^2}{A}}}$$

$A = 12750$

$n = 256$

$\bar{d} = 5.8043$ (from measurements)

$\bar{\delta} = 1/2\sqrt{(12750/256)} = 3.5286$

$s_e = 0.26136/\sqrt{(256^2/12750)} = 0.11528$

$Z = (5.8043 - 3.5286)/0.11528 = 19.74$

Critical Z ($\alpha = 0.05$) $= 1.96$

Calculated Z exceeds critical Z and therefore reject H_0; data are not random.

$R = 5.8043/3.5286 = 1.6449$

so data tend significantly towards regularity.

For the bedding plane in Fig. B7.1.1(b)

Test and H_0 as above.

$A = 13200$

$n = 52$

$\bar{d} = 8.5173$

$\bar{\delta} = 1/2\sqrt{(13200/52)} = 7.966$

$s_e = 0.26136/\sqrt{(52^2/13200)} = 0.57746$

$Z = (8.5173 - 7.966)/0.57746 = 0.9547$

Critical Z ($\alpha = 0.05$) $= 1.96$

Calculated Z does not exceed critical Z and therefore fail to reject H_0; data compatible with random point distribution.

Overall conclusions

The results shown in Boxes 7.1–7.3 indicate a clear difference between the two assemblages of trace fossils: those in Fig. B7.1.1(a) are clearly non-random in the direction of uniformity or even regularity; those in Fig. B7.1.1(b) are random. This probably reflects different strategies of resource exploitation and/or different types of organism.

As a useful index, the nearest-neighbour statistic $R = \bar{d}/\bar{\delta}$ may be quoted. This varies from 0 (all points coincident) to 2.15 (perfect equilateral spacing).

Testing for anisotropy

Isotropic point distributions show no directional preference in point arrangement: in anisotropic distributions, the probability of encountering other points starting from any given point is not equal in all directions. This may apply over short or long distances, giving different types of anisotropy: see Fig. 7.1.

Anisotropy due to strain

When an isotropic distribution is strained, such as a deformed rock in which grain centres constitute the points, the resulting anisotropy can be demonstrated by investigating the directions between the centre of a grain and the centres of adjoining grains: directions parallel to the principal stress will be recorded less frequently. However, this relationship only applies with respect to adjoining grains. Graphical (e.g. Fry's method) and other techniques for specialised analysis of this type of anisotropy have been developed largely by structural geologists: refer, for example, to Ramsay and Huber (1983).

Lineations

Anisotropy due to linear arrangement of points is of more general geological interest. It may, for example, be found that volcanoes in an area occur preferentially on linear trends along tectonic features. Artificial sampling schemes may also need to be tested for linear trends: if such trends coincide with linear geological features, the sample will be biased.

Linear trends can readily be detected by calculating or measuring bearings between all pairs of points in the chosen area. Higher frequencies will be observed of directions in which linear trends are oriented. Such data, then, can be analysed using the test for uniform distributions of oriented data given in Section 5.2.4. However, the frequency distributions of inter-point orientation data is clearly dependent on the shape of the area confining the points. A null hypothesis of isotropy only converts to an expected uniform distribution if the area is circular. An elongate area will inevitably tend to produce higher frequencies in the direction of the elongation, and even a square area involves preferred orientations between opposite corners. A general method for any shape of area requires Monte Carlo simulation to specify confidence intervals and is consequently quite involved: refer to Lutz (1986).

7.2

GRAPHICAL DISPLAY OF SPATIAL DATA

The usual requirement in analyses of spatial data is to use values of a measured variable, which we will symbolise by z, taken from discrete sample locations called control points, to produce a model or estimate of the values of the variable over the whole area. Typical geological z variables for which this may be required are concentrations of elements in stream sediments in a geochemical survey, depths to a formation boundary in an oilfield and mean grain size in a sandstone. We may investigate statistical trends in the data by trend surface analysis (Section 7.3) or we may merely estimate grid values (Section 7.4); both methods may require presentation of results in the form of a three-dimensional (3D) surface, in which every x, y coordinate or grid reference is associated with a unique point on the surface with a calculated z value (envisaged as altitude) corresponding to the analysed variable (Fig. 7.2). Graphical portrayal of such surfaces is clearly an important part of the procedure, and although it is the final stage, following on from the application of the methods described in the next two sections, we here describe graphical display methods first as they will be used to illustrate what follows. This section, then, assumes that all required z values have already been calculated.

In order to define a smooth surface in principle we need to specify z values for an infinite number of x, y coordinates. This is impossible in practice, so we use the next best thing: a large number of closely spaced points arranged on a grid. The impression or illusion of smoothness can be achieved by sufficient close and numerous points. Clearly, the result will be better with a finer grid resolution, but this means more computer time for

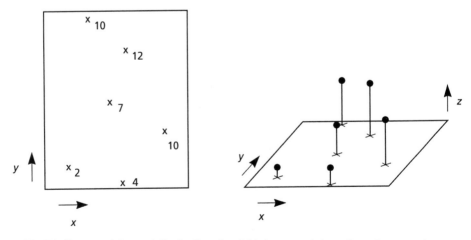

Fig. 7.2 Concept of the spatially distributed variable in two and three dimensions: x and y are the geographic coordinates, z is the spatially distributed variable.

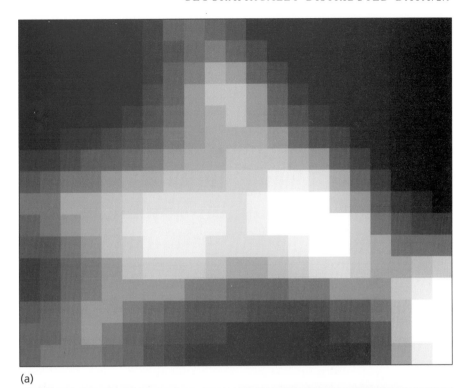

(a)

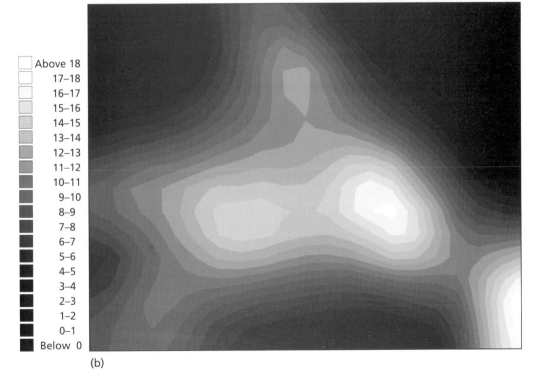

☐	Above 18
☐	17–18
☐	16–17
☐	15–16
☐	14–15
☐	13–14
☐	12–13
☐	11–12
☐	10–11
☐	9–10
☐	8–9
☐	7–8
☐	6–7
☐	5–6
☐	4–5
☐	3–4
☐	2–3
☐	1–2
☐	0–1
☐	Below 0

(b)

Fig. 7.3 Styles of graphical portrayal of surfaces. (a) Grey-level coding of grid squares; (b) isoline; (c) isoline with grey-level coding; (d) perspective construction. All plots are based on the same grid resolution.

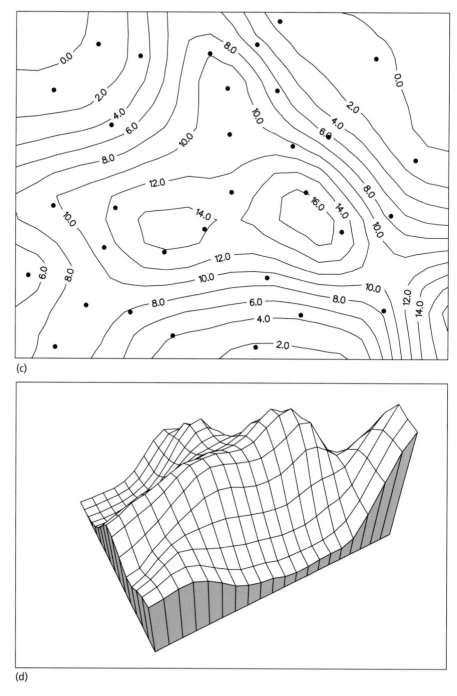

(c)

(d)

Fig. 7.3 *Continued*

calculation of z values, so the choice of grid spacing is a compromise decided by the computer power available and the user's requirements. Starting, then, from the required number of gridded x, y, z values, graphical display can be done by raster-based colour coding, 3D projection or isolines: refer to Fig. 7.3.

7.2.1 Raster-based colour coding

The simplest approach is to convert z values to a range of appropriate colours. With each grid square assigned a colour, this will yield an undesirably blocky result if there are relatively few grid cells (Fig. 7.4). However, the visual appearance will be good if the grid resolution is equal to the maximum resolution of the device. In the case of graphics screens, this may mean 256×256 or more pixel values to estimate. Hence this approach has the disadvantage of requiring a great deal of computer time and memory, but, having calculated the values, such data can be incorporated with great effect into geographic information systems (GIS) for comparison with other types of data, for example from satellite images.

7.2.2 Three-dimensional projections

7.2.2.1 Types of projection

In plan view, the array of x, y grid points at which z values are estimated can be connected with straight lines to form a grid. In three dimensions, we can imagine this grid in a distorted form with elevations from a basal plane related to the z value. This is often known as a wireframe model of the data. It can be graphically constructed in two dimensions by the use of standard projections; these are defined by equations relating the three x, y, z data dimensions to the two X, Y dimensions of the piece of paper or computer screen.

Isometric projection

Here the x and y grid lines run at 30° angles from horizontal, but with similarly proportioned scales. The equations for calculating the new two-dimensional coordinates x, y are:

$$X = k_1 (x \cos 30° - y \cos 30°)$$
$$Y = k_1 (x \sin 30° + y \sin 30°) + k_2 z$$

where k_1 and k_2 are constants chosen to produce appropriate scaling. See Fig. 7.5.

Formation thickness

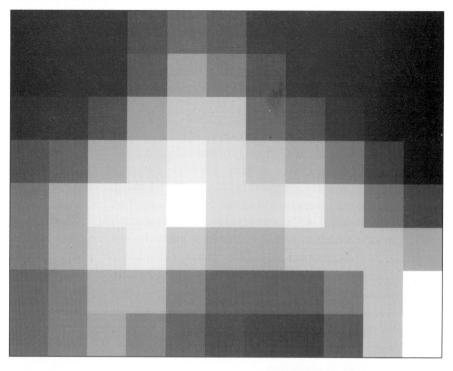

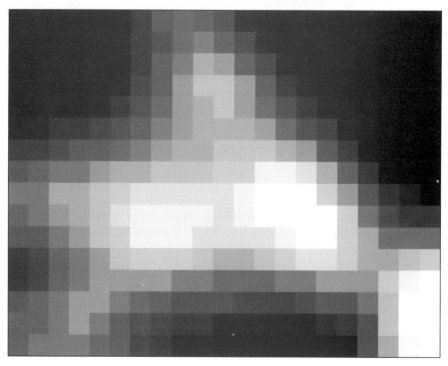

Fig. 7.4 Grey-level coding of gridded z values, using the data on thickness of an alluvial sandstone (Appendix 3.9). As more grid values are calculated, the visual impression improves.

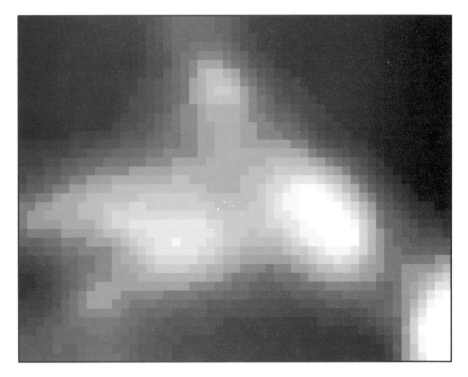

Fig. 7.4 *Continued*

Oblique projection

Here the x grid lines are horizontal and the y grid lines are at 45°. The shortening effect of perspective is accommodated by halving the scale of y relative to x:

$$X = k_1 x + (k_1/2)y \cos 45°$$
$$Y = k_2 z + (k_1/2)y \sin 45°$$

See Fig. 7.5.

Clearly, the equations for the oblique and isometric projections could be presented in a simplified form, but the versions shown here would facilitate alteration of the specified angles. This would give the impression of a change in the viewpoint, but the perspective would be unnatural.

Perspective construction

Many software packages now offer perspective views as the default construction. True perspective differs from the isometric and oblique constructions in that the grid lines are not parallel but converge towards peripheral vanishing points, and increments of x, y and z are not constant along the

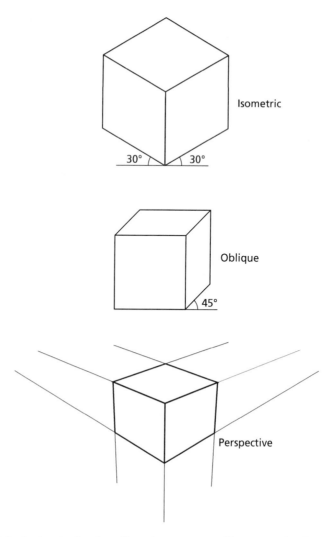

Fig. 7.5 Methods of projecting three dimensions on to two. The construction lines on the perspective diagram meet at vanishing points.

lines but decline with distance from the viewer. The equations are complex and allow great flexibility in choice of the viewing position, in terms of distance, elevation and direction. See Figs 7.5 and 7.6.

7.2.2.2   **Choice of projection**

Perspective construction is intuitively natural and attractive but has some important drawbacks.
1   The flexibility in the choice of apparent viewing position allows the user to customise the appearance of the surface. This reduces objectivity: it is possible to enhance (or hide from view) specific features of a surface.

Furthermore, a diversity of viewpoints makes comparisons between related surfaces difficult.

2   The complexity of the geometrical construction causes great difficulty in retrieving actual z values at a point given x, y grid references. This can be done quite readily from isometric and oblique surfaces using a ruler.

Crudely, then, the compromise is between intuitive appreciation and scientific rigour. Geologists are specialists at conceptualising in three dimensions, and a useful rule is to use the intuitive perspective approach when the surface is real, for example a formation boundary in an oilfield, so that our experience of interpreting landscapes can be effective. However, when the surface is imaginary, for example where it conveys the results of a geochemical survey, there is nothing to be gained from simulating real perspective.

7.2.2.3   **Graphical enhancements**

The basic wireframe model is a 'see-through' construction, exactly as if it were made only of wires. The visual three-dimensionality of this can be improved by portrayal as a surface of opaque 'tiles'. This construction is often achieved on the computer screen by filling in the grid quadrilaterals starting from the apparent rear of the surface and working forwards. The intuitive appreciation of the topography can be further enhanced by simulating differential illumination from a specific direction. This is done by calculating the angle between the 'rays' of the simulated light source and the pole to the plane of the tile, and arranging a proportional darkness of colour of the tile.

The difficulty of retrieving values from perspective surfaces, mentioned in the previous section, can be partly overcome by colour coding the tiles according to the z value, but this leaves jagged quadrilateral edges around each colour category. An increasingly common graphical method is to construct smooth isolines on to the desired projection. Section 7.2.3 describes how such contour lines can be calculated on the map projection; each such calculated set of x, y coordinates and z contour value can be transformed using one of the 3D projections to draw the isolines superimposed on, or instead of, a wireframe-type surface.

7.2.3   **Isoline construction**

Isolines are lines joining points having equal values of some property. Geologists are familiar with topographic contours, which are isolines of altitude, and it is easy to apply this experience to interpretation of isolines of geological attributes such as depth, temperature, porosity or geochemical concentration. Indeed, the ease with which geologists can visualise the shape of surfaces on the basis of isoline plots, based on experience of topographic maps, is the main advantage of this style of portrayal.

Formation thickness (m)

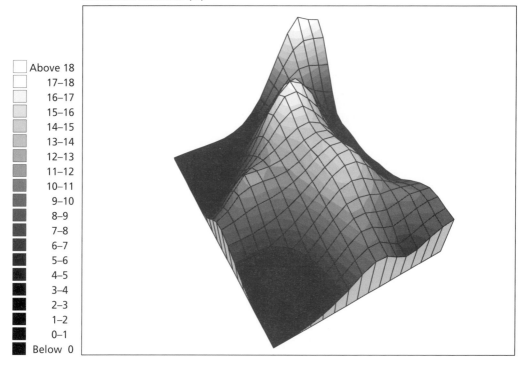

	Above 18
	17–18
	16–17
	15–16
	14–15
	13–14
	12–13
	11–12
	10–11
	9–10
	8–9
	7–8
	6–7
	5–6
	4–5
	3–4
	2–3
	1–2
	0–1
	Below 0

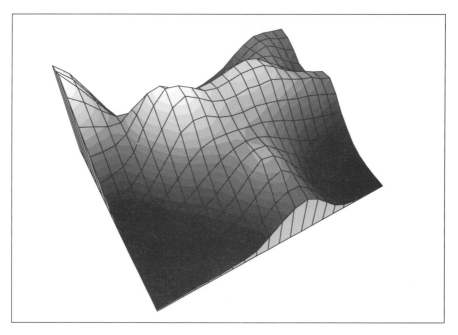

Fig. 7.6 Four different perspective views of the same surface. The appearance can alter dramatically.

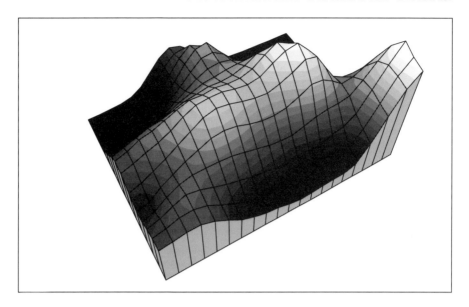

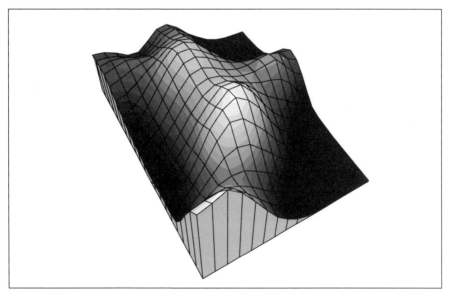

Fig. 7.6 *Continued*

Isolines can appear incidentally on raster-based plots (Section 7.2.1) where different-coloured classes meet; however, the principle of isoline construction is to avoid calculating unnecessary point values, but instead to find only a selection of points at specified z values. No value is calculated for points which fall between isolines, and knowledge about such points must be inferred by the user.

7.2.3.1   **Computer algorithms**

Individual grid square construction

Starting from a grid with known z values at the grid intersections, or nodes, the simplest algorithm is to take each grid square one at a time and to find points on the sides of the square with values equal to the required isoline values. Like values are simply joined together by straight lines. The points on the sides are found by simple linear interpolation (Section 6.3.1) between the node values (see Fig. 7.7). This divides the square into polygons, which, if desired, can be readily filled with colour. As the construction is completed for successive grid squares, the isolines emerge as continuous lines.

This approach has the disadvantage of being wholly unsuitable for vector-based pen plotters, as each isoline will be clumsily constructed from a large number of separately drawn segments. Furthermore, smoothing of the angular isolines cannot be calculated.

Threading isolines through the grid

It is more intuitive, but more algorithmically complex, to construct one complete isoline at a time, rather than one grid square at a time. This is done by tracing a specific isoline across one grid square, using interpolation as above, and then applying the same process to an adjacent cell, chosen according to which boundary the first isoline segment crosses. This can be repeated until the isoline joins up or meets the map edge. The result is that the coordinates through which the isoline passes are retrieved in a systematic

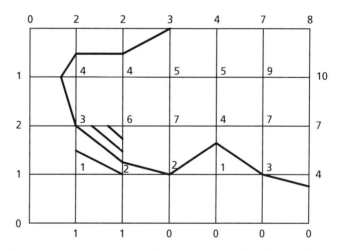

Fig. 7.7 Isoline construction based on a grid of calculated z values. Two methods are used: (a) each contour can be iteratively traced through the grid, like the contour for value 3 shown here; (b) all the required contour segments in each grid cell can be constructed, one grid cell at a time: this is shown for one grid cell here.

sequence, and a pen plotter can trace through these to produce a neat graphical plot. Furthermore, the sequence can be used for smoothing by splines.

7.2.3.2   Isoline smoothing: splines

Isolines constructed using the methods described above are composed of straight-line segments spanning grid squares, and the result will obviously look more unnaturally angular with a coarse grid. Often, increasing the grid resolution can give a reasonable illusion of rounding (Fig. 7.8), but true curvature can only be achieved by splines. Splines are somewhat similar to regression lines, except that they pass smoothly and exactly through any number of points in sequence. If an equation is required to pass exactly through a series of points, two points will constrain a linear, three points a quadratic and four points a cubic equation. In practice, isoline splines are often cubic and are calculated for every set of four sequential points (so that there is a three-point overlap between adjacent sets of four). The initial result has different curves for each set of four points, with three alternative

Formation thickness (m)

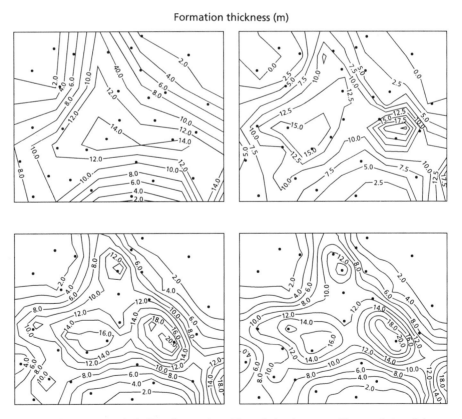

Fig. 7.8 Improvement in isoline plots as the grid resolution increases. The angularity of the isolines diminishes as the grid cells become smaller and the line segments shorter.

curves between each pair. The complete spline solution involves weighting these to form a complete single curve with a continuous derivative; this means that the curvature does not change abruptly (Fig. 7.9).

Splined isolines are primarily cosmetic: they do not add information or improve the surface estimate. However, it is certainly likely that the 'true' surface is smoothly curved so it could be argued that the splined version must be more accurate! Occasionally, however, bad spline algorithms are encountered which allow the spline to bow out between data points, producing a spurious sinuosity. Spline algorithms are mathematically quite complex but are standard on nearly all modern contouring software packages.

7.3 TREND SURFACE ANALYSIS

Trend surface analysis is simply a type of multiple regression (see Sections 3.3 and 8.2) in which the independent variables are based on the x, y coordinates or grid references. The method seeks to obtain a best-fit surface defined by an equation describing the z variable as a function of geographic position. Such a surface is called a trend surface and is constrained to be planar or geometrically curved. It is important to appreciate that this surface is not necessarily a good estimate of the distribution of the variable: the purpose is not primarily to produce a picture but to test statistical hypotheses such as: does the percentage of cadmium in soils increase from north to south? and: does the formation top tend to be higher in the middle of the map? These are posed against the ubiquitous null hypothesis that there is no trend.

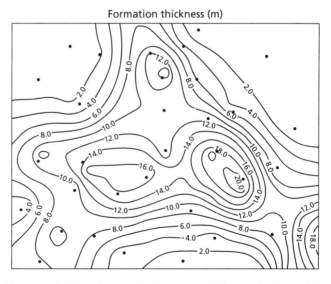

Formation thickness (m)

Fig. 7.9 Isolines smoothed by splines, using the same grid data as the bottom left plot in Fig. 7.8. It must be remembered that the apparent increase in quality is artificial.

A succession of increasingly complex forms of equations are available. It is common practice to start from the simple first-order trend equation and then work through successively higher orders, testing at each stage to see if there is any significant improvement in fit. This procedure can find the best trend model for the data, but we will find that there are some theoretical objections.

7.3.1 First-order (linear) trend surfaces

The equation for a plane defined in (x, y, z) space is:

$$z = b_0 + b_1x + b_2y$$

In linear trend surface analysis, we have only two independent variables, x and y (the geographical coordinates), and z is a dependent variable (the spatially distributed variable of interest). Given values of x, y and z at control points, we have a model for the data:

$$z = \beta_0 + \beta_1x + \beta_2y + \varepsilon$$

where ε is the error in z due to poor fit of the trend surface to the data.

We use multiple regression techniques to find values of b_0, b_1 and b_2 (estimates of the corresponding β values) so that the error is minimised. These values will then define an inclined plane which describes the linear spatial trend, if any, of z values. The highest and lowest points will inevitably be on the edges and, particularly, the corners of the map, and may indicate to an explorationist the direction for further investigation; first-order surfaces are likely to indicate the direction of provenance of grains or elements. In the context of data on subsurface formation boundaries, the first-order surface will convey the regional structural trend. The inclined plane, though, is useless as a description of the data unless the trend is statistically significant. Statistical testing proceeds, as with ordinary regression, by analysis of variance (ANOVA) (see Section 3.3.1). Trend surface analysis can be executed in statistics packages simply as a multiple regression with the special set of independent variables.

7.3.2 Second-order (quadratic) trend surfaces

In this category of trend surface we allow the surface to be curved, but only with one sense of curvature (convex or concave) and with simple parabolic cross-sections. The model is:

$$z = \beta_0 + \beta_1x + \beta_2y + \beta_3x^2 + \beta_4y^2 + \beta_5xy + \varepsilon$$

Compared with the linear case, we have extra terms for the squared and xy variables.

The model allows high or low points within the perimeter of the map area, and so is most useful in geology for focusing attention on to areas with

Analysis of variance for first-order trend surface

H_0: $\beta_1 = 0$ and $\beta_2 = 0$
H_1: $\beta_1 \neq 0$ and/or $\beta_2 \neq 0$

The ANOVA table follows.

Source	Sum of squares	Degrees of freedom	Mean square	F-test
First-order trend	SS_R	2	$MS_R = SS_R/2$	MS_R/MS_D
Deviation	SS_D	$n - 3$	$MS_D = SS_D/(n - 3)$	
Total	SS_T	$n - 1$		

n, sample size.

The SS values, plus the β coefficients and the rest of the ANOVA table, are normally provided by statistics packages.

If the calculated F ratio exceeds the critical value from tables (Appendix 2.5, using degrees of freedom $v = 2$, $n - 3$) we reject the null hypothesis and conclude that there is a spatial trend. The spatial trend is expressed by the regression equation and can be illustrated as an isoline or other spatial plot.

Worked example: see Boxes 7.4 and 7.5.

Warnings

1 The same constraints pertain as for ordinary classical regression: see Section 3.3.1.

2 The trend surface is not intended to be a good estimate of the distribution of the variable: it merely describes the trend.

3 If the statistics do not lead to rejection of H_0, the trend surface conveys no useful information.

4 The lack of a first-order trend does not preclude the possibility of higher-order trends.

significantly greater values. With formation boundary data, second-order trends may describe domes and basins. However, this and higher orders of equation become less useful for simple hypothesis testing because the alternative hypotheses are complex, involving a wide range of alternatives. Rejection of H_o may be due to a significant high or low, but may also be due merely to the data having a particularly geometrically curved trend in one part of the map area.

If the first-order trend for a data set is significant, the second-order trend will inevitably also be significant, as it includes the first-order terms. The ANOVA below includes a test for the significance of the improvement of fit of second order over first order. If the first-order test has already been done, it is only the test for improvement that is required.

Box 7.4 Trend surfaces

Control points with formation thickness (m)

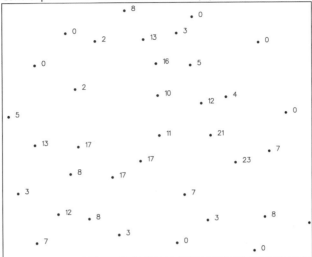

First order trend surface

Fig. B7.4.1

Appendix 3.9 gives data at 36 control points (boreholes) on thickness of an alluvial sandstone formation. Are there any spatial trends in the data? Isoline plots of first- to fourth-order trend surfaces are shown in Fig. B7.4.1, together with the location of and z values at the control points. These show heterogeneities, but these need to be assessed for significance before interpretation: see Boxes 7.5–7.7.

continued on p. 294

Box 7.4 *Continued*

Second order trend surface

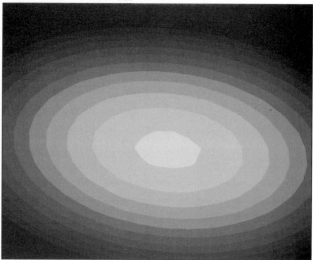

Third order trend surface

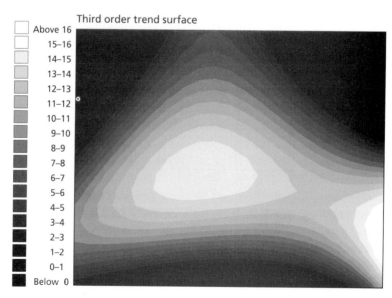

	Above 16
	15–16
	14–15
	13–14
	12–13
	11–12
	10–11
	9–10
	8–9
	7–8
	6–7
	5–6
	4–5
	3–4
	2–3
	1–2
	0–1
	Below 0

Fig. B7.4.1 *Continued*

Box 7.4 *Continued*

Fourth order trend surface

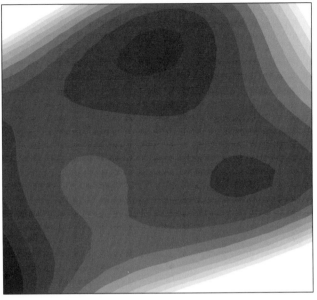

Fig. B7.4.1 *Continued*

Box 7.5 Worked example: ANOVA for first-order trend surface

Using the data set on sandstone formation thickness, a multiple regression procedure on a statistical software package produced the equation:

$$z = 8.99 + 0.094x - 0.255y$$

This is represented by the first-order surface in Box 7.4 and gives the following ANOVA.

ANOVA for significance of first-order trend surface

H_0: β_1 and $\beta_2 = 0$
H_1: β_1 and/or $\beta_2 \neq 0$

Source	Sum of squares	Degrees of freedom	Mean square	F ratio
First-order trend	59.74	2	29.87	0.64
Deviation	1537.01	33	46.58	
Total	1596.75	35		

Box 7.5 *Continued*

$R^2 = 0.037$

From tables (Appendix 2.5), critical value of F at $\alpha = 0.05$ and 2, 33 degrees of freedom is 3.3. The calculated F ratio does not exceed the critical value, so we fail to reject the null hypothesis and conclude that there is no first-order trend. In this context, this means that there is no systematic tendency for the sandstone thickness to increase or decrease in any one direction. The first-order surface shown in Box 7.4 contains no useful information.

Analysis of variance for second-order trend surface

Test 1:

H_0: β_1, β_2, β_3, β_4 and $\beta_5 = 0$
H_1: one or more of the above β coefficients $\neq 0$

Test 2:

H_0: β_3, β_4 and $\beta_5 = 0$
H_1: one or more of the above β coefficients $\neq 0$

The ANOVA table follows. The SS values and the β coefficients are normally provided by statistics packages.

Source	Sum of squares	Degrees of freedom	Mean square	F-test
Second-order trend	SS_{R2}	5	$MS_{R2} = SS_{R2}/5$	Test 1: MS_{R2}/MS_{D2}
First-order trend	SS_{R1}	2		
Increase of second over first	$SS_{R2-1} = SS_{R2} - SS_{R1}$	3	$MS_{R2-1} = SS_{R2-1}/3$	Test 2: MS_{R2-1}/MS_{D2}
Second-order deviation	SS_{D2}	$n - 6$	$MS_{D2} = SS_{D2}/(n - 6)$	
Total	SS_{T2}	$n - 1$		

n, sample size.

If the calculated F ratio for test 2 exceeds the critical value from tables (with degrees of freedom $v = 3$, $n - 6$), we reject the null hypothesis and conclude that there is a second-order spatial trend.

The spatial trend is expressed by the regression equation and can be illustrated as an isoline or other spatial plot.

Worked example: see Boxes 7.4 and 7.6.

Warnings (In addition to the warnings listed for first-order trends in Section 7.3.1.)

1 It has been demonstrated that the shape of second- and higher-order surfaces can be influenced by uneven control point distributions: control points should be very approximately uniformly distributed.
2 Second- and higher-order trends are capable of developing extreme but spurious gradients, especially when extrapolated beyond the control points.

Box 7.6 Worked example: ANOVA for second-order trend surface

Using data and procedures as for Box 7.5, the equation is:

$$z = -3.22 + 1.33x + 2.73y - 0.0535x^2 - 0.178y^2 - 0.0269xy$$

(see Box 7.4 for shape of surface).

Analysis of variance for significance of improvement of second- over first-order trend surface

H_0: β_3, β_4 and $\beta_5 = 0$
H_1: one or more of the above β coefficients $\neq 0$

Source	Sum of squares	Degrees of freedom	Mean square	F ratio
Second-order trend	446.88	5	89.38	
First-order trend	59.74	2		
Increase of second over first	387.14	3	129.05	3.367
Second-order deviation	1149.87	30	38.33	
Total	1596.75	35		

$R^2 = 0.280$

From tables (Appendix 2.5), critical value of F at $\alpha = 0.05$ and 3, 30 degrees of freedom is 2.92. The calculated F ratio exceeds the critical value, so we reject the null hypothesis and conclude that there is a second-order trend. Referring to the map in Box 7.4, this means that the sandstone thickness tends to be greater in the middle of the area.

7.3.3   **Third-order (cubic) and higher-order trend surfaces**

Third-order trend surfaces are allowed one change in the sense of curvature (convex to concave) in any cross-section, analogous to the cubic polynomial curve (Section 3.3.1.2), and successive orders of trend are allowed additional such increments of flexibility. The model for the third order is:

$$z = \beta_0 + \beta_1 x + \beta_2 y + \beta_3 x^2 + \beta_4 y^2 + \beta_5 xy + \beta_6 x^3 + \beta_7 y^3 + \beta_8 x^2 y + \beta_9 xy^2 + \varepsilon$$

You will notice that the kth order of trend adds $k + 1$ terms to the equation for order $k - 1$. The terms can be summarised as in Table 7.3 (discounting the constant term). The number of terms progresses 2, 5, 9, 14, 20, 27 . . . , with the number of terms for the kth order given by $(k^2 + 3k)/2$.

Third or higher orders of trend surface are complex and there are corresponding problems with conceptualising what exactly the alternative hypothesis is that is being tested. It may be useful to use second-order surfaces to define parts of an area with significantly higher values, because it is easy to imagine geological processes that can produce this type of pattern, for example: a salt dome producing tectonic high; a patch reef producing a local concentration of bioclasts; an igneous intrusion producing geochemical high. Natural principles tend to involve linear or quadratic terms, such as the inverse square law which determines diffusive effects. It becomes less likely that geological processes will produce variables having distributions modellable by surfaces with the complexity of higher orders of trend.

Table 7.1

Order	Terms
1	$x \quad y$
2	$x \quad y \quad x^2 \quad y^2 \quad xy$
3	$x \quad y \quad x^2 \quad y^2 \quad xy \quad x^3 \quad y^3 \quad x^2 y \quad xy^2$
4	$x \quad y \quad x^2 \quad y^2 \quad xy \quad x^3 \quad y^3 \quad x^2 y \quad xy^2 \quad x^4 \quad y^4 \quad x^3 y \quad xy^3 \quad x^2 y^2$
5	$x \quad y \quad x^2 \quad y^2 \quad xy \quad x^3 \quad y^3 \quad x^2 y \quad xy^2 \quad x^4 \quad y^4 \quad x^3 y \quad xy^3 \quad x^2 y^2 \quad x^5 \quad y^5 \quad x^4 y \quad xy^4 \quad x^3 y^2 \quad x^2 y^3$
etc.!	

General analysis of variance for higher orders of trend surface

Test 1:

H_0: all β coefficients $= 0$
H_1: one or more of the β coefficients $\neq 0$

Test 2:

H_0: all the $(k + 1)$ additional β coefficients for order k over order $(k - 1) = 0$
H_1: one or more of these β coefficients $\neq 0$

Source	Sum of squares	Degrees of freedom	Mean square	F-test
kth-order trend	SS_{Rk}	$v_k = (k^2 + 3k)/2$	$MS_{Rk} = SS_{Rk}/v_k$	Test 1: MS_{Rk}/MS_{Dk}
$(k - 1)$th-order trend	SS_{Rk-1}			
Increase of kth over $(k - 1)$th	$SS_{RI} = SS_{Rk} - SS_{Rk-1}$	$k + 1$	$MS_{RI} = SS_{RI}/(k + 1)$	Test 2: MS_{RI}/MS_{Dk}
kth-order deviation	SS_{Dk}	$n - v_k - 1$	$MS_{Dk} = SS_{Dk}/(n - v_k - 1)$	
Total	SS_{Tk}	$n - 1$		

n, sample size.

If the calculated F ratio for test 2 exceeds the critical value from tables (with degrees of freedom $v = k + 1$, $n - v_k - 1$), we reject the null hypothesis and conclude that there is a kth-order spatial trend.

The spatial trend is expressed by the regression equation and can be illustrated as an isoline or other spatial plot.

Worked example: see Boxes 7.4 and 7.7.

Warnings (In addition to the warnings listed for second-order trends in Section 7.3.2.)

1 Higher orders of trend surface almost invariably produce extreme gradients and hence extreme values at edges and corners (and sometimes even between control points) which bear no relation to real trends in the data. These are known as edge effects and also occur in surface estimates (Section 7.4.2.3) where local trends are extrapolated without control (Fig. 7.10a).

2 Higher-order trends should not normally be used as it is very unlikely that the trend surface equations will have any relevance in describing the causative geological processes (Fig. 7.10b).

3 As the number of terms in the equation approaches the number of data points, a near perfect fit is likely. This will say nothing about the real statistical trends in the data. This is clear for lower orders – for example, a first-order equation with two terms will give an exact fit to three control points – and is equally true (but less intuitive) for higher orders.

4 High-order trend surface equations involve raising x and y values to high powers. As x and y may be grid references expressed in large numbers, the terms in the equation can become very large indeed! The z variable, though, is likely to comprise small values (say, <10), and the balancing of the equation is effected by very large numbers being multiplied by very small coefficients! This being the case, precision in a large string of numbers is required in computation, and it is sometimes found that the software and/or computer is inadequate. Fortunately, the result is usually unambiguously nonsensical! This can be avoided to an extent by local rescaling of grid references.

TP4(a)

7.3.4   **Trend surface residuals**

The difference between the trend surface and the actual control point z value is a measure of the error of the surface and is known as the residual. The distribution of the magnitude of residuals over an area can be investigated by interpolating an estimated residual surface, using the techniques to be described in Section 7.4. This has two uses.

1   The distribution can be assessed for homoscedasticity and autocorrelation, as is done for ordinary bivariate regression (Section 3.3). This achieves two objectives: (a) the technical validity of the regression can be appraised: the residuals should be homoscedastic but not autocorrelated; and (b) patterns of autocorrelation may indicate trends which could be modelled by higher orders of trend surface.

2   There may be some specialised circumstances in which the residuals are of greater geological interest than the trends. For example, a first- or second-order trend surface may include only mundane information about regional trends, whereas the interest lies in local anomalies. Local anomalies appear much more clearly after the trend has been removed, i.e. in a map of residuals. For example, residuals have been used to find reefs (potential oil reservoirs) from data giving depth to the overlying cap rock. Large-scale folding constitutes the trend which obscures the pattern of reefs, and the use of residuals has the effect of unfolding the strata.

7.4   **SURFACE ESTIMATION**

7.4.1   **Introduction**

When a geologist is confronted with values of a variable at various specified points scattered over a map, the commonest requirement is the production of an image of a 3D surface representing the likely distribution of the variable throughout the area. We have seen in Section 7.2 how the surface can be displayed once it is defined by values at grid points: this section addresses the problem of how those grid point values are calculated.

Geologists are usually more concerned that the surface is plausible rather than rigorous: the purpose is to aid visualisation. With such a loose criterion of acceptability, it is no surprise that a wide range of methods have been used, and it is unusual to find two software products which use the same method. The choice of best method may vary according to the type and nature of the data. The only surface estimation technique which is often claimed to be fundamentally superior and rigorous is kriging: this is described separately in Section 7.4.3, along with a critical review of these claims.

Surface estimation routines generally involve two assumptions in order to produce a realistic estimate: (a) continuity and uniqueness; and (b) spatial autocorrelation.

Box 7.7 Worked example: ANOVA for third- and fourth-order trend surfaces

Using data and procedures as for Box 7.5, the equation for the third-order surface is:

$$z = 1.81 - 1.16x + 4.43y - 0.029x^2 - 0.697y^2 + 0.562xy + 0.0055x^3$$
$$+ 0.0198y^3 - 0.0288x^2y - 0.0018xy^2$$

(see Box 7.4 for shape of surface).

ANOVA for significance of increase in fit for third- over second-order trend surfaces

H_0: β_6 to β_9 all $= 0$
H_1: at least one of β_6 to $\beta_9 \neq 0$

Source	Sum of squares	Degrees of freedom	Mean square	F ratio
Third-order trend	947.52	9	105.28	
Second-order trend	446.88	5		
Increase of third over second	500.64	4	125.16	5.012
Third-order deviation	649.23	26	24.97	
Total	1596.75	35		

$R^2 = 0.593$

From tables, critical value of F at $\alpha = 0.05$ and 4, 26 degrees of freedom is 2.74. The calculated F ratio exceeds the critical value, so we reject the null hypothesis and conclude that there is a third-order trend. However, the shape of the surface (Box 7.4) is sufficiently complex to make it difficult to identify which aspects of the formation thickness variation are significant! It appears that the third-order terms are describing a triangular pattern, with a ridge of high value to the east-southeast.

The equation for the fourth-order surface is:

$$z = 9.1 - 0.69x - 2.83y - 0.17x^2 + 0.75y^2 + 1.01xy + 0.0052x^3 - 0.084y^3$$
$$- 0.0009x^2y - 0.0935xy^2 + 0.00031x^4 + 0.00298y^4 - 0.0021x^3y$$
$$+ 0.00156xy^3 + 0.00263x^2y^2$$

(see Box 7.4 for shape).

continued on p. 302

Box 7.7 *Continued*

Test for significance of increase in fit for fourth- over third-order trend surfaces

H_0: β_{10} to β_{15} all $= 0$
H_1: at least one of β_{10} to $\beta_{15} \neq 0$

Source	Sum of squares	Degrees of freedom	Mean square	F ratio
Fourth-order trend	1045.11	14	74.65	
Third-order trend	947.52	9		
Increase of fourth over third	97.59	5	19.52	0.743
Fourth-order deviation	551.64	21	26.27	
Total	1596.75	35		

$R^2 = 0.655$

   From tables, critical value of F at $\alpha = 0.05$ and 5, 21 degrees of freedom is 2.68. The calculated F ratio does not exceed the critical value, so we fail to reject the null hypothesis and conclude that the fourth-order trend is not a significant improvement over the third-order trend. The fourth-order surface shown in Box 7.4 is no more useful than the third-order surface.

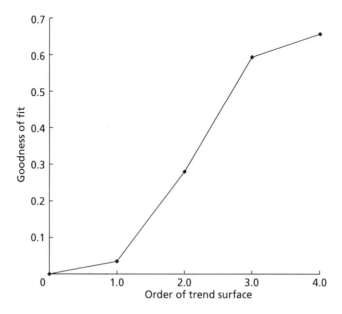

Fig. B7.7.1 *Continued*

Box 7.7 *Continued*

Summary of Boxes 7.4–7.7

We can view the increase in fit as the order is increased by plotting R^2 against order (Fig. B7.7.1).

The goodness of fit inevitably improves as we add more terms to the equation, but the improvement is only substantive in the second and third orders. This gives visual support to the statistical result that the best model is the third-order trend surface.

Continuity and uniqueness

The surface being estimated is assumed to be continuous, with no gaps or breaks, and must have a single value at each and every coordinate. In the case of data on depths to structural surfaces, this criterion prohibits data involving repetition due to thrusting, and precipitous breaks or gaps due to faulting in general will be smoothed off in the estimated surface. Special techniques for faulted surfaces do exist, but these are rather contrived. A more problematic contravention of continuity occurs in stream sediment data: geochemical characteristics of a stream sediment are sourced from within the stream catchment area, and the watershed is a barrier preventing continuity. Furthermore, stream flow imposes an asymmetry on element distribution: control points upstream should have the greatest influence on an estimate. Accurate estimation of surfaces for such geochemical data clearly requires consideration of topography and drainage pattern; this is a complex problem for which developments in GIS should provide solutions.

Spatial autocorrelation

In order to estimate the value of a variable at a new point from the known values at the control points, we must assume that there is some relationship between the two types of value: they must not be completely independent. The concept of dependence/independence in this context can best be imagined in terms of relative smoothness, and this can be quantified, as with time series data, by using the autocorrelation coefficient (refer to Section 6.3). Consider an extreme example: suppose we wish to produce an estimate of the altitude of a point on the surface of an alpine topography (Fig. 7.11). If the known values at control points are 20 km away, then they will be independent and useless in producing our estimate: they will not be able to predict whether our estimated point is on a peak or a valley bottom. In this case, the surface is not sufficiently smooth relative to the distance to control points, or, in technical terms, is not autocorrelated over distances comparable to control point spacing. In contrast, if the control points are 100 m away

Control points with chromium (PPM)

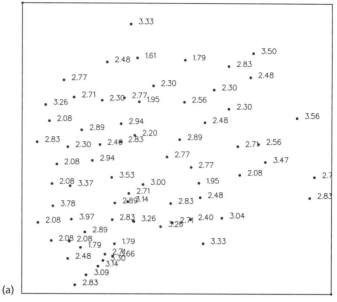

(a)

First order trend surface

(b)

Fig. 7.10 (a–e) Edge effects on trend surfaces of data on chromium from the Carswell area (Appendix 3.14). The third- and fourth-order surface show extreme gradients where unconstrained by control points in the northwest, northeast and southwest.

Second order trend surface

(c)

Third order trend surface

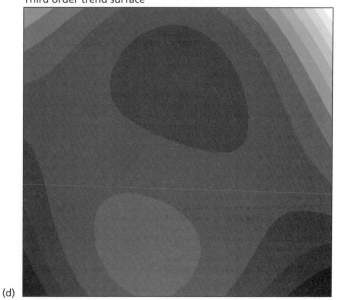

(d)

Fig. 7.10 *Continued*

Fourth order trend surface

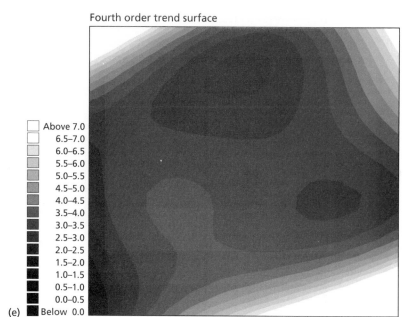

☐	Above 7.0
☐	6.5–7.0
☐	6.0–6.5
☐	5.5–6.0
☐	5.0–5.5
☐	4.5–5.0
☐	4.0–4.5
☐	3.5–4.0
☐	3.0–3.5
☐	2.5–3.0
☐	2.0–2.5
☐	1.5–2.0
☐	1.0–1.5
☐	0.5–1.0
☐	0.0–0.5
☐	Below 0.0

(e)

Third order trend surface

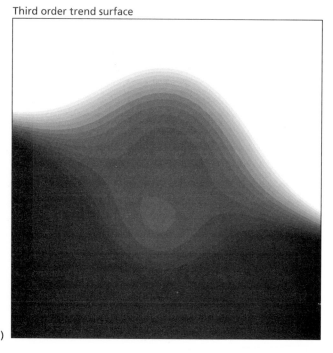

(f)

Fig. 7.10 *Continued* (f,g) This shows an extrapolation of the third- and fourth-order trends to a larger peripheral area. The extrapolation is taken to a ridiculous degree: the black areas are effectively at zero ppm (if the log transform wasn't used, these areas would have negative ppm) and the white areas extrapolate to beyond 1 000 000 ppm!

Fourth order trend surface

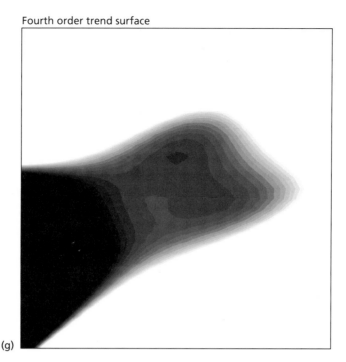

(g)

Fig. 7.10 *Continued*

from the point to be estimated, there will be some similarity or dependence between them and they will be usable in arriving at the estimate. The surface is smooth relative to the control point spacing: it is autocorrelated over that distance.

In this simple scenario, it would be intuitively clear what an appropriate density of control points would be, as the whole surface can be seen in advance. This may also be true in many geological problems; for example, in an investigation of subsurface formation boundaries in an oilfield we might be able to anticipate a certain degree of severity of folding and faulting and choose the control point spacing so as to be sufficiently close to be able to define it. However, in many geological jobs, such as geochemical surveys, the smoothness of the data may not be easy to anticipate.

It must be emphasised that an estimated surface based on control points which are too widely spaced will be completely without foundation and may be totally misleading: great care must be taken. It is important to get independent evidence of smoothness by calculating autocorrelations or semivariances (see Section 7.4.3). The control point spacing must be less than the distance at which the autocorrelation function declines to near zero or the semivariance reaches the sill value. This should ensure that at least two or three control points will contribute to most estimations, but there will always be a likelihood that parts of an area will remain not estimatable due to locally sparse control points.

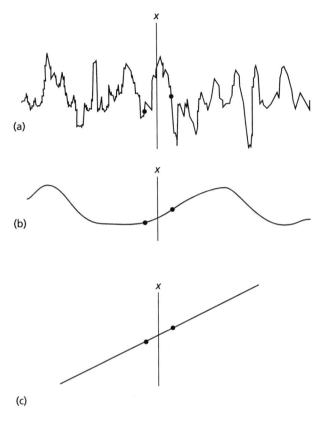

Fig. 7.11 Cross-sections through hypothetical surfaces: the dots indicate control points, and *x* marks a location for which we need an estimate of the surface. (a) Very little smoothness: estimate will be totally unreliable; (b) smooth surface: good estimation possible; (c) perfect smoothness: estimation can be perfect. In practice, if we do not know the smoothness of the surface in advance, we can't attach any reliability to the estimate.

7.4.2   General surface estimation techniques

Surface estimation techniques need to combine: (a) a method of selecting the set of control points which are to be used in estimating a new grid point value; with (b) a method of calculating the estimate from the data values at that set of control points.

7.4.2.1   Selecting the control points

The neighbourhood criterion

This technique derives from the concept of autocorrelation distance mentioned above. A radius is specified which determines the circular neighbourhood around the point to be estimated. Only control points within this neighbourhood are used in calculating the estimate (Fig. 7.12). Although the autocorrelation distance provides the rationale, in practice the autocorrelation is seldom calculated: the radius is often just guessed by the user or, more often, a default value provided by the software is used. It should not be forgotten that too large a radius will include fundamentally independent control points and the resulting estimate will include a degree of arbitrariness.

Triangulation

Any scatter of control points can be tessellated by triangles with control points at all apices (although tessellation to the edge of the map area requires addition of extra 'pseudo-points' on the edges). The result of this is that every point in an area is within or on the edge of a triangle. The estimation of a new grid point value can then be done using only the control points at the apices of the triangle within which it lies, or, more commonly and for a more accurate result, these three plus the three extra points at the apices of the three adjoining triangles (Fig. 7.13).

There are an almost infinite number of different ways of obtaining such a tessellation. The standard and best method is Delauney triangulation, in which the acuteness of triangles is minimised. Delauney triangulation requires the operation of a computer algorithm which is fortunately widely available in spatial analysis packages.

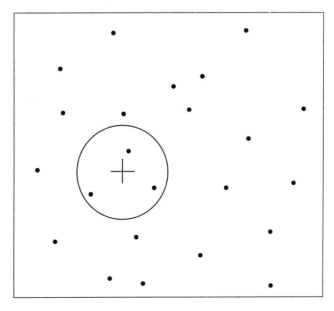

Fig. 7.12 Selection of control points (dots) by constructing a circle around the point to be estimated (cross). The radius should be equal to the distance of autocorrelation. If this is small, not enough control points will be usable in the estimation.

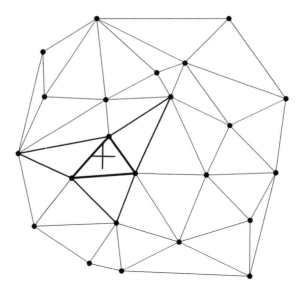

Fig. 7.13 Selection of control points (dots) by triangulation. We use the three control points at the apices of the triangle surrounding the point to be estimated (cross), or these plus the three extra apices of the other adjoining highlighted triangles.

7.4.2.2   **Calculating the estimate**

Inverse distance weighted averaging

A point estimate calculated as a simple average of all the chosen control point values will clearly be unsatisfactory: further points will be weighted equally to nearby points, so all points estimated from the same set of control points will have the same value, giving a blocky result. A simple solution to this is to weight the control point values according to the inverse of the distance to the grid point being estimated. For example, if three control points are to be used and these have values z_1, z_2 and z_3 and are distances d_1, d_2 and d_3 from the estimated point (Fig. 7.14), the estimated value z' is calculated by

$$z' = \frac{(z_1/d_1) + (z_2/d_2) + (z_3/d_3)}{(1/d_1) + (1/d_2) + (1/d_3)}$$

This is simply a weighted average of z_1, z_2 and z_3, in which the z values are multiplied by the weights, summed, then divided by the sum of the weights to remove bias. (Note that if the weights were all 1 this would be equivalent to an ordinary average.) The general equation for n control points is:

$$z' = \frac{\sum_{i=1}^{n} (z_i/d_i)}{\sum_{i=1}^{n} (1/d_i)}$$

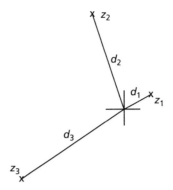

Fig. 7.14 If we are to estimate the z value at a point from three surrounding points with values z_1, z_2 and z_3, the values are allocated weights based on the distances d_1, d_2, and d_3.

The use of inverse distance is fairly arbitrary, and the relative weight applied to near and far points can be varied by using other distance transformations, e.g. $1/d^2$ or $1/d^{1/2}$.

Local trends

Inverse distance-weighted average methods will inevitably produce an estimate which is within the range of the contributing control point values, but the true value need not be so restrained: see Fig. 7.15. A better estimate in such cases would make use of extrapolated local trends. A first- or second-order trend surface may be calculated on the basis of the contributing control point values, and the value on the trend surface at the estimated grid point can then be calculated and used as the estimate. Figure 7.15(b) demonstrates that a second-order trend surface is best for this.

A variety of this method which makes better use of first-order trends calculates local trend surfaces centred on each contributing control point. These surfaces are extrapolated to give values at the point to be estimated, and this set of values can be inverse-distance-weighted to produce the final estimated value.

Three-dimensional splines

In Section 7.2.3 on isoline construction, it was noted that splines can be used to join points together into a smooth line. This method can be extended into three dimensions, so that points are joined on a smooth surface. For computational simplicity, each part of the splined surface is calculated from a fixed number of local control points. Typically, six adjacent control points are linked by a third-order (cubic) polynomial spline. The final smooth surface then gives estimates at all points.

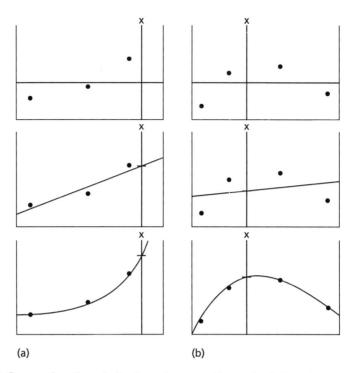

Fig. 7.15 Cross-sections through situations where we wish to estimate the value at a point x on the basis of control points (dots). In (a), the estimation requires extrapolation; in (b) interpolation. In the top diagrams, the simple average is used; the middle diagrams use linear trends; and the bottom diagrams use quadratic trends. The average estimates are restricted to the range of data values; the linear trend fails to find the high in (b); the quadratic trend gives a reasonable result.

7.4.2.3   **Choice of technique and appraisal**

The available computer software for surface estimation makes use of various permutations of the above techniques. Most common are linkages of the neighbourhood concept with local trend surfaces and/or inverse distance weighting, and linkages of Delauney triangulation with cubic splines (Box 7.8).

In general, methods based on triangulation seem to yield poorer results with geological data. This may be because triangulation methods were devised by cartographers for topographic contouring, where control points can be carefully chosen on the visible surface during surveying. In geology, the 'surface' is invisible a priori and the positioning of the control points is independent of its features. Consequently, the positioning of an estimated point within a particular triangle connecting three control points will be to an extent arbitrary.

It is important to be aware of the method used by your software, as each method can produce is own particular brand of artefact. The user should always be wary of such spurious effects so that they do not influence the geological interpretation, especially in exploration. All too often, unfortunately, even exhaustive study of the software manual reveals insufficient detail about the technique used!

7.4.2.4   **Artefacts and spurious effects**

Edge effects (Box 7.9)

1   Extreme, unrealistic values are common at map edges when local second- or (especially) third-order polynomial surfaces have been used and become overextrapolated where unconstrained beyond the peripheral control points.
2   Flat areas will occur between the peripheral control points and the edges of the map if inverse distance techniques are used without local trends.
3   Triangulation methods require estimation of values at 'pseudo-points' around the map border. These estimates sometimes appear to be poor, and this affects all subsequent stages of surface estimation in the peripheral areas. Most commonly, these areas appear too flat.

Bull's-eye effect – contours circling control points (Box 7.9)

1   Where simplistic inverse distance techniques are used, control points will always be the locus of any local 'highs' or 'lows'. This is simply because, as noted above, inverse distance estimates cannot exceed the range of contributing control point values, so the locally highest and lowest estimated values will be the ones nearest the locally highest and lowest control points. Obviously, in geological surveys, the real highs and lows are likely to be between control points.
2   In nearly every case, an inclined linear 'ridge' or 'valley' on the real surface will be estimated as a series of local 'highs' and 'lows', each centred around a control point. This arises because 'ridge top' and 'valley bottom' estimates are accurate only at positions coinciding with control points; elsewhere the estimate is influenced by laterally disposed, less extreme control points. It might be thought that local quadratic trends would correctly estimate the ridge or valley trend, but in practice there are seldom sufficient control points within the neighbourhood to accurately define the structure. This problem can only be overcome by development of a technique for recognising the linear trend in advance of the estimation procedure.

Honouring the control points (Box 7.9)

Surface estimation algorithms based on simple inverse distance weighting or splines always produce surfaces giving estimates at control points which are equal to the known true values: this is known as honouring the control points. However, some methods include extra smoothing functions which

Box 7.8 Example: surface estimation procedures

The data used here are the same as for Boxes 7.5–7.7, given in Appendix 3.9, giving thickness (m) of a sand formation at various boreholes. The result of three interpolation algorithms is shown to indicate the type and degree of difference that can typically be attributed to the algorithm.

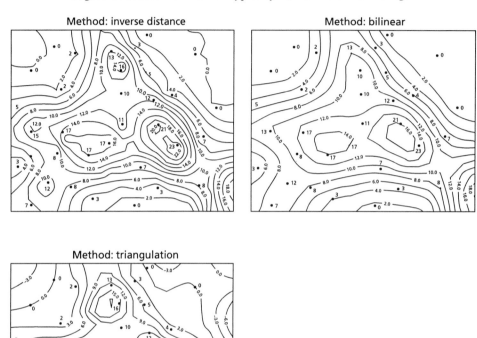

Fig. B7.8.1

The simple inverse distance method has produced a surface in which the highs and lows are constrained to be at control points (or, to be precise, the grid nodes nearest to control points).

The 'bilinear' method here is based on an inverse distance method but the initial estimate is smoothed by using local trends. The overall form of the surface looks reasonable, but the control points are not honoured, so the estimate is demonstrably inaccurate.

The triangulated surface has some unnatural-looking irregularities and some dubious, unsupported features around the edges.

Using geological intelligence, we might interpret the dendritic pattern as

Box 7.8 *Continued*

alluvial or deltaic, and we would expect that the highs (points of greatest thickness) on the surface would be linear 'ridges'. All of these surface estimates have broken the ridges into separate peaks: this is a parsimonious but unintelligent interpretation of the data. The computer is not programmed with geological intelligence, and we must not be misled into thinking that these peaks are real and need to be interpreted geologically or even drilled for oil!

Box 7.9 Artefacts in surface estimates

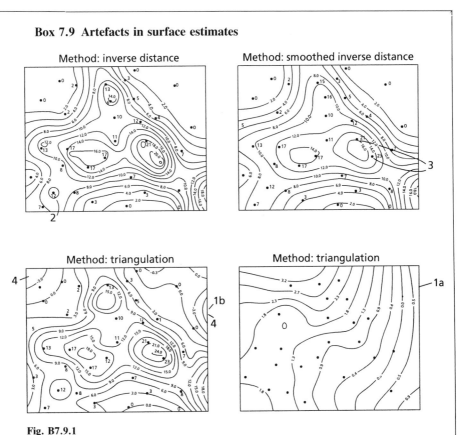

Fig. B7.9.1

1 Edge effects. 1a extrapolation of local trends where unconstrained by control points; or 1b poor pseudo-point estimation in triangulation.
2 Bull's eye effect (Fig. B7.9.1b): due to failure to extrapolate trends beyond the range of the control point values, or to failure to model as quadratic 'ridge' or 'valley'.
3 Failure to honour control points: caused by smoothing.
4 Implausible estimates: extrapolation to negative ppm (corrected by using log transform).

result in estimated values at control points differing from the known true value. It might be thought that an estimated surface with incorrect estimates at every point for which real values are known would be fundamentally useless! This is a reasonable opinion where the control point z values have little or no error attached, such as borehole depth data, but if the z variable involves measurement error or has high inherent local variance, such as permeability, smoothing off potentially extreme values is likely to give a better picture of the surface (except exactly at the control points!).

Implausible estimates (Box 7.9)

It is not at all uncommon for surface estimates of geochemical survey data to include regions with estimated negative quantities! This results from extrapolation of local and unrepresentative trends, plus the inability of the software to recognise the absurdity. This can be avoided by using log-transformed data, which will both decrease the gradients and ensure that estimates of zero amounts or less are impossible (as the log of zero is $-\infty$): see Fig. 7.16.

7.4.2.5 Testing surface estimates

Most surface estimation methods produce surfaces which inevitably go through all the control points. Unlike trend surface analysis, we cannot assess significance by using a measure of deviation of the data from the model. With most surface estimates, at all the known data points the fit is

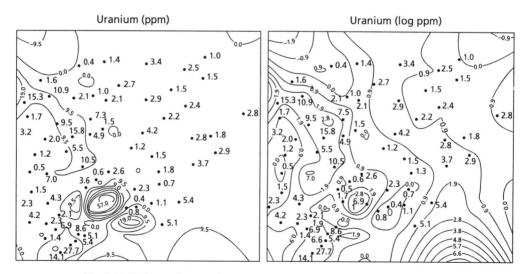

Fig. 7.16 Surface estimates of uranium in the Carswell area (Appendix 3.14). When expressed in ppm, the extreme value near the mine in the central southwest induces an extreme local gradient, causing a 'ghost' low with negative ppm estimates. This problem does not occur with the log-transformed data.

perfect, and everywhere else the fit is unknown! Surface estimates aim to provide useful predictions about the areas between control points, so the obvious way to assess them is to check the accuracy of predictions by obtaining new data. New data, however, may be expensive to obtain; indeed, avoidance of excessive data collection may be the motivation behind surface estimation.

Fortunately, there is a devious way around this problem. The ability of a surface estimation method working on, say, 30 control points to predict the value at a 31st point, is essentially the same as its ability to predict a 30th from 29. We can, then, hold one control point back from the estimation procedure and use it to assess the method's predictive capability. This can be done iteratively with a number of different points in turn, so that we can calculate statistics (e.g. standard errors and paired sample t-tests) for the difference between the predicted and actual values. Such statistics will be pertinent for assessing the reliability of genuinely new point estimates. This approach is known as a cross-validation or jack-knife technique; the idea can be generally used in statistics for assessing predictions.

7.4.3   ## Kriging

Kriging is a different and special way of making estimates of spatially distributed values from point values. Its advocates have developed a distinct culture of terminology and methodology known as *geostatistics*. The term geostatistics *sensu stricto* does not mean just any statistical method applied to geological data; it specifically pertains to the particular methods to be described in this section. The term is sometimes erroneously used in the broad sense; the term Matheronian geostatistics can be used to avoid ambiguity (Matheron, together with Krige, were the main pioneers of the method). Kriging was developed for the very specific application of predicting gold reserves in mines in South Africa from borehole information, but is becoming increasingly used in hydrocarbon exploration and other geological industries.

The following are the key concepts and jargon involved.

Regionalised variable

A regionalised variable has values that are spatially continuous: all points in an area, in principle, have a unique value. The values are not random, but neither are they exactly describable by any geometric function. A crucial aspect of the non-randomness of regionalised variables is that values are to some extent smooth (refer back to Fig. 7.11). A regionalised variable may consist of a drift component and residuals.

Drift

The drift is the component of a regionalised variable resulting from average and trend effects. The drift component may be described by equations such as those used to describe trend surfaces, except that drift may be recognised on a more local scale. First-, second- and third-order trends may be used.

Residuals

The residuals are simply the difference between the calculated drift and the actual data values.

Stationarity

A variable is stationary if there is no significant drift within a specified area. Residuals are, by definition, stationary.

Semivariogram

The semivariogram is a graph that describes the properties of a regionalised variable and is used to find the weightings to be attached to various control point values in order to make estimates at new points.

7.4.3.1   **Semivariance**

Semivariance describes the relatedness of control point values which are specified distances apart. It is ideally calculated from data along a regular traverse or grid.

Semivariance

$$\gamma_h = \frac{\sum\limits_{i=1}^{n-h} (z_i - z_{i+h})^2}{2(n-h)}$$

where γ_h is the semivariance for a distance h, n is the number of points in the traverse and z_i is the value of the geological variable at the ith point along the traverse.

The semivariance is calculated for various distances and the semivariogram is constructed simply as a graph of γ_h against h (Fig. 7.17). Where h is zero, pairs of z values at a distance h apart are identical and the semivariance clearly equals zero. As h increases, the similarity between the pairs of z values is likely to decrease and the semivariance increases. (This is similar in concept to autocorrelation, Section 6.3.3.1, except that autocorrelation at

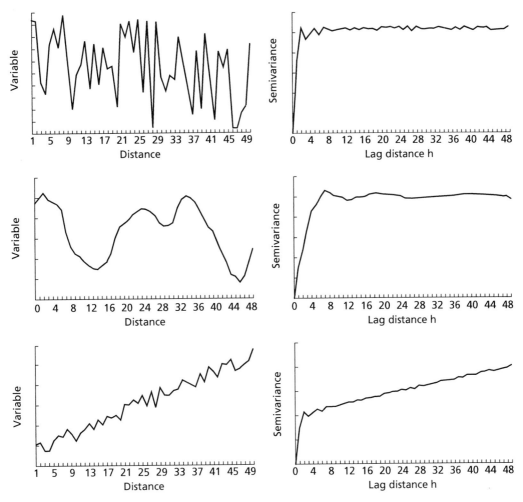

Fig. 7.17 Examples of data series and their semivariograms. The random sequence in (a) gives a semivariogram with a short range; the range is larger with smoother data (b). In (c), the trend results in a continuous increase in semivariance.

zero distances is 1 and values decline at higher distances.) At a certain value of h, we reach a point at which the semivariance ceases to increase: this is the range value. Points that are a distance apart that exceeds the range value are spatially independent: they are beyond the distance within which the inherent smoothness of the regionalised variable confers a degree of similarity. Often, the semivariogram levels off beyond the range at a sill value: the semivariance at the sill is equal to the variance. The only factor affecting the relationship between points that distance apart is the overall variance.

Semivariance can be calculated in any direction, and it is routine to obtain four profiles (north–south, east–west and the two diagonals). This allows the anisotropy of semivariance to be constrained as an ellipse.

Anisotropy will inevitably result from geological heterogeneities such as stratigraphic variation and effects of tectonic stress. For example, the direction having the shortest range is likely to be parallel to the maximum shortening direction in deformed rocks.

7.4.3.2 **Punctual kriging**

The semivariogram fits in well with the ideas on spatial autocorrelation and choice of control points in Sections 7.4.1 and 7.4.2.1: we now have a means of quantifying the concepts introduced there.

Choice of control points to be used in a new estimate

In kriging, we only use control points that are at a distance less than the range from the point at which the estimate is required. The circle of radius equal to the range around a point is said to enclose the neighbourhood.

Calculating the estimate

In Section 7.4.2.1, we noted that inverse distance weighting could be used to find the weightings to be attached to control point values, but the choice of inverse distance was arbitrary. The portion of the semivariogram within the range shows the degree of relatedness of points h apart, so these are more satisfactory. Weights attached to points a distance h from the point to be estimated, then, are based on the semivariance for that h. In practice, the weights are derived from a semivariogram model: an idealised curve of standard form which should convey the trend of the data points.

The calculation of the estimates can be done by a fairly complex matrix formulation; this will not be discussed here, and it remains well hidden from the user in the computer package! If this procedure is applied to the raw control point data, the procedure is known as punctual kriging.

A consequence of the matrix solution in kriging is the derivation of an estimate of the error associated with each calculation. So, for each estimated point value, we have a value called the estimation variance. It is normal to present maps of the estimation variance along with the maps of the estimated z values, thus showing the user where the estimates are unreliable. (See Box 7.10.)

7.4.3.3 **Universal kriging**

Punctual kriging only uses the semivariogram information to estimate surfaces from control point values. Semivariograms are efficient ways of describing the degree of smoothness of stationary regionalised variables, but, if there is a spatial trend or drift (i.e. it is non-stationary), the semivariogram result will contain the compounded effects of smoothness plus drift. Furthermore, semivariances will not be useful for predicting drift in the vicinity of the point to be estimated. Punctual kriging should only be used on drift-free, genuinely stationary data.

The problem of accounting for drift is solved by universal kriging. Here we include extra terms in the matrix formulation to describe the drift within the neighbourhood being considered. These terms are like those for trend surfaces (Section 7.2) and likewise may be linear, quadratic or higher-order (though no more than quadratic is usually advisable). Universal kriging can be a problematic procedure (see Section 7.4.3.4), but for data with simple trends we can obtain a reasonable result by calculating the trend surface for the whole area, punctually kriging the residuals, then adding the trend to the kriged result: see Box 7.11.

7.4.3.4   Critique of kriging

Punctual kriging can be seen as a more rigorous form of the simple weighted averaging procedures described in Section 7.4.2, and as such deserves attention. However, the problem of accounting for local trends remains serious.

It has been argued that universal kriging involves theoretical difficulties that negate its apparent ability at dealing with drift. The criticism is of circularity in procedure: the semivariogram provides information on the range and the drift is calculated for the set of points within the range. However, the data are not stationary until the drift is subtracted, and the semivariogram is not valid unless the data are stationary. So we need to use the semivariogram to find the drift, but we need to subtract the drift before calculating the semivariogram!

In practice, the semivariogram used is a model loosely based on data, the drift is found and subtracted, and then the semivariogram is recalculated for comparison. This becomes a highly interactive and rather subjective procedure: there is no single unique solution.

7.5   ANALYSIS OF FRACTAL DIMENSION

A fractal is a structure that exhibits scale invariance: it looks similar at a range of magnifications. This may be a property both of geometrically regular objects such as snowflakes and irregular shapes such as coastlines: it is claimed that the irregularities of a coast outline on a map projection are similar in type to its irregularities on meso- and microscales. The contortion of a coastline on a map causes the line to occupy the map plane to a larger extent than does a straight line. As the straight line has a dimension of 1 and the plane has a dimension of 2, the coastline can be said to have a fractional dimensionality between 1 and 2: this is its fractal dimension. Similarly, a topography has a fractal dimension between 2 and 3: the Netherlands or the American Midwest would near 2; the Himalayas would work out to be somewhat greater! Any definable object has a fractal dimension and the

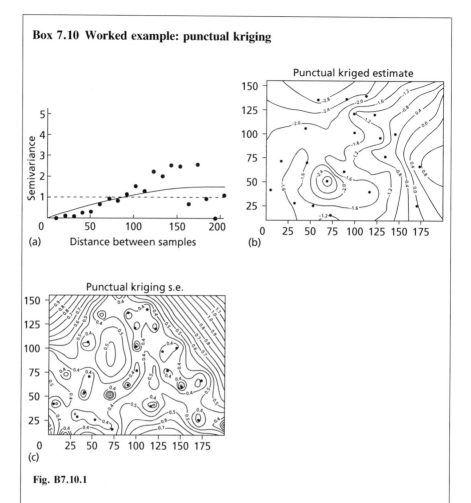

Box 7.10 Worked example: punctual kriging

Fig. B7.10.1

Using the data on grain size (phi units) of an alluvial sandstone formation (Appendix 3.9), we obtain the semivariogram in Fig. B7.10.1(a). A semivariogram model with rather poor fit has been used. The systematic increase in semivariance from lag of 0 to 150 is indicative of a large-scale trend or drift in the data. This casts suspicion on the use of punctual kriging in this case, and explains why a normal semivariogram model with reasonable fit could not be found: the data aren't stationary. The scatter of the data points at higher distances merely reflects that there are few pairs of points with that amount of separation. The horizontal dashed line gives the variance.

Use of the semivariogram model produced the surface estimate and error surface shown in Fig. B7.10.1(b,c). Standard errors are inevitably larger further from the control points (dots).

Box 7.11 Worked example: universal kriging

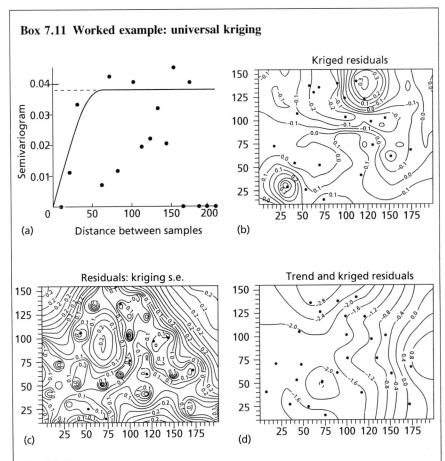

Fig. B7.11.1

Using the same data as Box 7.10, it was found that a first-order trend surface provided a reasonable fit to the data. The residuals from the trend surface, then, could be regarded as closer to stationary. The semivariogram of the residuals is shown in Fig. B7.11.1(a). Although the data points are quite scattered, this is a more typical semivariogram and the model shown was used for kriging.

The kriged residuals and error plots obtained were as shown in Fig. B7.11.1(b,c). The standard errors are much less than those shown in the punctually kriged alternative in Box 7.10: this is because the punctual semivariogram included redundant information on the trend.

Adding the kriged residuals to the first-order trend restores the two components of the surface (Fig. B7.11.1d). This can be taken to be a superior surface estimate.

usefulness of this for characterising geological properties is being explored. A method for finding fractal dimensions between 1 and 2 is given in Box 7.12. However, it is clear that the fractal dimension is only an efficient descriptor for entities with genuinely fractal properties: a sine wave has a calculable fractal dimension but is better characterised by its wavelength and amplitude.

It has been argued that fault planes may have fractal properties, and that this could produce a fractal pattern of seismic activity, with dimensionality

Box 7.12 Calculation of fractal dimension

Grids of various cell sizes are superimposed over the feature: in this case the feature is a stylolite and we show, as examples, the grids with cell sizes of 50 and 5 units, in a rectangle with length 200 units (Fig. B7.12.1).

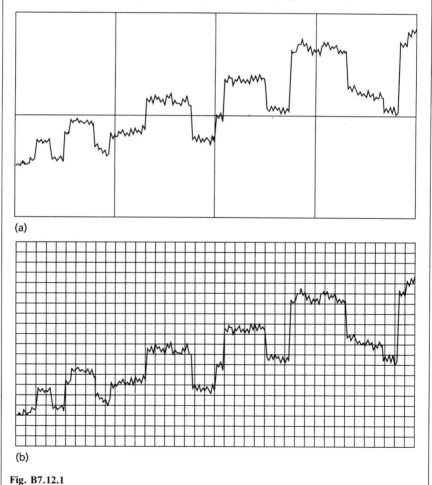

(a)

(b)

Fig. B7.12.1

Box 7.12 *Continued*

For each grid, count the number of cells intersected by the feature. In this case:

Cell size	Cell count
200	1
100	4
50	6
25	14
12.5	33
5	99
2.5	224

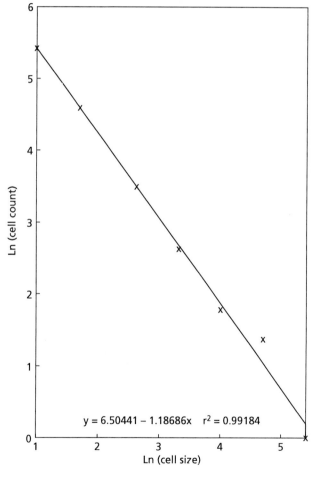

$y = 6.50441 - 1.18686x \quad r^2 = 0.99184$

Ln (cell count) vs Ln (cell size)

Fig. B7.12.2

continued on p. 326

Box 7.12 *Continued*

A graph of

ln(cell count) vs. ln(cell size)

yields an approximately linear scatter. The gradient of the best-fit line gives the fractal dimension (if we ignore the minus sign) (Fig. B7.12.2). In this case, the fractal dimension is 1.187.

It is easy to work out that, using this method, a straight line yields a dimension of 1 and the whole plane gives a dimension of 2.

The goodness of fit or correlation coefficient could be used as an index of the suitability of the fractal dimension as a descriptor: a genuine fractal structure should show a goodness of fit very near to 1.

---

between 0 and 1 (0 would be a single event; 1 would be continuous activity). Also, fractal dimensionalities between 2 and 3 are being used to model the distribution of porosity and permeability in reservoirs.

FURTHER READING

Cadigan R.A. (1962) A method for determining the randomness of regionally distributed quantitative geologic data. *Journal of Sedimentary Petrology*, **32**(4), 813–818.

Lutz T.M. (1986) An analysis of the orientations of large scale crustal features: a statistical approach based on aerial distribution of point-like features. *Journal of Geophysical Research*, **91**(B1), 421–434.

Ramsay J.G. and Huber M.I. (1983) *The Techniques of Modern Structural Geology*, Vol. 1. Academic Press, London.

sources of methods for point distribution analysis.

Watney W.L. (1985) Resolving controls on epeiric sedimentation using trend surface analysis. *Mathematical Geology*, **17**, 427–451.

Whitten E.H.T. and Bayer R.E. (1964) Process–response models based on heavy mineral content of the San Isabel granite, Colorado. *Bulletin of the Geological Society of America*, **75**, 841–862.

case studies of trend surface analysis.

Jones T.A., Hamilton D.E. and Johnson C.R. (1992) *Contouring Geological Surfaces with the Computer*. Van Nostrand, New York.

an up-to-date text.

Clark I. (1979) *Practical Geostatistics*. Applied Science Publishers, Barking.

for details of kriging methodology.

Bardossy A. and Bardossy G. (1984) Comparison of geostatistical calculations with the results of open pit mining at the Iharkut bauxite district, Hungary: a case study. *Mathematical Geology*, **16**, 173–192.

possibly the most accessible kriging case study.

Philip G.M. and Watson D.F. (1986) Matheronian geostatistics — *quo vadis? Mathematical Geology*, **18**, 93–117.

a scathing criticism of kriging which prompted an intemperate correspondence in subsequent issues of the journal!

8 Multivariate Methods

8

8.1 INTRODUCTION

8.1.1 Why use multivariate methods?

Multivariate methods allow the simultaneous analysis of any number of variables. Earlier in the book we have introduced methods for one (Chapter 2) and two (Chapter 3) variables, but we have only used as many as three variables simultaneously in the special case of spatial data (Chapter 7). The particular problem inherent in multivariate analysis is conceptualisation and graphical representation: our appreciation of univariate, bivariate and spatial data is based on constructed (or imagined) one-, two- or three-dimensional diagrams. If we have, say, 10 variables to analyse, the corresponding construction must involve 10 dimensions. This is impossible to draw or imagine, except in an abstract way, and herein lies the problem. The reader will be relieved to discover that one of the functions of multivariate methods is to reduce the dimensionality of the data to an imaginable and plottable two or three dimensions! Furthermore, although multivariate methods are designed to deal with multidimensional complexity, they also work on two- or three- dimensional data, so we can draw perfectly valid diagrams to illustrate what a particular method is doing.

Multivariate methods are important because virtually all geological data are inherently, naturally, multivariate: this is necessary to formulate a complete numerical description of a geological object, whether a rock, a bag of sediment or a fossil. An igneous rock may be described by a long list of percentages of phases, oxides or elements, a sedimentary rock by percentage of various clast types and grain size frequencies, and a fossil by measurements of all its component parts. Even with more 'remotely sensed' data, geological characteristics are resolved by collecting many measurements: responses at different spectral bands in satellite images, or a diversity of electric logs from down-hole recording devices.

It is clear that many variables are needed to fully characterise the geological object, but under what circumstances are they all needed for simultaneous analysis using a multivariate method? This question is best answered by reversing it; it is the use of univariate or bivariate methods that should need special justification – what is the justification for *not* using the bulk of the geological information? These are some reasonable justifications.

1 Logistical: data for only one or two variables are available or obtainable within the pertaining constraints.

2 Geological: only one or two variables may be relevant due to the specified scope of the study, i.e. the project may focus on the variable rather than the

geological objects. This may be particularly true in resource appraisal, where only, say, distribution of copper around a mine or permeability in a hydrocarbon reservoir is economically relevant.

3 Empirical: it may be found that, for a simple job of hypothesis testing, for example in a two-sample t-test, the use of only one variable may be sufficient to obtain a rejection of the null hypothesis (H_0). If two populations can be shown to be significantly different using just one variable, then they will inevitably prove to be significantly different if a multivariate test is used. But the converse does not apply: it is not at all unlikely that a multivariate test will result in rejection of H_0 even if univariate tests on all the individual variables fail to reject H_0.

4 Mathematical: the simultaneous analysis of multiple variables only provides extra information if variables are correlated with each other: this produces the special and important structure of the data scatter in multivariate space. If there are no significant correlations, variables might as well be dealt with separately, but this is seldom the case with real geological data.

There may be a number of more subtle but genuine reasons for focusing on just one or two variables, but it is still frequently the case that multivariate methods are overlooked because of ignorance or blind adherence to traditional procedures.

The potential of using multivariate data can be appreciated by referring back to Chapter 3: there we saw how the bivariate approach revealed properties of a sample which were not apparent using one variable at a time; the multivariate approach has correspondingly greater power.

It is essential to begin the analysis of multivariate data simply, using dot or box-and-whisker plots and bivariate plots and determining sample means, variances and correlations. These activities can help to guide more sophisticated analyses, if they are needed, and to reveal errors of data recording and entry.

A further warning is needed: multivariate analysis requires large samples. Just as two observations on a pair of variables are sure to give a correlation coefficient of 1 (as two points fix a straight line), so multivariate data with few observations on many variables will give exciting but misleading results!

8.1.2   **Notation**

We must first introduce some notation and describe generalisations of models and statistics which we use in multivariate analysis. Although the mathematics of matrices and vectors is not essential in principle for the study of multivariate methods, the compact notation can simplify presentation considerably and we shall use it for that purpose. A brief treatment of matrix algebra is given in Appendix 1.

Data matrix

The data, consisting of measurements on p variables made on each of n items, are set out in an $n \times p$ matrix, in which each row represents one observation. Denoting the variables by $X_1, \ldots, X_p$ and the value of X_i in the ith observation by x_{ij}, the data matrix is written

$$X = \begin{pmatrix} x_{11} & x_{12} \ldots x_{1p} \\ x_{21} & x_{22} \ldots x_{2p} \\ \vdots & \vdots \quad \vdots \quad \vdots \\ x_{n1} & x_{n2} \ldots x_{np} \end{pmatrix}$$

The n values of variable X_j can be written as a column vector

$$x_j = \begin{pmatrix} x_{1j} \\ x_{2j} \\ \vdots \\ x_{nj} \end{pmatrix}$$

and the ith observation is represented by a row vector

$$x_j^T = (x_{i1} \quad x_{i2} \ldots x_{ip})$$

Note that geochemical data are sometimes presented in the opposite way to that described above: the percentages of oxides found in a single specimen (i.e. a single observation) are printed in a column.

Appendix 1 shows how vectors can be related to points in space. A single observation on p variables can be treated as a point in p-dimensional space, and the n values observed on a single variable as a point in n-dimensional space. Both ways of regarding the data will be used extensively. For example, we shall form clusters of observations which are close to each other in p-dimensional space and of variables which are close in n-dimensional space.

While, in general, the p variables are correlated, the observations must be independent in what follows.

8.1.3   Choice of method

Multivariate methods include some which are merely multivariate versions of univariate or bivariate techniques, plus others which deal specifically with the properties and problems that emerge with large numbers of variables. The following is a general guide to the purposes of the main types of techniques.

Multivariate generalisations of univariate statistics

For making statistical statements about population parameters: most often for testing for equivalence of population means.

Example: to test the null hypothesis that the compositions of granites from two separate intrusions are the same.

Multiple regression

Finding and testing a 'best-fit' equation relating one dependent variable to any number of independent variables.

Example: finding an equation relating sandstone permeability to a number of wireline logs, for use in prediction.

Discriminant functions

Finding the equation for a line which best separates two (or more) user-defined subgroups within the data set, and allocating new data to one or other of the groups on this basis.

Example: discriminating between areas of high and low gold resource potential on the basis of geochemical, geophysical and remotely sensed data.

Principal components and factor analysis

Finding the directions of maximum variance in the data, using these to ordinate data in one, two or three dimensions and interpreting them as factors influencing the data.

Example: obtaining a scatter plot to show the relationships between fossil specimens defined by large numbers of morphological measurements.

Similarity coefficients and cluster analysis

Quantifying similarity between pairs of objects or observations, and using this to produce an empirical classification.

Example: obtaining a classification of a suite of sedimentary rocks on the basis of sediment components.

8.2 GENERALISATIONS OF UNIVARIATE AND BIVARIATE STATISTICS

Although this section deals with only basic statistics on multivariate data, it is perhaps the most difficult part of Chapter 8 and could be omitted by less ambitious students!

8.2.1 Joint probability distributions

In considering probability distributions of single random variables, we introduced the idea of a probability density function (p.d.f.) $f(x)$ from which we could find the probability of obtaining values of X in any given interval.

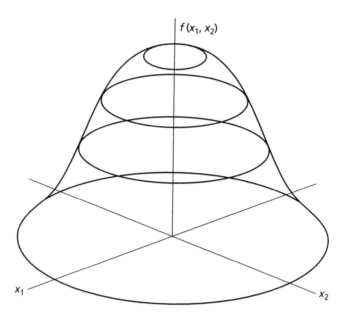

Fig. 8.1 The concept of the bivariate frequency distribution. This idea can be extrapolated for situations of higher dimensionality.

When two or more variables are involved, we can find p.d.f. $f(x_1, \ldots, x_p)$ of the p variables: this could be used to calculate probabilities of the values of the variables lying in specific multidimensional intervals, defined in terms of each of p variables. Such functions are called joint p.d.f.s and the variables are said to have joint distributions. When $p = 2$ the density can be represented in three dimensions.

Figure 8.1 is a representation in two dimensions of the three-dimensional density of the bivariate normal distribution. The surface is described by a function which takes into account correlations between the pair of variables as well as mean values and variances. We can imagine versions of this diagram for larger numbers of variables.

8.2.1.1 Parameters of multivariate normal distributions

The joint normal density of p variables has the following parameters:

the means $\mu_1, \ldots, \mu_p$ of the p variables
their p variances $\sigma_1^2, \ldots, \sigma_p^2$
their $p (p - 1)/2$ covariances $\sigma_{12}, \ldots, \sigma_{1p}, \ldots, \sigma_{p-1,p}$

It is convenient to display these parameters in arrays and we define the following vector and matrices.

The mean vector

$$\mu = \begin{pmatrix} \mu_1 \\ \mu_2 \\ \vdots \\ \mu_p \end{pmatrix}$$

The covariance matrix

$$\Sigma = \begin{pmatrix} \sigma_{11} & \sigma_{12} & \cdots & \sigma_{1p} \\ \sigma_{21} & \sigma_{22} & \cdots & \sigma_{2p} \\ \vdots & \vdots & \vdots & \vdots \\ \sigma_{p1} & \sigma_{p2} & \cdots & \sigma_{pp} \end{pmatrix}$$

where, for consistency with the other elements in the matrix, the variances have been denoted by $\sigma_{11}, \ldots, \sigma_{pp}$. Note that they constitute the leading diagonal of the covariance matrix.

Two functions of a covariance matrix will be used to summarise the total variation in the p variables.

1 The sum of the variances; it is given by the sum of the elements in the leading diagonal of the covariance matrix; it is called the trace of the matrix and written $\text{tr}(\Sigma)$.

2 The determinant of the covariance matrix, $|S|$, which is called the generalised variance.

The matrix is symmetric: both σ_{12} and σ_{21} represent the covariance between variables X_1 and X_2, for example.

When all the variables are uncorrelated (and, as we are discussing parameters of normal distributions, therefore independent), the covariance matrix is diagonal:

$$\Sigma = \begin{pmatrix} \sigma_{11} & 0 & 0 & \cdots & 0 \\ 0 & \sigma_{22} & 0 & \cdots & 0 \\ \vdots & \vdots & \vdots & \vdots & \vdots \\ 0 & 0 & 0 & \cdots & \sigma_{pp} \end{pmatrix}$$

It will frequently be more appropriate to use correlations than covariances and these too are set out in a matrix:

$$\rho = \begin{pmatrix} 1 & \rho_{12} & \cdots & \rho_{1p} \\ \rho_{21} & 1 & \cdots & \rho_{2p} \\ \vdots & \vdots & \vdots & \vdots \\ \rho_{p1} & \rho_{p2} & \cdots & 1 \end{pmatrix}$$

where $\rho_{ij} = \sigma_{ij}/\sqrt{(\sigma_{ii}\sigma_{jj})}$. Note that the correlation matrix is simply a covariance matrix which has been derived from standardised variables.

The 1s in the leading diagonal represent the correlation of each variable with itself. Estimated correlation matrices are usually presented without these elements, and, as this matrix is also symmetrical, it is usual to display only the elements below the leading diagonal (only the lower triangle).

When the variables are uncorrelated, the correlation matrix is diagonal; in fact, it is an identity matrix.

8.2.1.2 Estimates of the parameters

The elements of these arrays are estimated in the same ways as the corresponding parameters in univariate and bivariate statistics.

Means

The means $\mu_1, \ldots, \mu_p$ are estimated by the sample arithmetic means

$$\bar{x}_1 = 1/n\Sigma x_{i1}$$
$$\vdots \qquad \vdots \quad \vdots$$
$$\vdots \qquad \vdots \quad \vdots$$
$$\bar{x}_p = 1/n\Sigma x_{ip}$$

and they will be arranged in a vector

$$\bar{x}^T = (\bar{x}_1 \, \bar{x}_2 \ldots \bar{x}_p)$$

Variance and covariance

Unbiased estimates of the sample variances are provided by

$$s_j^2 \text{ or } s_{jj} = \frac{1}{n-1}\Sigma(x_{ij} - \bar{x}_j)^2$$

where the sum is the corrected sum of squares of values of X_j. Similarly, the covariances between pairs of variables X_j, X_k are estimated by

$$s_{kj} = \frac{1}{n-1}\Sigma(x_{ij} - \bar{x}_j)(x_{ik} - \bar{x}_k)$$

where the sum is the corrected sum of products of values of X_j, X_k.

These parameters are set out in the sample covariance matrix

$$S = \begin{pmatrix} s_{11} & s_{12} & \cdots & s_{1p} \\ s_{21} & s_{22} & \cdots & s_{2p} \\ \vdots & \vdots & \vdots & \vdots \\ \vdots & \vdots & \vdots & \vdots \\ s_{p1} & s_{p2} & \cdots & s_{pp} \end{pmatrix}$$

Two measures of the variation in data on p variables are used:

total variance $= s_{11} + s_{22} + \ldots + s_{pp}$
the generalised variance $= |\Sigma|$

The first of these will be of particular importance in the sections dealing with principal components analysis (PCA) and the second appears in tests for the equality of covariance and correlation matrices.

Correlation coefficient

The sample correlation coefficients between variables X_j, X_k are calculated from

$$r_{jk} = \frac{\Sigma(x_{ij} - \bar{x}_j)(x_{ik} - \bar{x}_k)}{\sqrt{\{\Sigma(x_{ij} - \bar{x}_j)^2(x_{ik} - \bar{x}_k)^2\}}}$$

$$= \frac{s_{ij}}{\sqrt{\{s_{jj}s_{kk}\}}}$$

Covariances and correlations are calculated from corrected sums of squares (CSS) and corrected sums of products (CSP), so it is helpful to define the sums of squares and products matrix (SSPM).

$$\begin{pmatrix} \text{CSS}(x_1) & \text{CSP}(x_1, x_2) \ldots \text{CSP}(x_1, x_p) \\ \text{CSP}(x_2, x_1) & \text{CSS}(x_2) \quad\quad \ldots \text{CSP}(x_2, x_p) \\ \vdots & \vdots \quad\quad \vdots \quad\quad \vdots \\ \text{CSP}(x_p, x_1) & \text{CSP}(x_p, x_2) \ldots \text{CSS}(x_p) \end{pmatrix}$$

The estimated covariance matrix is calculated from

$$S = \frac{\text{SSPM}}{n - 1}$$

8.2.1.3 **Some important properties of the multivariate normal distribution**

1 If variables $X_1, \ldots, X_p$ have a joint normal distribution then any linear compound of the variables

$$Y = a_1X_1 + a_2X_2 + \ldots + a_pX_p$$

has a univariate normal distribution. This is a most useful property because much of our analysis will require the construction of linear compounds in order to obtain univariate indices.

Note that the compound may be written in terms of vectors

$$Y = a^TX$$

where a^T and x are the vectors $(a_1, \ldots, a_p)$ and $(X_1, \ldots, X_p)$.

2 A set of q linear compounds

$$Y_1 = a_{11}X_1 + \ldots + a_{1p}X_p$$
$$\vdots \quad\quad \vdots \quad\quad \vdots \quad\quad \vdots$$
$$Y_q = a_{q1}X_1 + \ldots + a_{qp}X_p$$

of joint normal variables has a joint normal distribution. In matrix notation this may be written

$$Y = AX$$

where A is the $q \times p$ matrix

$$\begin{pmatrix} a_{11} & a_{12} \ldots a_{1p} \\ a_{21} & a_{22} \ldots a_{2p} \\ \vdots & \vdots \quad \vdots \quad \vdots \\ a_{q1} & a_{q2} \ldots a_{qp} \end{pmatrix}$$

and $\mathbf{X}$ is again the vector of variables.

These new variables will have their own mean vector and covariance and correlation matrices; their relationship to the parameters of the original distribution is described below.

3 Uncorrelated normal random variables are independent. In PCA we shall seek independent linear compounds of the original variables and we shall see that this requires us to find the values $a_{11}, \ldots, a_{qp}$, which produces a covariance matrix in which all the off-diagonal elements are zero.

4 Multivariate generalisations of the central limit theorem have been established so that, if we have large samples, the sample means tend to have a joint normal distribution.

8.2.1.4   **Linear compounds**

Linear compounds are important in multivariate analysis because we often wish to compute simple, meaningful indices from several variables.

Suppose that $X_1, \ldots, X_p$ have a multivariate distribution with mean vector $\mathbf{\mu}^T$ and covariance matrix Σ and that we form a linear compound

$$Y = a^T X$$
$$= a_1 X_1 + \ldots + a_p X_p$$

Then the mean of Y is

$$\mu_Y = a^T \mu$$
$$= a_1 \mu_1 + \ldots + a_p \mu_p$$

and the variance is

$$\mathrm{Var}(Y) = \mathrm{Var}(a_1 X_1 + \ldots + a_1 X_p)$$
$$= a_1^2 \, \mathrm{Var}(X_1) + \ldots + a_p^2 \, \mathrm{Var}(X_p) + 2a_1 a_2 \mathrm{cov}\,(X_1, X_2) + \ldots +$$
$$2a_1 a_p \mathrm{cov}\,(X_1, X_p) + \ldots 2a_{p-1,\,p} a_p \mathrm{cov}\,(X_{p-1}, X_p)$$

which may be written more conveniently as

$$\mathrm{Var}(Y) = a^T \Sigma a$$

Further, if the variables $X_1, \ldots, X_p$ are multivariate normal then Y has a normal distribution.

Worked example: see Box 8.1.

In PCA we shall obtain sets of linear compounds $Y_1, \ldots, Y_q$ with specified properties:

Box 8.1 Worked example: linear compounds

Heights, lengths and widths of a species of fossil have mean values equal to 4.6, 9.4 and 2.8 cm respectively; the covariance matrix is

Height   1.2
Length   0.2   2.3
Width   0.6   0.7   1.9

Find the mean and variance of the linear compound

$Y = 0.6 \times$ height $+ 0.2 \times$ length $+ 0.3 \times$ width

Solution

The mean of the compound is

$\mu_Y = 0.6 \times 4.6 + 0.2 \times 9.4 + 0.3 \times 2.8$
$\quad = 5.48\,\text{cm}$

The variance is

$$\text{Var}(Y) = (0.6 \quad 0.2 \quad 0.3) \begin{pmatrix} 1.2 & 0.2 & 0.6 \\ 0.2 & 2.3 & 0.7 \\ 0.6 & 0.7 & 1.9 \end{pmatrix} \begin{pmatrix} 0.6 \\ 0.2 \\ 0.3 \end{pmatrix}$$

Premultiplying the matrix by the row vector gives

$$(0.94 \quad 0.79 \quad 1.07) \begin{pmatrix} 0.6 \\ 0.2 \\ 0.3 \end{pmatrix}$$
$$= 1.043\,\text{cm}^2$$

$Y_1 = a_{11}X_1 + a_{12}X_2 + \ldots + a_{1p}X_p = a_1^T X$
$Y_2 = a_{21}X_1 + a_{22}X_2 + \ldots + a_{2p}X_p = a_2^T X$
$\vdots \qquad \vdots \qquad \vdots \qquad \vdots \qquad \vdots \qquad \vdots$
$Y_q = a_{q1}X_1 + a_q\,22X_2 + \ldots + a_{qp}X_p = a_q^T X$

These will have some joint distribution with a mean vector and covariance matrix whose diagonal elements are the variances of the compounds $a^T X$ and whose off-diagonal elements are their covariances. The covariance matrix is obtained by combining calculations for individual compounds in

$$\Sigma_y = \begin{pmatrix} a_1^T \\ a_2^T \\ \vdots \\ a_q^T \end{pmatrix} \Sigma(a_1 \, a_2 \ldots a_p)$$

$\quad = A\Sigma A^T$, say

Box 8.2 Worked example: linear compounds

A second compound of the height, length and width of the species of fossil referred to above was formed, using weights 0.1, 0.7 and 0.4. Find the covariance matrix for the two linear compounds.

Solution

The transforming matrix is

$$A = \begin{pmatrix} 0.6 & 0.2 & 0.3 \\ 0.1 & 0.7 & 0.4 \end{pmatrix}$$

so the required covariance matrix is

$$\begin{pmatrix} 0.6 & 0.2 & 0.3 \\ 0.1 & 0.7 & 0.4 \end{pmatrix} \begin{pmatrix} 1.2 & 0.2 & 0.4 \\ 0.2 & 2.3 & 0.7 \\ 0.6 & 0.7 & 1.9 \end{pmatrix} \begin{pmatrix} 0.6 & 0.1 \\ 0.2 & 0.7 \\ 0.3 & 0.4 \end{pmatrix}$$

After multiplying the first two matrices, this gives

$$\begin{pmatrix} 0.94 & 0.79 & 0.95 \\ 0.50 & 1.91 & 1.29 \end{pmatrix} \begin{pmatrix} 0.6 & 0.1 \\ 0.2 & 0.7 \\ 0.3 & 0.4 \end{pmatrix} = \begin{pmatrix} 1.007 & 1.021 \\ 1.021 & 1.895 \end{pmatrix}$$

Note how in this case the matrix has been reduced in dimensions to 2×2 by the multiplication process; the number of rows (and columns) of the final covariance matrix will be the same as the number of compounds we choose to construct.

---

where **A** is the matrix formed from the row vectors of weights $a_1^T, a_2^T, \ldots, a_p^T$.

Worked example: see Box 8.2.

In practice, the elements of the vectors $a_1, \ldots, a_p$ will be chosen in ways which suit particular purposes: in PCA they will be chosen to maximise variances of compounds and in factor analysis they are selected to produce a small number of new variables which explain observed correlations. For the purposes of discrimination or testing hypotheses about mean vectors they are chosen to maximise distances.

8.2.2   **Generalisations of univariate and bivariate inference**

In this section we shall extend the work on confidence limits for means and tests on means and correlations. The procedures are discussed in statistical texts in terms of two principles, which do not always lead to the same procedures or conclusions.

1 Comparing likelihoods under null and alternative hypotheses.

2   The union–intersection principle by which we calculate linear compounds which maximise the discrepancy between hypothesis and observation.

Rather than examine these principles in detail we shall quote results and rationalise them heuristically. In particular, the idea of standard distance is useful.

8.2.2.1   Tests on mean vectors

Hotelling's T^2

In univariate statistics the t statistic used for testing a hypothesis about a single mean, the quantity

$$t = \frac{\bar{x} - \mu_0}{s/\sqrt{n}}$$

may be regarded as the distance of the sample mean from μ_0, measured in standard errors of the sample mean; and similarly the statistic used to compare two means,

$$t = \frac{|\bar{x}_1 - \bar{x}_2|}{s\sqrt{\{(1/n_1) + (1/n_2)\}}}$$

$$= \frac{|\bar{x}_1 - \bar{x}_2|\sqrt{\{n_1 n_2/(n_1 + n_2)\}}}{s}$$

is the distance between the sample means, measured in standard errors of the difference in sample means.

In what follows we form linear compounds of the variables to produce new scalar variables. Using the results in Section 8.2.1.3 we obtain standard distances of the form

$$\frac{|a^T \bar{x} - a^T \mu_0|}{s_a}$$

and

$$\frac{|a^T \bar{x}_1 - a^T \bar{x}_2|}{s_a}$$

where $s_a = a^T S a$ and the vectors **a** are chosen to maximise the distances.

Test on a single vector, μ

This test is useful when we wish to compare a data set with some standard. The data might be dimensions of a number of specimens of a fossil, for example, and we may wish to see how they compare with the mean dimensions of a particular species established in type descriptions. We shall take it that the covariance matrix is unknown and estimate it from the sampled data.

Then it would be of interest to test the null hypothesis

$$H_0: \mu = \mu_0$$

against

$$H_1: \mu \neq \mu_0$$

where μ is the mean vector of the dimensions of the new specimens and μ_0 is the standard vector of means. In fact, we compare the sample mean $\bar{Y}$ of a linear compound $a^T x$ with that of the standard mean, $a^T \mu_0$, choosing weights which maximise the distance

$$\frac{a^T (\bar{x} - \mu_0) \sqrt{n}}{\sqrt{\{a^T S a\}}}$$

where S is the estimate of the covariance matrix, obtained by dividing the matrix of corrected sums of squares and products by $(n - 1)$.

The required weights are proportional to

$$S^{-1}(\bar{x} - \mu_0)$$

and the maximised distance (called the Mahalanobis distance) can be shown to be

$$D_p = \sqrt{\{n(\bar{x} - \mu_0) S^{-1} (\bar{x} - \mu_0)\}}.$$

When the null hypothesis is true, the square of this statistic has a distribution which is related to F:

$$\frac{(n - p)D_p^2}{(n - 1)p} \simeq F_{p,n-p}$$

Accordingly, we accept the null hypothesis at the α level of significance unless

$$\frac{(n - p)D_p^2}{(n - 1)p} > F_{\alpha;p,n-p}$$

Worked example: see Box 8.3.

Tests on two mean vectors

We may have data from two sources and wish to compare the mean vectors by testing the hypothesis

$$H_0: \mu_1 = \mu_2$$

against the alternative

$$H_1: \mu_1 \neq \mu_2$$

As before, we shall form linear compounds $a^T \bar{x}_1$, $a^T \bar{x}_2$ of the sample

Box 8.3 Worked example: Hotelling T^2

This test is often presented in the following form: reject the null hypothesis that $\mu = \mu_0$ at the $100\alpha\%$ level of significance when

$$T^2 \geqslant \frac{(n-1)p}{n-p} F_{\alpha; p, n-p}$$

where T^2 is a quantity called Hotelling's T^2 and is related to the F distribution by

$$\frac{(n-p)}{(n-1)p} T^2 \simeq F_{p, n-p} \quad \text{so that } T^2 \text{ is the } D_p^2 \text{ in the text}$$

It is then called Hotelling's T^2 test.

Twenty observations were made on each of two variables X_1, X_2, representing short and long axis lengths of clasts in mm. The sample means were $\bar{x}_1 = 3.85$, $\bar{x}_2 = 5.40$ and the sample covariance matrix was

$$S = \begin{pmatrix} 0.7586 & 0.2183 \\ 0.2183 & 1.3492 \end{pmatrix}$$

Test the null hypothesis that the data came from a (bivariate normal) distribution with mean vector $\mu^T = (4.0 \quad 6.0)$.

$$7(\bar{x} - \mu)^T S^{-1} (\bar{x} - \mu) = 7(-0.15 \quad -0.60) \begin{pmatrix} 1.3826 & -0.2237 \\ -0.2237 & 0.7774 \end{pmatrix} \begin{pmatrix} -0.15 \\ -0.60 \end{pmatrix}$$

$$= 1.89 \text{ to } 2 \text{ D}$$

Now

$$F_{0.05; 2, 18} = 3.55 \quad \text{(Appendix 2.5)}$$

and

$$\frac{(n-1)p}{n-p} F_{0.05; 2, 18} = 7.49$$

There is no evidence against the null hypothesis that the mean vector is $\mu^T = (4.0 \quad 6.0)$.

means and test for the equality of the corresponding compounds of the true means.

We shall only consider cases where the covariance matrices for the two populations or distributions are equal but unknown. The test works well even when the covariance matrices are unequal, provided that the sample sizes are roughly equal.

The pooled estimate of the common covariance matrix is

$$S = \frac{(n_1 - 1)S_1 + (n_2 - 1)S_2}{n_1 + n_2 - 2}$$

where S_1, S_2 are the estimated covariance matrices from the two samples and n_1, n_2 are the sample sizes.

The standard distance between the sample means is

$$\frac{|\bar{x}_1 - \bar{x}_2|}{s_a}$$

where s_a is $a^T S a$, and it is maximised when a^T is the vector

$$S^{-1}(\bar{x}_1 - \bar{x}_2)$$

The distance is then

$$D_p = (\bar{x}_1 - \bar{x}_2)S^{-1}(\bar{x}_1 - \bar{x}_2)$$

Again, the square of the maximised distance is, when the true means are equal, related to the F distribution:

$$\frac{n_1 + n_2 - p - 1}{p(n_1 + n_2 - 2)} \cdot \frac{n_1 n_2}{n_1 + n_2} D_p^2 \sim F_{p,\, n1\, n2-p-1}$$

The procedure is to accept that the mean vectors are equal unless the value of this test statistic exceeds $F_{\alpha;p,n1+n2-p-1}$.

Simultaneous confidence intervals for differences between means of all the variables should calculated and it can be shown that the limits are

$$a^T (\bar{x}_1 - \bar{x}_2) + \sqrt{s_a \cdot \frac{n_1 + n_2}{n_1 n_2(n_1 + n_2 - p - 1)}} \; F\alpha;\, p,\, n_1 + n_2 - p - 1$$

for all linear compounds $a^T(\mu_1 - \mu_2)$. Individual confidence limits for differences in means of individual variables can again be obtained by choosing a^T to be of the forms

$(1, 0, 0, \ldots, 0)$
$(0, 1, 0, \ldots, 0)$

but it is possible for all of them to include zero even when the data cast doubt on the null hypothesis.

When the a^T vector which maximises the distance is used, however, at least one interval will not include zero, as expected from the result of the test.

8.2.2.2 Tests on covariance and correlation matrices

The most important cases are the following.
1   A test for a lack of correlation between all pairs of variables. (If there are no significant correlations, multivariate techniques should not be used.)
2   A test for the equality of two or more covariance or correlation matrices.

Test for lack of correlation

If no pairs of variables are correlated, the correlation matrix R is a $p \times p$ identity matrix:

$$\begin{pmatrix} 1 & 0 & 0 & \ldots & 0 \\ 0 & 1 & 0 & \ldots & 0 \\ \vdots & \vdots & & & \vdots \\ 0 & 0 & 0 & \ldots & 1 \end{pmatrix}$$

The test derived by considerations of likelihood leads to the test statistic

$$-n\ln|R|$$

If R is an identity matrix, its determinant is equal to 1 and the logarithm of this is zero. Values of $|R|$ in the region of 1 produce small values of the statistic, suggesting that the true correlations may be zero. For large samples, the distribution of the statistic is approximately chi-squared, with $p(p-1)/2$ degrees of freedom.

Test for the equality of two or more covariance matrices

The type of function used to assign objects to particular groups will depend on whether or not the covariance matrices of the groups are equal, and tests for the equality of mean vectors depend upon the assumption that the covariance matrices are equal.

The null hypothesis is that, in k groups,

$$\Sigma_1 = \Sigma_2 = \ldots = \Sigma_k$$

and the alternative is that at least two of the matrices differ. The test, derived by likelihood considerations, requires the calculation and comparison of estimates of the covariance matrices under the two hypotheses.

When the matrices differ each is best estimated separately in the usual way:

$$S_i = \frac{\text{Matrix of corrected sums of squares and products}}{n_i - 1}$$

When all the matrices are equal, the pooled estimate of the common matrix is

$$S = \frac{(n_1 - 1)S_1 + \ldots + (n_k - 1)S_k}{n_1 + \ldots + n_k - k}$$

A test statistic involving generalised variances $|S_i|$ is given by

$$(n_1 + \ldots + n_k - k)\ln|S| - \{(n_1 - 1) \ln|S_1| + \ldots + (n_k - 1)\ln|S_k|\}$$

If the generalised variances were exactly equal the sum in braces would be the same as the first term. Low values of the statistic are therefore consistent with the matrices being equal. For large values of n_i, the statistic comes from a distribution which is approximately chi-squared, with $(k - 1)p(p - 1)/2$ degrees of freedom, but small sample sizes tend to exaggerate the effect of the sum in braces. There is a scaling factor, Box's modification, which allows the test to be applied with smaller n.

8.3 MULTIPLE LINEAR REGRESSION

In Chapter 3 we examined the dependence of a single dependent variable on one other explanatory or regressor variable. It is usual, however, for the dependence to be on two or more regressors. For example, the quality of a crude oil may depend on age, depth, temperature at which it formed and many other variables. In this section we shall discuss the setting up and testing of models for such dependence.

The ideas of many of the tests will be familiar from Section 2.6; analysis of variance plays an important part.

Among the possible objectives of the procedures are the following.

1   Summarising large amounts of data, as when contours are expressed as functions of distances from a reference point (see trend surface analysis, Section 7.3).

2   Understanding the processes which result in observed phenomena, for example hydrocarbon maturity as a function of age, depth, temperature, etc.

3   Estimating quantities which are difficult to measure but which are related to others which are readily determined, such as prediction of porosity from wireline logs.

Multiple regression is not strictly a multivariate technique because we are dealing with only one dependent variable, but it is usually included in books and chapters on multivariate analysis. The computations are very complex and always performed on a computer. We shall not give details of them but describe the statistics produced by statistical packages and explain their uses.

8.3.1 Typical regression models

Regression models may take the following forms:

$$Y = \beta_0 + \beta_1 x_1 + \beta_2 x_2 + \varepsilon$$
$$Y = \beta_0 + \beta_1 x_1 + \ldots + \beta_k x_k + \varepsilon$$
$$Y = \beta_0 + \beta_1 x + \beta_2 x^2 + \varepsilon$$
$$Y = \beta_0 + \beta_1 x_1 + \beta_2 x_2 + \beta_3 x_1 x_2 + \varepsilon$$

where the ε terms are random variables with mean equal to zero and variance σ^2 which is the same for all values of the regressor variables. An important condition is that they should be uncorrelated.

$\beta_0, \beta_1, \ldots, \beta_k$ are called partial regression coefficients and β_i represents the change in Y produced by a unit change in X_i when all the others remain constant. In real life, however, all the other regressors will change!

These models are all linear in the partial regression coefficients β and we shall not attempt to deal with non-linear models.

In Section 3.4.1.1, we saw that simple linear regressions could be drawn in two dimensions: one for the dependent variable and the other for the regressor; the regression line was chosen to minimise the sum of squared deviations from the data points to the line, measured parallel to the Y axis. The first model given above describes points in three-dimensional space and we find the best-fitting plane through them; the second and fourth represent points in hyperspace (more than three dimensions) and our task is to fit hyperplanes through them. It is impossible to imagine figures in four or more dimensional space but many familiar theorems, notably Pythagoras's theorem, apply in these dimensions. The planes are chosen to minimise the sum of squared deviations from the data points to the plane or hyperplane.

The third model is an example of a polynomial regression. Although it represents a quadratic, it is still linear in the partial regression coefficients and is therefore a linear model.

8.3.2   Data matrices and vectors of data and coefficients

It is convenient to think of the data on the independent and regressor variables set out in a table, called the data matrix, as in Table 8.1.

To avoid excessive writing we use matrix notation. A matrix in which the column of y values is replaced by a column of 1s and the other columns remain as above is represented by an upper-case letter, X.

When the values of Y are needed, we represent them in a vector and the partial regression coefficients and error terms are also referred to in terms of vectors:

$$Y = \begin{pmatrix} y_1 \\ y_2 \\ \vdots \\ y_n \end{pmatrix} \quad X = \begin{pmatrix} 1 & x_{11} & x_{12} & \ldots & x_{1k} \\ 1 & x_{21} & x_{21} & \ldots & x_{2k} \\ \vdots & \vdots & \vdots & & \vdots \\ 1 & x_{n1} & x_{n2} & \ldots & x_{nk} \end{pmatrix}$$

$$\beta = \begin{pmatrix} \beta_0 \\ \beta_1 \\ \vdots \\ \beta_k \end{pmatrix} \quad \varepsilon = \begin{pmatrix} \varepsilon_1 \\ \varepsilon_2 \\ \vdots \\ \varepsilon_n \end{pmatrix}$$

Table 8.1

Y	X_1	X_2	.	.	.	X_k
$\begin{pmatrix} y_1 \\ y_2 \\ \vdots \\ y_n \end{pmatrix}$	$\begin{matrix} x_{11} \\ x_{21} \\ \vdots \\ x_{n1} \end{matrix}$	$\begin{matrix} x_{12} \\ x_{22} \\ \vdots \\ x_{n2} \end{matrix}$	.	.	.	$\begin{matrix} x_{1k} \\ x_{2k} \\ \\ x_{nk} \end{matrix}$

Then the linear models may be written

$$y = X\beta + \varepsilon$$

8.3.3 Estimates of the parameters

As in simple linear regression, the principle of least squares is used: we choose the hyperplane which minimises the sum of squares of the residuals. The variation, measured by the total corrected sum of squares of the Y values, is broken down into a part, called the regression sum of squares, which is systematic, and another which is regarded as random, just as we did with data from designed experiments (Fig. 8.2).

It can be shown that the estimators are unbiased. They are, in general, correlated with each other. The fitted regression model is written in matrix notation:

$$\hat{y} = X\beta$$

Residuals are held in the vector

$$z = y - \hat{y}$$

and the residual sum of squares is simply the sum of squares of these values. The variance is estimated by

$$s^2 = \frac{\text{Residual SS}}{n - k - 1}$$

where k is the number of regressor variables in the model.

The calculations are set out in an analysis of variance table, Table 8.2, in which k is the number of regressors and N is the total number of observations.

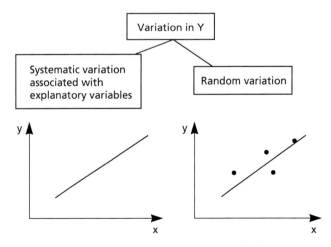

Fig. 8.2 Decomposition of variation in a dependent variable into its components.

Table 8.2 Analysis of variance table for multiple regression

Source of variation	d.f.	Sum of squares	Mean square	Mean square ratio, F
Regression on $X_1, \ldots, X_k$	k	Regression SS	Regression SS/k	Regression MS/ Residual MS
Residual	$n - k - 1$	Residual SS	Residual SS/$n - k - 1$	—
Total	$n - 1$	CSS (Y)	—	—

8.3.4

Tests of a model

We test H_0: $\boldsymbol{\beta} = 0$, i.e. all the partial regression coefficients $\beta_1, \ldots, \beta_k$ are zero, against H_1: $\boldsymbol{\beta} \neq 0$, i.e. at least one of the coefficients is not zero.

When the null hypothesis is true and the random terms have a normal distribution the mean square ratio has an F distribution with parameters k, $n - k - 1$; when at least one of the partial regression coefficients differs from zero, the distribution is a modified version of F (called non-central F) and we tend to get high values of the mean square ratio.

Box 8.4 Worked example: multiple regression

What, if any, is the relationship between permeability and the other petrophysical parameters of the Sherwood sandstone (Appendix 3.10)?

From a statistics package, the regression equation is

Permeability = 0.16 + 50.6 porosity − 331 mat. cond. − 0.179 TFF − 0.453 ind. pol.

Using:

H_0: regression coefficients are zero
H_1: at least one coefficient is non-zero

Analysis of variance

Source	d.f.	SS	MS	F
Regression	4	234.922	58.731	8.18
Error	35	251.179	7.177	
Total	39	486.101		

$R^2 = 48.3\%$

The calculated F (8.18) exceeds the critical F from Appendix 2.5 at $\alpha = 5\%$ and 4, 35 degrees of freedom, so we reject the null hypothesis. There is some relationship between permeability and the other variables, but this result doesn't tell us which variables contribute significantly to this.

As before, the expectation of the regression mean square is

$\sigma^2 +$ a positive term caused by systematic variation

The test consists of comparing the mean square ratio with $F_{\alpha;k,n-k-1}$, where α is the chosen significance level.

Worked example: see Box 8.4.

Individual coefficients may be tested, assuming that the variable concerned is the last one added to the model, but, unless the contributions of the regressor variables are independent, the importance of any one of them — as measured by changes in the regression sum of squares — depends upon what other variables are already in the model. We shall not, therefore, deal with this test.

8.3.5 Criteria for choice of regression model

We have seen how to test a given regression model but it is important to develop methods of choosing sensible models to test. If there are only two possible independent variables, we have three models to try (using x_1, x_2, and x_1 with x_2); if there are three variables there are seven alternatives; with four there are 15 options, and so on. We now describe some of the most important.

Examine all possible regressions

Some criteria for selecting models are as follows.

1 Coefficient of multiple determination, R_p^2. This is the proportion of the total corrected sum of squares which is accounted for by the regression sum of squares.

$$R_p^2 = \frac{\text{Regression sum of squares}}{\text{Total corrected sum of squares}}$$

The value increases when a variable is added to the model until all variables are included, but there are diminishing returns.

2 Make the mean square error small. The addition of a variable may actually increase the mean square error: for a decrease to occur, the re-

Table 8.3 Effect of addition of variable on ANOVA for multiple regression

Source of variation	d.f.	Sum of squares	Mean square
Regression on X_1, X_2, X_3	3	81.9723	27.32
Residual	26	146.1668	4.86
Regression on X_1, X_2, X_3, X_4	4	84.3624	21.09
Residual	25	143.7767	5.75
Total	29	228.1391	

duction in the error sum of squares must be at least equal to the previous mean square to allow for the lost degree of freedom. An example is given in Table 8.3.

3 Adjusted R_p^2. This is described as the coefficient of multiple determination adjusted for the degrees of freedom:

$$\bar{R}_p^2 = 1 - \frac{n-1}{n-p}(1 - R_p^2)$$

$$= 1 - \frac{\text{Residual MS}}{\text{Total MS}}$$

It increases as the the number of regressors increases, provided that the decrease in error sum of squares is enough to compensate for the loss of a degree of freedom. Note that the method is equivalent to reducing the residual mean square.

4 The C_p statistic. This is used to enable us to choose the model which produces least bias. It is defined to be

$$C_p = \frac{\text{Residual SS }(p)}{\text{Residual MS (Full model)}} - n - 2p$$

It can be shown that when there is no bias the expectation of the statistic is p, so we plot values of C_p against p. Values which lie near the line from the origin to the point obtained by including all variables are then showing little bias.

Stepwise regression

1 Examine all simple linear regressions and choose the one which produces the highest significant mean square ratio.

2 For each of the remaining variables, calculate the correlation with Y, allowing for the fact that the first variable is in the model (this is called the partial correlation). This is equivalent to computing

$$F_j = \frac{\text{Residual SS }(\beta_j | \beta_0, \beta_1)}{\text{Residual MS }(x_j, x_1)}$$

where the numerator is the decrease in residual sum of squares caused by the inclusion of X_j when X_1 is already in the model and the denominator is the residual mean square with both X_1 and X_2 in the model.

The variable X_j which gives the highest value of this statistic is entered into the model, provided that F_j exceeds some threshold value, F_{In}, chosen to make the probability of acceptance low (for example, 0.05 or less) when X_j provides no significant, further information about Y.

3 The variable chosen at stage **1** is then tested to see if it should be removed or retained, by computing

$$F_1 = \frac{\text{Residual SS } (\beta_1 \mid \beta_0, \beta_2)}{\text{MSE } (x_1, x_2)}$$

and comparing it with a threshold value F_{Out}: if $F_1 < F_{Out}$, then x_1 is deleted. Note F_{Out} must be less than or equal to F_{In}. The two values are usually chosen to be equal.

4   And so on – until no more variables qualify for admission and none is deleted. At each stage every previously chosen variable is tested but only one, if any, is deleted.

Worked example: see Box 8.5.

Box 8.5 Worked example: stepwise regression

Using the same data as for Box 8.4, which variables contribute significantly to the regression equation?

Using a statistics package to obtain a stepwise regression, we obtain:

	Step	
	1	2
Constant	−12.778	−7.874
Porosity	66	58
Mat. cond.		−210
R^2	36.15	46.17

This has used a forwards approach and is interpreted as follows.

Step 1:

Permeability $= -12.778 + 66$ porosity
Goodness of fit $= 36.15\%$

Porosity has the main influence on permeability.

Step 2:

Permeability $= - 7.874 + 58$ porosity $- 210$ mat. cond.
Goodness of fit $= 46.17\%$

The regression is significantly improved by the inclusion of matrix conductivity.

No other variables significantly improve the goodness of fit. If we wish to predict permeability from data such as these, we would use just porosity and matrix conductivity, but the goodness of fit is quite low, so the reliability of the estimates would not be good.

Forward selection

This follows the same path as stepwise regression except that there is no testing of previously chosen variables to see how they are affected by later ones.

Backward elimination

All candidate regressors are included and the one with the lowest partial correlation coefficient with Y, when all the rest are allowed for, is deleted if the partial F statistic is less than F_{Out}. This process is continued until a model is fitted with no variable having a partial F statistic less than F_{Out}.

Note There is, in general, no single best model: you may have to choose from several which appear to do the job reasonably. The methods described above do not always produce the same 'final' model.

8.3.6   ## Examination of adequacy

It is important to test for conditions such as that of no correlation between error terms. This can be done by the following.

1   Using the methods described for simple linear regression analysis (Section 3.4.1).

2   Test for autocorrelation between residuals (Durbin–Watson test). This test requires special statistical tables, but test results may be produced by statistical software. It is based on the correlation between successive residuals, and has a null hypothesis of $\rho = 0$.

8.3.7   ## Multicollinearity

This is said to occur when some of the regressor variables are highly correlated. It can produce some very difficult problems in regression analysis.

1   The variances and covariances of the estimators of the partial regression coefficients become large; this implies that their precision is low.

2   When there are two regressors and they are highly correlated, the partial regression coefficients become equal in size but opposite in sign, no matter what the values of β_1, β_2 are.

3   The computer is unable to compute certain matrices properly and will produce meaningless or very unreliable results.

Most computer software for multiple regression will give warnings when multicollinearity is serious.

8.3.8   **Influential values: Cook's distance**

Sometimes observations are a long way from the others in hyperspace and may have a large influence on the model produced. If they turn out to be erroneous, they should be removed, but often they are providing useful information and then they should be retained. They may be detected by using a quantity known as Cook's distance, which is the square of the distance between $\hat{\beta}$ based on all n observations and $\hat{\beta}$ based on the data with the ith observation removed. A value of Cook's distance which exceeds 1 indicates an influential observation.

8.4   **DISCRIMINANT ANALYSIS**

8.4.1   **Principles and applications**

Discriminant analysis is used to assign objects to one of two or more established groups. For example, we may have a specimen of rock and want to know if it should be classified as granite, granodiorite or diorite; or the problem may be to assign a fossil to one of several species. Initially we need to have 'training group' data for each of the categories that we wish to separate, so we might need initial representative data from each of the groups granite, granodiorite and diorite, or from each species. We aim to find discriminant functions: these are vectors in the directions of optimal separation between the groups. If we plot the data from the established groups along a discriminant function they plot as separately as is possible, and the positions of new, unclassified points can be used to allocate them to one or other of the groups. It is important to realise that the vector between group means will not necessarily achieve this: we need to consider the covariance structure of the groups (see Fig. 8.3).

The crucial aspect of discriminant functions is that they are determined by the user's choice of training group. These are chosen so that the discriminant function emphasises the geological property of interest. Furthermore, we don't need any prior information about which variables may have discriminating ability, so we can include variables on a trial basis. Often in applied geology the fundamental geological property of interest is economic status of a potential resource. We can set up a control group to represent known economically viable areas (e.g. near working mines) and another for known barren areas, and include in the analyses information from geological mapping, geochemical surveys and remote sensing. The result will indicate to us which variables are useful signatures of an economic deposit, and will allow us to calculate an index of economic potential for new areas.

The ideas behind the technique can be illustrated by reference to an artificial, univariate case.

A specimen of a fossil is to be assigned to one of two species, A or B, on the basis of its length only. The lengths of both species have normal

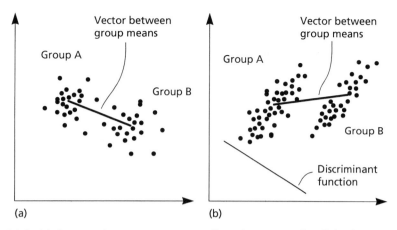

Fig. 8.3 In (a) the vector between group means allows the groups to be efficiently separated and would be nearly parallel with the discriminant function vector. In (b), the covariance structure results in the vector between group means being a poor discriminator.

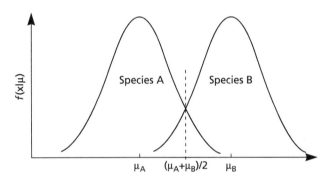

Fig. 8.4 The partition between two groups is provided by $(\mu_A + \mu_B)/2$.

distributions with the same variance σ^2 but different means μ_1, μ_2 (Fig. 8.4). In the absence of any other information the probabilities of the species belonging to A and B are the same.

It would be sensible to assign the specimen to species A if the length x is closer to μ_A than to μ_B, i.e. if

$$x - \mu_B > \mu_A - x$$

or

$$x > (\mu_A + \mu_B)/2$$

Another way of regarding the rule is to say that the specimen should be assigned to A if

$$f(x|\mu_A) > f(x|\mu_B)$$

where $f(x|\mu_A)$, $f(x|\mu_B)$ are the densities of populations A and B or, in terms of likelihoods,

$$L(\mu_A|x) > L(\mu_B|x).$$

If we knew that species A was comparatively rare, however, we would move the dividing line nearer to μ_A, requiring more convincing evidence from the length before assigning the specimen to the rarer species. Denoting the prior probabilities of the specimen being of species A and B by $\Pr(A)$ and $\Pr(B)$, the rule would be to assign the specimen to A if

$$L(\mu_A|x)\Pr(A) > L(\mu_B|x)\Pr(B)$$

i.e.

$$\frac{L\,(\mu_A|x)}{L\,(\mu_B|x)} > \frac{\Pr(B)}{\Pr(A)}$$

and to B otherwise.

In practice, it is easier to deal with logarithms of these functions, so we shall rewrite the inequality

$$\ln\{L(\mu_A|x)\} - \ln\{L(\mu_B|x)\} > \ln k$$

where $k = \ln\{\Pr(B)/\Pr(A)\}$.

These ideas are now applied to multivariate data, arguing by analogy with the univariate case.

8.4.2   **Use of p variables in discriminating between two populations**

Suppose that an object is known to belong to one of two populations whose mean vectors are μ_1 and μ_2 and whose covariance matrix is Σ. We use a linear compound of the variables to produce an index from a univariate normal distribution, as we did when testing hypotheses about vector means, and the compound used is again that one which maximises the distance between the means of the populations: the weights are given by

$$a^T = (\mu_1 - \mu_2)^T\Sigma^{-1}$$

In practice, the weights must be estimated by

$$(\bar{x}_1 - \bar{x}_2)^T S^{-1}$$

where $\bar{x}_1$, $\bar{x}_2$ and S are estimates of the two mean vectors and the common covariance matrix of the two populations, calculated from previously gathered data. These weights define the discriminant function. For the moment we shall take it that the variables used are suitable for discriminating between the two populations and leave tests of this assumption to the next section. Given a new observation x, we compute its discriminant function score by simple matrix multiplication:

$$(\bar{x}_1 - \bar{x}_2)^T S^{-1} x$$

and compare the value with the discriminant function scores for the two sample mean vectors:

$$(\bar{x}_1 - \bar{x}_2)^T S^{-1} \bar{x}_1$$

and

$$(\bar{x}_1 - \bar{x}_2)^T S^{-1} \bar{x}_2$$

We assign the individual to the group with mean vector μ_1 if the value of the computed discriminant function score is closer to the first of these values than to the second, and to the second group otherwise. Algebraically, we assign the new specimen to the group with mean vector μ_1 if

$$(\bar{x}_1 - \bar{x}_2)^T S^{-1} (x - \bar{x}_1) < (\bar{x}_1 - \bar{x}_2)^T S^{-1} (\bar{x}_2 - x)$$

It can be shown that when the prior probabilities $\Pr(A)$, $\Pr(B)$ of the specimen belonging to populations A, B are taken into account the criterion for placing a new item in population A becomes

$$(\bar{x}_1 - \bar{x}_2)^T S^{-1} x - (\bar{x}_1 - \bar{x}_2)^T S^{-1} (\bar{x}_1 + \bar{x}_2)/2 > \ln\{\Pr(B)/\Pr(A)\}$$

Worked example: see Box 8.6.

There is, however, a serious risk of misclassifying specimens when the prior probabilities of their membership of populations is not taken into account. Unfortunately, it can be extremely difficult to obtain the necessary information required to calculate the probabilities, and perhaps the best safeguard against serious error in this respect is the scientist's knowledge of his or her subject. In the rest of this chapter it will be assumed, for simplicity, that all the prior probabilities are equal.

8.4.3

Assessment of the linear discriminant function

We now turn to three questions about the function used in practice.

1 Are the chosen means of the parent populations sufficiently different to provide worthwhile discrimination?

2 If they are, could we omit some variables and still be able to discriminate between the populations? This is an important question because the use of fewer variables will lessen the time and effort spent on measurement, analysis and computation.

3 How can we compare the performance of potential discriminator functions?

Test for worthwhile discrimination

This is the same as the test for a significant difference between sample means, described in Section 8.2: if there is no difference between the mean

Box 8.6 Worked example: discrimination

Data gathered on the lengths, widths and heights of two species of brachiopod gave the following sample mean vectors and estimates of covariance matrices.

Mean vectors

Species *A*   Species *B*

$$\begin{matrix} l \\ w \\ h \end{matrix} \begin{pmatrix} 3.3 \\ 2.7 \\ 1.7 \end{pmatrix} \qquad \begin{matrix} l \\ w \\ h \end{matrix} \begin{pmatrix} 2.8 \\ 2.5 \\ 2.0 \end{pmatrix}$$

Covariance matrices

Species *A*:   Species *B*:

$$\begin{matrix} l \\ w \\ h \end{matrix} \begin{pmatrix} 0.0432 & & \\ 0.0176 & 0.0212 & \\ 0.0137 & 0.0125 & 0.0203 \end{pmatrix} \qquad \begin{matrix} l \\ w \\ h \end{matrix} \begin{pmatrix} 0.0396 & & \\ 0.0230 & 0.0190 & \\ 0.0168 & 0.0147 & 0.0109 \end{pmatrix}$$

Sixty specimens of species *A* were examined, 18 of species *B*. It is thought that the abundance of apecies *B* is about four times that of *A*. A new specimen was found to be 3.2 cm long, 2.6 cm wide and 1.9 cm high. Calculate a linear discriminant function and use it to allocate the new specimen to *A* or *B*, first ignoring the information about relative abundances and then taking it into account.

The estimates of the covariance matrices are similar so we may combine them to produce a pooled, unbiased estimate *S* of the common covariance matrix:

$$S = (59S_1 + 17S_2)/(59 + 17)$$

$$= \begin{pmatrix} 0.0424 & 0.0188 & 0.0144 \\ 0.0188 & 0.0207 & 0.0130 \\ 0.0144 & 0.0130 & 0.0182 \end{pmatrix}$$

Inversion of the matrix on a computer gives

$$S^{-1} = \begin{pmatrix} 40.5592 & -30.2544 & -10.4805 \\ -30.2544 & 110.1740 & -54.7609 \\ -10.4805 & -54.7609 & 102.3526 \end{pmatrix}$$

The discriminant function is then

$$(\bar{x}_1 - \bar{x}_2)^T S^{-1} x = (0.5 \quad 0.2 \quad -0.3) \begin{pmatrix} 40.5592 & -30.2544 & -10.4805 \\ -30.2544 & 110.1740 & -54.7609 \\ -10.4805 & -54.7609 & 102.3526 \end{pmatrix} \begin{pmatrix} x_1 \\ x_2 \\ x_3 \end{pmatrix}$$

$$= 17.37x_1 + 23.34x_2 - 46.90x_3$$

Box 8.6 *Continued*

There is a heavy weighting of the height here and it does seem sensible to give higher weights to variables which have lower variances and have well-separated means in the population.

For the new specimen the value of the discriminant function is

$17.37 \times 3.2 + 23.34 \times 2.6 - 46.90 \times 1.9 = 27.16$

Now the value of the discriminant index

$(\bar{x}_1 - \bar{x}_2)^T S^{-1} (\bar{x}_1 + \bar{x}_2)/2$

is

$17.37 \times 3.25 + 23.34 \times 2.60 - 46.90 \times 1.95 = 25.68$

and, as the new specimen's value of the discriminant function is greater than this, we assign it to species A, ignoring the belief that the abundance of B is about four times that of A.

Let us now use that information. Substituting in

$(\bar{x}_1 - \bar{x}_2)^T S^{-1} x - (\bar{x}_1 - \bar{x}_2)^T S^{-1} (\bar{x}_1 + \bar{x}_2)/2 > \ln\{\Pr(B)/\Pr(A)\}$

we get $27.16 - 25.68 = 1.48$ on the left-hand side and $\ln(4) = 1.39$ on the right-hand side. In this case, we arrive at the same conclusion whether or not we take the relative abundances into account.

vectors then clearly we cannot discriminate between the populations on that basis. We therefore compute the test statistic

$$\frac{(n_1 + n_2 - p - 1)}{p(n_1 + n_2 - 2)} \cdot \frac{n_1 n_2}{n_1 + n_2} (\bar{x}_1 - \bar{x}_2)^T S^{-1} (\bar{x}_1 - \bar{x}_2)$$

and accept that the chosen variables are useful for discrimination if the value exceeds $F_{\alpha;p,n1+n2-p-1}$. It does not mean that they are the best variables, however.

Retention and discarding of variables

When the test which we have just described indicates that the chosen set of variables produces some discrimination between populations, it means only that at least some members of the set are useful; some may be of no help – or of no additional help when others have been included – and these should be discarded. Considerable work has been done on this subject and we shall describe only one method, which is related to methods of choosing subsets of variables in multiple regression.

Suppose that the complete list of variables is $X_1, \ldots, X_p$ and that we wish to test the pth variable to see whether or not it should be retained. The

Mahalanobis distances D_p^2 and D_{p-1}^2, with and without this variable, are computed and the change $D_p^2 - D_{p-1}^2$ in distance is tested for significance by using the statistic

$$\frac{(n_1 + n_2 - p - 1)\,(D_p^2 - D_{p-1}^2)}{\{(n_1 + n_2 - 2)\,(n_1 + n_2)/n_1 n_2\} + D_{p-1}^2}$$

Box 8.7 Worked example: important variables in discriminant function analysis

Does length play an important part in the discrimination between species A and B whose data are given in Box 8.6?

When all three variables are used for discrimination, the Mahalanobis distance

$$D_3 = (0.5\ 0.2\ -0.3)S^{-1}\begin{pmatrix} 0.5 \\ 0.2 \\ -0.3 \end{pmatrix} = 27.447$$

where S^{-1} is the inverse of the unbiased 3×3 covariance matrix.

When length is omitted, the covariance matrix for height and width is

$$S_2 = \begin{pmatrix} 0.0207 & 0.0130 \\ 0.0130 & 0.0182 \end{pmatrix}$$

and its inverse is

$$S_2^{-1} = \begin{pmatrix} 87.6095 & -62.5782 \\ -62.5782 & 99.6438 \end{pmatrix}$$

The new Mahalanobis distance is

$$(0.2\ \ -0.3)\begin{pmatrix} 87.6095 & -62.5782 \\ -62.5782 & 99.6438 \end{pmatrix}\begin{pmatrix} 0.2 \\ -0.3 \end{pmatrix}$$

which is

$$(36.2954\ \ -42.4088)\begin{pmatrix} 0.2 \\ -0.3 \end{pmatrix}$$

or 19.985.

The reduction in Mahalanobis distance is therefore $27.447 - 19.985$, i.e. 7.462. To see if the reduction is significant we calculate

$$\frac{(n_1 + n_2 - p - 1)(D_p^2 - D_{p-1}^2)}{(n_1 + n_2 - 2)(n_1 + n_2)/n_1 n_2 + D_{p-1}^2} = \frac{(60 + 18 - 3 - 1) \times 7.462}{(60 + 18 - 2)\,78/480 + 19.985}$$
$$= 17.1$$

The 5% point of F with parameters 1 and 74 is about 4 so the reduction in distance is certainly significant. Length is valuable for discriminating between the two species.

where the Mahalanobis distances are of the form

$$(\bar{x}_1 - \bar{x}_2)^T S^{-1}(\bar{x}_1 - \bar{x}_2)$$

When the reduction in distance is not significant, the statistic comes from the F distribution with 1 and $n_1 + n_2 - p - 1$ degrees of freedom. If the computed value exceeds $F_{\alpha;1,n1+n2-p-1}$ then the pth variable is retained because its removal would reduce the distance significantly.

Worked example: see Box 8.7.

Probability of misclassification

The value of a discriminator depends upon the proportions of objects which are correctly assigned. The theoretical treatment of this problem becomes very complex when, as is usually the case, the parameters of the populations are not known. In practice, the probabilities may be assessed by assigning cases which are already known. This method gives over-optimistic assessments when the weights in the linear compound are calculated from data which include the cases to be assigned, and two methods of overcoming this difficulty are as follows.

1 Part of the available data is used to calculate weights and the remaining objects are assigned by means of the resulting discriminant function. The proportion of misclassifications in each direction is recorded. A large data set is required.

2 Assign each object in turn, computing weights from the remaining ($n - 1$) observations. This obviously requires a considerable amount of computing time as the parameters of the distributions must be estimated n on each occasion.

The importance of the probabilities of misclassification will depend upon the context. In some studies the overall probability may be paramount but in others the penalty for misclassifying an object from one population may be far greater than that for the opposite error. For example, it could be far more costly to exploit a mineral deposit which is not commercially viable than to leave one unexploited which could have produced a valuable yield. A company has been known to attempt to exploit a region for oil and produced only water. Unfortunately, we cannot consider the solution of such problems here but it is important to be aware of their existence.

Assumptions required

The assumptions which have been made in this description of discriminant analysis are:

1 the data come from multivariate normal distributions; and
2 the covariance matrices are equal.

The method works well even when the data are not from normal distributions but the statistical tests for the retention or discarding of variables are not reliable. When the covariance matrices are not equal the discriminant function includes quadratic as well as linear terms and it is often more difficult to interpret in physical terms. Furthermore, the analysis may be more sensitive to departures from the normal distribution than is the case when linear discrimination functions are used. Provided that large samples are used, the test of a function is not very badly affected by unequal covariance matrices.

8.5   EIGENVECTOR METHODS

8.5.1   Principal components analysis

8.5.1.1   Introduction

Suppose that we have a bivariate data scatter ordinated against two perpendicular axes representing variables with a fairly strong linear correlation (Fig. 8.5). We might have similar variances for each variable. If we were to find a vector which passed along the 'long axis' of the data scatter, plus a second vector perpendicular to the first, we would have a new pair of perpendicular axes that could provide an alternative reference frame for the data scatter. The new axes are different in that the amount of variance in the direction of the first axis is maximised and variance along the second axis is minimised. The useful properties of the result are that we can reduce dimensionality of the data scatter in an optimal way by using the first axis only, and we may be able to make special interpretations about the reason for the elongation of the data scatter in this direction. This may seem mundane with a two-dimensional example, but the method is extremely useful for viewing and understanding high-dimensionality data scatters.

PCA is a technique for finding such linear compounds of correlated variables. These linear compounds are called principal components (PCs) and they have two useful properties.

1   In general, most of the total variance of the p variables in a data set can be accounted for by a comparatively small number k of the new variables. We say that the dimensionality of the data is reduced from p to k.

2   They are uncorrelated, which facilitates examination and analysis of the data.

There are two important implications of these remarks.

1   If the original variables are uncorrelated there is no point in applying the technique: always look at the correlation matrix and scatter diagrams before performing PCA.

2   When the number of variables is small – three or four, perhaps – the dependence can usually be resolved by visual inspection of the correlation matrix and again PCA is not necessary.

Consider the artificial correlation matrix based on 30 observations on five variables which have been labelled 1 to 5:

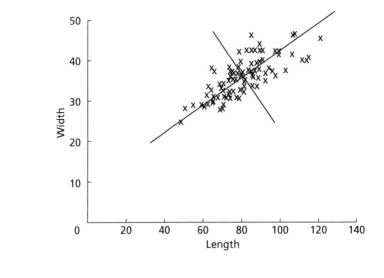

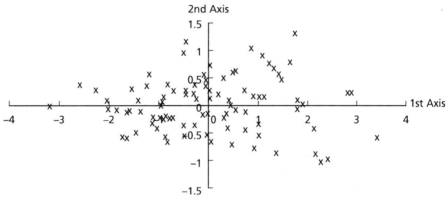

Fig. 8.5 The bivariate scatter of dimensions of echinoids (data as for Fig. 3.1) shown in (a) can be replotted against the new axes (b). In this case, axis 1 corresponds to size and axis 2 to shape.

$$
\begin{array}{c c c c c}
 & 1 & 2 & 3 & 4 \\
2 & 0.85 & & & \\
3 & 0.14 & 0.21 & & \\
4 & 0.23 & 0.19 & 0.90 & \\
5 & 0.78 & 0.95 & 0.25 & 0.32
\end{array}
$$

It is easy to see that the variables are in two groups, particularly if the order of variables is changed:

$$
\begin{array}{c c c c c}
 & 1 & 2 & 5 & 4 \\
2 & 0.85 & & & \\
5 & 0.78 & 0.95 & & \\
4 & 0.23 & 0.19 & 0.32 & \\
3 & 0.14 & 0.21 & 0.25 & 0.90
\end{array}
$$

The variables 1, 2 and 5 form one group; the other group consists of variables 3 and 4. The example is artificial but the procedure used is worth carrying out in practice.

8.5.1.2 Geometrical interpretation

We may regard n observations on p variables as n points in p-dimensional space. We try to define a space in fewer dimensions to explain nearly all the variation.

The process is equivalent to rotating the axes in p-dimensional space in directions such that projections of the points on them are spread out to the maximum extent. This idea is illustrated for two variables in Fig. 8.5(a), where the first axis has been rotated to pass through the data points. The origin of the axes has been moved to the point $(\bar{x}_1, \bar{x}_2)$ but this does not affect the covariance or the PCs. Notice that the second axis has been rotated through the same angle in order to preserve the right angle between axes. Rotations which maintain right angles between all the axes are said to be orthogonal. Notice how in Fig. 8.5(b) the variables show little or no correlation when referred to the new axes.

We shall now assume that the origin of axes has been moved to the point in p-dimensional space which corresponds to the sample means. Further, we shall refer only to the sample covariance and correlation matrices, since the 'true' matrices are never known in practice.

8.5.1.3 Definition of the components

The first PC, which we shall denote by Y_1, is that compound

$$Y_1 = a_{11}X_1 + a_{12}X_2 + \ldots + a_{1p}X_p$$

or

$$Y_1 = a_1^T X$$

which has the greatest possible variance.

The variance could be made arbitrarily large by choosing large values of the weights, so a restriction is imposed: the sum of squares of weights is constrained to be 1:

$$a_{11}^2 + a_{12}^2 + \ldots + a_{1p}^2 = 1$$

Subject to this condition, then, there is no other linear compound of the original variables which has a higher variance than Y_1. As we saw in Section 8.2.1.4, this variance can be written

$$\mathrm{Var}(Y_1) = a_1^T S_X a_1$$

where S_X is the sample covariance of the original variables.

The second PC is the linear compound

$$Y_2 = a_{21}X_1 + a_{22}X_2 + \ldots + a_{2p}X_p$$

which has the next highest variance, $a_2^T S_X a_2$. It can be shown that the components Y_1, Y_2 are uncorrelated.

The number of components which can be obtained is equal to the number of correlated variables in the data set. In practice, all the sample covariances (and correlations) are non-zero, so computation will produce p components from p variables.

Now the total variance of the components is the same as that of the original variables so that all components are needed to explain their total variance. It is usual, however, for most of the variance to be concentrated in the first few components and to this extent we can say that most of the variation in the data is explained by them.

8.5.1.4   **Computation of the weights (loadings) a_i**

The weights, commonly known as loadings, are derived from a covariance or a correlation matrix. They are affected by the scales of the variables, so if different types of variable are involved (for example, ppm silver, percentage organic material, pH) the correlation matrix would be appropriate. When the variables are measured on the same scale, the covariance matrix is often used because the results are easier to interpret and statistical tests are easier if they are at all appropriate. The weights obtained from the covariance and correlation matrices differ and it is not possible to compute weights from one type of matrix by transforming those from the other.

The set of p transformations may be combined in a single statement:

$$Y = AX$$

where A is a matrix whose rows are the vectors of weights

$$a_1^T = (a_{11} \ldots a_{1p})$$
$$\vdots$$
$$a_p^T = (a_{p1} \ldots a_{pp})$$

We have remarked that the new variables Y_i are uncorrelated and that their variances are given by $a_i^T S a_i$. The covariance matrix of the PCs is, then,

$$AS_X A^T = \begin{pmatrix} \text{Var}(Y_1) & 0 & \ldots 0 \\ 0 & \text{Var}(Y_2) & \ldots 0 \\ \vdots & \vdots & \vdots \\ 0 & 0 & \text{Var}(Y_p) \end{pmatrix}$$

or

$$AS_X A^T = S_Y$$

The transformation $Y = AX$ of the variables X has caused a transformation of their covariance matrix S_X to a matrix S_Y in diagonal form.

The computation of the weights and variances requires the use of a computer and it is important to choose reliable software. Some packages use mathematical rather than statistical terms for the vectors and matrices we have described here, making the output rather abstract; for that reason we explain them now.

In mathematical language, the rows of A (or equivalently the columns of A^T) are called eigenvectors of the matrix S_X and the elements in the leading diagonal of S_Y are called its eigenvalues. In statistical terms the elements of the eigenvectors are weights a_{ij} applied to the original variables and the eigenvalues are variances of the new, uncorrelated variables. PCA is an example of what are sometimes called eigenvector techniques.

Worked example: see Box 8.8.

8.5.1.5   **Principal component scores**

The n values of the new variables are computed by using the weights a_{ij} to calculate linear combinations of the values of the original variables in each observation: for the rth object, for example, the value of the kth PC Y_k will be

$$Y_{kr} = a_{k1}x_{k1} + a_{k2}x_{k2} + \ldots + a_{kp}x_{kp}$$

Box 8.8 Worked example: principal components analysis

Using the data on skull measurements from fossil oreodont mammals (Appendix 3.11), we obtain the following from a standard statistics package.

Eigenanalysis of the correlation matrix:

	PC1	PC2	PC3	PC4
Eigenvalue	3.4428	0.3898	0.1116	0.0557
Proportion	0.861	0.097	0.028	0.014
Cumulative	0.861	0.958	0.986	1.000

Loadings:

Variable	PC1	PC2	PC3	PC4
Brainw.	-0.497	-0.488	0.705	-0.135
Toothl.	-0.501	-0.468	-0.598	0.414
Bullal.	-0.519	0.290	-0.307	-0.743
Bullad.	-0.482	0.677	0.226	0.508

Box 8.8 *Continued*

PC1 is extremely strong (86.1% of total variance) and is decribed by loadings that are similar for each variable. It is clearly interpretable as

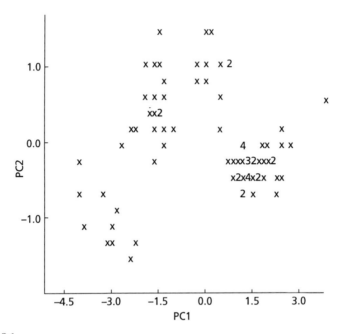

Fig. B8.8.1

due to intercorrelations resulting from size variation: larger skulls tend to have larger component parts. PC2 accounts for about three-quarters of the remaining variation and corresponds to a type of shape variation, in which the bulla has varying relative size. PC3 and PC4 are minor and may be picking up just random effects.

The PC scores plot for PC1 and PC2 is given in Fig. B8.8.1. The PC1 axis is now readable as size and PC2 as shape, so the scatter is a useful picture of the relationship between the two. Also, the scatter suggests the existence of a number of clusters, perhaps different species.

The values are called the kth PC scores and they constitute the new data. These would correspond to the coordinates in Fig. 8.5(b), and indeed it is routine to produce scatter plots of PC scores, particularly PC1 vs. PC2. For example, see Box 8.8.

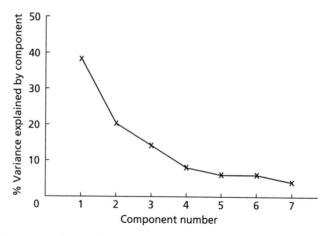

Fig. 8.6 A 'scree' diagram, showing the decline in variance explained with increasing principal component number.

8.5.1.6 **Reduction of dimensionality of the space**

Normally, some components will be found to contribute little to the total variance. The implication is that the values of the component scores are roughly constant and that their dimensions may be ignored – in effect, their values are replaced by their means, which we arranged to be zero. In this sense the dimensionality of the space is reduced, simplifying the analysis and interpretation of the data. When k components are omitted, only the first $(p - k)$ columns of the transforming matrix A are used to compute component scores. The components are still expressed as compounds of the original p variables at this stage but methods of reducing the number of variables are also available. Indications of these are given in Section 8.5.1.8.

Among the common criteria used for omitting components are the following.

1 Increase in the percentage of the total variance explained when the next most important component is included. If k components have been included and the addition of the next does not affect the proportion of total variance significantly, that component and the remainder are omitted. Judgement of the significance of the increase is often subjective and may be aided by a plot of the cumulative proportion against the number of components, illustrated in Fig. 8.6. It is thought that there is a tendency to retain too many components when this criterion is used.

2 The proportion of total variance explained. It has been suggested that enough components should be included to explain a high proportion (90%) of the total variance. This criterion may lead to the omission of too many components.

3   A component is omitted if its variance is less than the average of all p components (less than 1 when the correlation matrix is used).
4   Approximate equality of the last k eigenvalues (variances). If the variances of the last k components are equal the variation in them is described as spherical or isotropic because none contains more information than any other. As the important aspects of the data are summarised in the first $(p - k)$ components the last k are omitted.

8.5.1.7   **Correlations between variables and components**

The square of the sample correlation coefficient between variable X_i and PC Y_j can be shown to equal

$$r_{ij}^2 = a_{ij}^2 \cdot \mathrm{Var}(Y_j)/s_{ii}$$

where s_{ii} is the sample variance of variable X_i (which will be, when the correlation matrix is used to produce PCs, equal to 1). This shows that the squared correlation is a measure of the proportion of the variance of X_i which is explained by the jth PC. We shall return to this point in the next section.

8.5.1.8   **Retention and discarding of variables**

When several variables are investigated together, a large quantity of data is required. If some of the variables studied can be omitted the amount of computation and material will be reduced, and PCs have been used for this purpose.

Frequently, a component corresponding to a small eigenvalue – that is, a component with small variance – is considered to be of little importance. A variable with a high weighting on this component may therefore be declared redundant. A procedure for reducing the number of components used from p to k which is based on such reasoning is as follows: discard the variable with the highest absolute value of the coefficients in the component (assuming that it has not been removed at an earlier stage), and then repeat for each of the $(p - k)$ smallest eigenvalues.

The components to be considered for this treatment can be chosen according to their contributions to the total variance. When the components are derived from a correlation matrix, the total variance is equal to the number of variables, since each variable contributes 1 to the sum. It has been suggested that PCs whose contributions fall below 0.7 should be omitted (subject to leaving at least four components).

8.5.1.9   **Some common geological interpretations of principal components**

(refer to Box 8.9)

Box 8.9 Examples of geological interpetations of PC eigenvalues and loadings

1 Principal components analysis of soil geochemistry data from Masson Hill, Derbyshire (Appendix 3.4). The soil overlies dolomitised limestone and basalt with mineralised veins (first six PCs only shown).

	PC1	PC2	PC3	PC4	PC5	PC6
Eigenvalue	4.9035	1.4766	1.1953	0.6156	0.3732	0.1734
Proportion	0.545	0.164	0.133	0.068	0.041	0.019
Cumulative %	54.5	70.9	84.2	91.0	95.2	97.1

Loadings:

Variable	PC1	PC2	PC3	PC4	PC5	PC6
Zn	−0.325	0.429	0.221	−0.003	0.461	0.585
Pb	−0.419	−0.151	0.067	−0.109	−0.230	0.261
Cd	−0.309	0.317	0.476	0.083	0.279	−0.619
Mg	0.064	0.650	−0.296	−0.596	−0.262	−0.006
Ca	−0.365	0.026	−0.389	0.437	−0.028	0.135
Cu	−0.360	−0.327	−0.033	−0.505	0.096	−0.255
Ag	−0.385	−0.331	−0.084	−0.316	0.110	0.105
Sr	−0.286	0.178	−0.628	0.213	0.141	−0.332
Ba	−0.357	0.145	0.280	0.187	−0.739	−0.033

PC1 is very strong and has similar loadings on all but one of the variables: it may be interpretable as a dilution effect. High positive PC1 scores indicate low amounts of all of the variables (except Mg, which is neutral); this could be due to high quantities of a further major component − perhaps SIO_2 or organic material. Negative PC scores will indicate less of the diluting component, and hence higher percentages of the listed elements. Alternatively, Mg may be neutral because it occurs equally in veins and in the host dolomite and basalt. Samples with negative scores, then, would have high amounts of all the other analysed components: all of these would be expected to be higher in veins.

PC2 points to an unexpected association of Ag and Cu, perhaps in higher-temperature veins, and of Zn, Cd and Mg. The interpretation is uncertain.

PC3 includes a correlation of Mg, Ca and Sr: this is likely to be the signature of the host carbonates.

2 Principal components analysis of measurements of forams (Appendix 3.12) (first six PCs only shown).

Box 8.9 *Continued*

	PC1	PC2	PC3	PC4	PC5	PC6
Eigenvalue	6.1684	1.3435	0.5182	0.4109	0.2771	0.1536
Proportion	0.685	0.149	0.058	0.046	0.031	0.017
Cumulative	0.685	0.835	0.892	0.938	0.969	0.986

Loadings:

Variable	PC1	PC2	PC3	PC4	PC5	PC6
Length	−0.327	−0.170	0.562	−0.221	−0.587	−0.360
Width	−0.384	−0.020	−0.048	−0.167	−0.101	0.634
Wdthlc	−0.394	−0.017	0.061	0.040	−0.129	0.315
Hghtlc	−0.372	0.021	0.083	0.468	0.249	−0.316
Hght2lc	−0.379	0.076	−0.012	0.402	0.203	−0.208
Diampro	−0.301	0.247	−0.729	−0.320	−0.267	−0.372
Breadth	−0.385	0.075	−0.075	0.136	0.051	0.255
Wdthfor	0.088	−0.765	−0.366	0.392	−0.341	0.039
Locfor	−0.247	−0.558	−0.001	−0.516	0.579	−0.146

PC1 is readily interpretable as size: higher positive PC1 scores will correspond with smaller specimens. Interestingly, 'Wdthfor', the width of the foramen, is size-independent.

PC2 is a shape factor: it seems that the two measurements of the foramen (width and location) vary to an extent independently of the other variables.

PC1 is only useful as a size index, but PC2 may contain taxonomically or stratigraphically useful information.

Geochemistry and petrology

1 PCs are the result of strong correlations between variables. If we are using compositional data expressed as percentage or ppm, then data are closed (Section 3.2.2) and it is inevitable that many induced correlations will result. PC1, then, is very often strong and entirely interpretable as the result of data closure. The PC loadings of trace and minor elements, oxides or phases will tend to be similar, and these are likely to be opposite in sign to the dominant components, e.g. silica. The positive correlation of minor elements is the result of their being equally diluted by the major component: the interpretation of PC1 is this dilution effect. Unfortunately, the closure effect is not linear, so it doesn't all come out in PC1: we can't interpret the other PCs as entirely due to real geological effects!

2 PCs may result from correlations of suites of variables repesenting a particular geological origin. Variables with similar loadings in the same PC may originate from mineralisation at the same temperature; from the same parent magma; from the same sedimentary environment; from the same provenance area, etc. Interpretation can only be done with knowledge of the geological system.

Palaeontology

1 Where data are measurements of various anatomical components on fossils, PC1 is almost inevitably interpretable as size: all variables tend to be positively correlated because a bigger organism is likely to have bigger parts! The PC1 score can be used as an index of size.

2 Once size is removed in PC1, PC2 and other PCs must represent aspects of shape, perhaps related to mode of life.

Wireline logs and remote sensing

These and other methods of automated data retrieval have the common attribute of recovering large amounts of information on various variables which have a subtle relationship with the geological properties. Information on rock type does occur in gamma, sonic, etc. electric logs, and in pixel values in certain bandwidths on satellite images; but the data need processing to enhance such information. It is often found that a geological attribute such as clay content will affect multiple variables similarly, and hence cause correlations. Consequently, it is routine procedure to investigate the PCs to find those related to the required geological property.

8.5.2 Factor analysis

We have seen in the previous section how PCs may have geological interpretations: each of such components can be taken to represent a geological factor. In a sense, PCs can be described as a method for factor analysis, but the term factor analysis should be reserved for methods involving rotation of eigenvectors. The principle is that we can represent the total variation in the data set by a small number (q, say) of geological factors, each manifested in a degree of correlation (elongation) in the data scatter. In factor analysis, we only find q vectors (whereas in PCA the number of PCs equals the number of variables, p), and it is found that the variance in those q directions can be maximised by rotating the axes away from the basic PC eigenvector solution. In the VARIMAX method, orthogonality of the axes is preserved, and rotation of the axis system is attempted to account for as much of the variance as possible. The OBLIQUE method allows the axes to rotate out of perpendicular: consequently they become, to an extent, correlated. It is argued that this is reasonable because real geological factors are likely to be correlated: grain size may be correlated with quartz percentage, water depth with current velocity, etc.

The idea of factor analysis is a reasonable one, but the method has been severely criticised on the following grounds.

1 There is no clear criterion for the number of factors that exist in the data: the number is chosen by the user. If we choose different numbers of factors, the factors found are different. So the result is dependent on the choice of the user, who may be influenced by preconceived ideas.

2 It is easy to systematically rotate one axis in two dimensions through 360° in order to find the direction of maximum variance, but it is difficult to systematically search all directions in multidimensional space! In practice, the user has some control over the rotation, and subjectivity can result.

3 As a result of points **1** and **2**, factor analysis can be a complicated procedure and it is more of an interactive modelling exercise. Modelling often suffers from uniqueness problems: it is difficult to know whether or not a result is the best or the only solution.

8.5.3   Eigenvectors of spherical data

Dip-and-strike or any other three-dimensional orientation data can be envisaged as points on a sphere: this concept is used in graphical analysis on Schmidt or other stereographic projections. Of the techniques introduced in Chapter 5, some can be modified by extension of the coordinate system to three dimensions by:

$$x = \cos u \sin \alpha$$
$$y = \cos u \cos \alpha$$
$$z = -\sin u$$

where u is plunge of the linear feature (or pole to planar feature) and α is plunge direction.

The mean direction (and dispersion) can be calculated from these three components in an analogous way to the mean for the circular case (Section 5.2.2), but there are problems with producing summaries of shapes of data distributions which don't exist in the circular case. Of particular concern is the scattering of data points along great circles of the sphere, typically resulting from cylindrical folding. The extent of this can be quantified by eigenvector analysis of the vectors defining the points on the sphere surface.

This can be done in two ways.

1 Representing each data point by itself and its duplicate on the opposite side of the sphere (with x, y and z values of opposite sign), followed by simple PCA of the variance/covariance matrix of the x, y and z variables.

2 A more complex approach in which the eigenvectors take directions of maximum moments of inertia.

In the case of method **1**, PC1 will find the centre of the data scatter (a good estimate of the mean direction) and PC2 will be at a perpendicular position, which, with PC1, defines the preferred great circle, if any. PC3 will be oriented away from the data scatter, and will define the pole of any preferred great circle: this would be the plunge of folded structures. The relative values of eigenvalues λ_1, λ_2 and λ_3 (of PC1, PC2 and PC3, respectively) are diagnostic (taking $=$ to mean similar):

$\lambda_1 = \lambda_2 = \lambda_3$: data scatter random: evenly distributed over sphere
$\lambda_1 \gg \lambda_2 = \lambda_3$: data scatter strongly focused on preferred value
$\lambda_1 = \lambda_2 \gg \lambda_3$: data evenly scattered along great circle: tight cylindrical folding
$\lambda_1 > \lambda_2 > \lambda_3$: data scattered around great circle but with preferred direction: typical open imperfect cylindrical folding

Using the moments of inertia method, the meanings of the first and third eigenvectors and eigenvalues are reversed.

8.6 SIMILARITY COEFFICIENTS AND CLUSTER ANALYSIS

8.6.1 Rationale

Most analyses undertaken are based on the assessment of a geological object in relation to some externally defined scale, such as a linear measurement scale on a ruler. This allows observations to be related to theories or laws which are defined in terms of these measurements, for example the grain size of a sandstone related to the depositional current velocity. However, such measurements can only indirectly provide a comparison between objects. The crucial aspect of an observation may not be that a measurement of rock A is x and of rock B is y, but that the difference between them is $x - y$. In this section, the basis of analysis will be measures of similarity or difference between objects; once these are calculated, the external frame of measurement is lost. This approach is important because so much of geology is, and should be, done on the basis of comparisons between objects. Examples are the following.

1 Stratigraphic correlation is based on similarities between successions; it doesn't matter what the absolute values are that may be used to describe a pair of correlating samples: all that matters is that the two samples are similar.

2 Decisions about which rock types to include in a geological unit for lithological mapping purposes, or which brachiopod specimens to include in the same species, are based on the criterion that the unit/species should comprise a set of rocks/specimens which are relatively similar to each other.

It is clear that similarity has to be assessed by considering multiple variables: it is unusual for one variable to be sufficiently diagnostic. Applications are based on the data structure conveyed in a matrix of similarity coefficients, giving the numerical similarity between all relevant pairs of objects.

Table 8.4 Symbols for m variables on two objects

	Variables				
	x_1	x_2	x_3	$\ldots$	x_m
Object A	x_{1A}	x_{2A}	x_{3A}	$\ldots$	x_{mA}
Object B	x_{1B}	x_{2B}	x_{3B}	$\ldots$	x_{mB}

8.6.2   Calculation of similarity

A suitable method of calculating numerical similarity between two objects, each described by multivariate data, can be applied to multiple pairs of objects to build up the required similarity matrix. Here we will describe methods for calculating similarity between two objects, say A and B, where the data for m variables are conveyed as shown in Table 8.4. The calculation can then be extended to any pair of a larger set of n objects.

There are three main categories of similarity measure.

1   Distance coefficients.
2   Correlation coefficients.
3   Association coefficients.

Association coefficients are restricted to binary data; the choice between distance and correlation coefficients for normal data is not straightforward!

8.6.2.1   Distance coefficients

If the data are imagined as a scatter of points in a multidimensional space defined by multiple variables, then it is easy to see how distance between two points can be used to measure similarity. Zero distance between two points clearly means that there are no numerical differences between the two objects; they are completely similar. Greater distances correspond to lesser similarity, so distance measures are often called dissimilarity, rather than similarity, coefficients.

Standardisation is normally applied prior to distance calculation; we do not, for example, want interpoint distances to be dominated by one variable which happens to be measured in ppm when the rest are percentages. Also, even when all units are the same, it is often the case that a small amount of variation in one variable is as geologically important as large variation in another.

Euclidean distance

The most intuitive distance measure is the direct staight line distance – the Euclidean distance. In two dimensions (i.e. with two variables x_1 and x_2) the

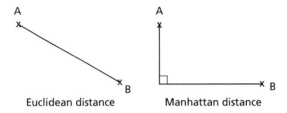

Euclidean distance   Manhattan distance

Fig. 8.7 The concepts of Euclidean and Manhattan distances between two points A and B.

distance d_{AB} between two points A and B is obviously the length of the hypotenuse of a triangle, where the sides are $(x_{1A} - x_{1B})$ and $(x_{2A} - x_{2B})$:

$$d_{AB} = \sqrt{(x_{1A} - x_{1B})^2 + (x_{2A} - x_{2B})^2}$$

(see Fig. 8.7).

In the multivariate general case (with m variables) this becomes as follows.

Euclidean distance coefficient

$$d_{AB} = \sqrt{\sum_{i=1}^{m}(x_{iA} - x_{iB})^2}$$

or, where we do not wish the measure to necessarily increase with more variables:

$$d_{AB} = \sqrt{\frac{1}{m}\sum_{i=1}^{m}(x_{iA} - x_{iB})^2}$$

Worked example: see Box 8.10.

Manhattan distance

Though intuitive, Euclidean distances are criticised on the grounds that the direct line approach underestimates joint differences. This is probably clearest in palaeontology: as a first hypothesis, differences in two shape characters of an organism should be regarded as due to two separate genetic changes, so the real difference between them is the sum of the differences, not the length of the hypotenuse. The Manhattan distance is the simple sum of absolute differences (Fig. 8.7).

Box 8.10 Distance coefficients

Appendix 3.13 shows data on the occurrence of five microfossil species (coded a to e) in 10 samples from borehole A and eight samples from borehole B. We wish to calculate the similarity between each sample in borehole A with each sample in borehole B in order to find possible stratigraphic correlations. What is, for example, the numerical similarity between sample $A4$ and $B2$, and between $A4$ and $B4$?

Standardised data:

	a	b	c	d	e
A4	−0.21	1.12	0.76	0.14	−0.93
B2	0.64	3.13	2.58	−0.15	−0.50
B4	−0.21	−0.39	0.76	−0.01	−0.07

Using Euclidean distance:

$$d_{AB} = \sqrt{\frac{1}{m}\sum_{i=1}^{m}(x_{iA} - x_{iB})^2}$$

we have:

$$
\begin{aligned}
d_{A4,\,B2} &= \sqrt{(((-0.21 - 0.64)^2 + (1.12 - 3.13)^2 + (0.76 - 2.58)^2} \\
&\qquad + (0.14 + 0.15)^2 + (-0.93 + 0.5)^2)/5) \\
&= 0.578
\end{aligned}
$$

and

$$
\begin{aligned}
d_{A4,\,B4} &= \sqrt{(((-0.21 + 0.21)^2 + (1.12 + 0.39)^2 + (0.76 - 0.76)^2} \\
&\qquad + (0.14 + 0.01)^2 + (-0.93 + 0.07)^2)/5) \\
d_{A4,\,B4} &= 0.348
\end{aligned}
$$

Using Manhattan distance:

$$d_{AB} = \frac{1}{m}\sum_{i=1}^{m}|x_{iA} - x_{iB}|$$

we have:

$$
\begin{aligned}
d_{A4,\,B2} &= ((0.64 + 0.21) + (3.13 - 1.12) + (2.58 - 0.76) \\
&\qquad + (0.14 + 0.15) + (0.93 - 0.5))/5 \\
&= 1.081
\end{aligned}
$$

and

$$
\begin{aligned}
d_{A4,\,B4} &= ((0.21 - 0.21) + (1.12 + 0.39) + (0.76 - 0.76) \\
&\qquad + (0.14 + 0.01) + (0.93 - 0.07))/5 \\
&= 0.502
\end{aligned}
$$

Both methods give the result that $A4$ has a greater similarity (lower distance coefficient) to $B4$ than to $B2$. This suggests that $A4$ is more likely to correlate stratigraphically with $B4$.

Manhattan distance coefficient

$$d_{AB} = \frac{1}{m} \sum_{i=1}^{m} |x_{iA} - x_{iB}|$$

Worked example: see Box 8.10.

The name derives from the constraints on routes between two points in Manhattan: only perpendicular sidewalks and elevators! The choice of Euclidean or Manhattan distance depends on the geologist's idea about the independence of variables in the causative process: do two differences really mean twice the difference, or just two linked consequences of one difference? It is difficult to prescribe a general rule of thumb.

8.6.2.2 **Correlation similarity coefficient**

Similarity conveyed by distance is not the only type of similarity that may be of interest. As we have seen (in Sections 3.2.2 and 8.5.1.9), two rocks with differing concentrations of a number of trace elements may only differ due to the amount of diluting quartz, and two fossils may be identical in shape but have greatly different measurements due to different size. We may wish to regard similarity in terms of ratios between different variable values. We can use the correlation coefficient (see Section 3.2.1.) in an unusual way for this purpose: instead of many objects and two variables, we have two objects but many variables. We can imagine the situation by a scatter plot where the axes are the objects and the data points are the variables.

Correlation similarity coefficient

$$r_{AB} = \frac{\sum_{i=1}^{m} (x_{iA} - \bar{x}_A)(x_{iB} - \bar{x}_B)}{s_A s_B (n - 1)}$$

where the means and standard deviations of objects A and B are calculated over all the variables.

Worked example: see Box 8.11.

Although standardisation is not so important with correlation coefficients, care should be taken that the result is not affected by outliers (as usual with correlation coefficients; see Fig. 3.3), due in this case to high or low values of one or two variables. Non-parametric correlation coefficients can be used as an alternative.

Box 8.11 Worked example: correlation similarity coefficient

Using the same data (samples A4, B2 and B4) as for Box 8.10, what are the corresponding correlation similarity coefficients?

Using Pearson's correlation coefficient, we obtain:

$r_{A4, B2} = 0.907$

$r_{A4, B4} = 0.166$

Here, a higher value means greater similarity, so we find that $A4$ is more similar to $B2$. This contradicts the results obtained from distance coefficients in Box 8.10. This result has occurred because $A4$ is more similar to $B4$ in terms of the magnitude of the values for species a to e, but it is more similar to $A2$ in terms of the relative proportions of the species. It is highly debatable as to which is the more useful result for the purpose of stratigraphic correlation.

8.6.2.3 **Association coefficients**

These are for use with binary data, and as such find use primarily in palaeontology, both for taxonomy, where the data are character presences/absences, and for biostratigraphy, where the data are species presences/absences. The data for two borehole samples may be, for example, as in Table 8.5.

The information available for comparison of A and B is limited, but clearly if the presences and absences are identical in A and B, we have perfect similarity, and, if there are no 1s or 0s corresponding between the two samples, we have minimum similarity. Unfortunately, there is no consensus as to how a scale should be established between these extremes, but all the available methods are based on a type of contingency table (Table 8.6). Association coefficients are calculated from a, b, c and d. An enormous number of varieties exist, designed to perform well according to various criteria. One example is presented here.

Jaccard association coefficient

$$J_{AB} = \frac{a}{a + b + c}$$

This doesn't consider that joint absences (d) indicate similarity. It is a simple proportion of the total variables which match.

Worked example: see Box 8.12.

Table 8.5 Binary data on species present in 2 samples

	Species*				
Sample	1	2	3	4	5
A	1	1	0	0	1
B	0	1	1	0	1

*1 indicates the species is present in the sample and 0 indicates absence. (For a critique of the use of such data, see Section 1.3.)

Table 8.6 Contingency table for comparison of 2 samples

		Sample B	
		Present	Absent
Sample A	Present	a	b
	Absent	c	d

a: Number of attributes present in both samples.
b: Number of attributes present in A, absent in B.
c: Number of attributes present in B, absent in A.
d: Number of attributes absent in both samples.

8.6.3   Stratigraphic correlation

8.6.3.1   The data and the similarity matrix

Conventionally, stratigraphic correlation is achieved by observing occurrences of index fossils (biostratigraphy) and/or subjective appraisals of lithological changes (lithostratigraphy). Recently, the use in hydrocarbon exploration of large numbers of species of microfossils with overlapping ranges and the awareness of the vagaries of first and last species occurrences have complicated the situation in biostratigraphy, while in lithostratigraphy it has become necessary to correlate using numerical values from wireline logs. A way of dealing with this is to correlate on the basis of overall similarity. The two sections or boreholes to be correlated are represented by discrete, closely spaced samples, and for each sample a value is recorded for each of a number of attributes which are potentially stratigraphically useful. These may be counts of microfossil species, sediment components or petrophysical measurements. It is hoped that any poorly stratigraphically constrained attributes will be outweighed by good information in a multivariate assessment of overall similarity. In the special case of biostratigraphy, it is possible to weight each fossil species by using an index known as the relative biostratigraphic value (RBV), based on the geographical extent, vertical range and facies dependence of a species – see Box 8.13.

   With each sample in each section (say, A and B) represented by appropriate multivariate data, we then use a similarity coefficient to assess similarity between each sample in section A and each sample in section B. If

---

Box 8.12 Worked example: association coefficients

If the data in Appendix 3.13 had been expressed in binary (presence/absence) form, the data for the samples used in Boxes 8.10 and 8.11 would have been:

A4 1 1 1 1 0
B2 1 1 1 1 1
B4 1 1 1 1 1

To obtain an association coefficient for the similarity between A4 and B2, we first make up the contingency table:

		Sample B_2	
		Present	Absent
Sample $A4$	Present	$a = 4$	$b = 0$
	Absent	$c = 1$	$d = 0$

Using the Jaccard coefficient:

$$J_{AB} = \frac{a}{a + b + c}$$

we obtain

$$J_{A4, B2} = 4/5 = 0.8$$

Note that the same coefficient value would have been obtained if we had compared A4 with B4: this illustrates the consequences of a paucity of information with binary data.

---

there are n samples from A and m from B, the resulting similarities can be expressed in an $n \times m$ matrix. This will contain a numerical appraisal of all possible correlations between the two sections. Worked example: see Box 8.14.

8.6.3.2 **Stratigraphic correlation from the similarity matrix**

A major constraint inherent in stratigraphy is that correlation traces must not cross (if we exclude the possibility of tectonic repetition). There are two main types of algorithm designed to work with this constraint: lateral tracing and slotting.

Lateral tracing

This method is simple and intuitive. The greatest similarity is found in the matrix and the corresponding pair of samples are correlated. Then,

Box 8.13 Relative biostratigraphic value (RBV)

The persistent difficulty in numerical approaches to stratigraphic correlation is to condense from the data available the stratigraphically useful information. With biostratigraphic data, the large number of variables involved allows the possibility of identifying specific variables (taxa) of stratigraphic worth; we can aim to weight the influence of each taxon accordingly.

Classically, the attributes of a good zone fossil are as follows.

1   Restricted stratigraphic range.
2   Wide geographical range.
3   Independence of facies.

These can be quantified by:

Vertical range V = (range occupied by taxon)/(total stratigraphic sequence being considered)

This can be measured in terms of thicknesses of strata or numbers of samples. V can be calculated from each of a number of different sections; the V value used is the highest of these.

Geographical range G = (number of sections in which taxon occurs)/(total number of sections investigated)

Facies independence F = (number of facies in which taxon occurs)/(total number of facies)

'Facies' can be defined however appropriate; often it is taken to convey just crude lithological differences.

For a good zone fossil, we should have low V, high G and high F. RBV indices have been designed to quantify this combination of attributes, for example McCammon's RBV:

$$RBV1 = F(1 - V) + (1 - F)G$$

and Brower's RBV:

$$RBV2 = (F(1 - V) + G(1 - V))/2$$

RBVs have a range between 0 (useless) to 1 (perfect zone fossil). Both McCammon's and Brower's RBVs have the property of giving V greatest influence. An unweighted RBV such as

$$RBV = (G + F + (1 - V))/3$$

would give a high RBV to a spatially and temporally ubiquitous species, which would be useless as a zone fossil.

Numerical biostratigraphy uses similarities between assemblages to obtain stratigraphically useful results. The concept of the RBV allows us to weight the contribution of each species to that measure of similarity. Association coefficients are the most often used similarity measures in this context; these can readily be adapted to incorporate the RBVs for each species. Equations such as the Jaccard coefficient:

Box 8.13 *Continued*

$S = a/(a + b + c)$

can be redefined so that:

a = sum of RBVs of species present in both samples
b = sum of RBVs of species present in A, absent in B
c = sum of RBVs of species present in B, absent in A
d = sum of RBVs of species absent in both samples

Box 8.14 The distance matrix for stratigraphic correlation

Using the Manhattan distance coefficient to obtain similarities between each sample in borehole A and each sample in borehole B for the data in Appendix 3.13, we obtain:

Borehole A	Borehole B							
	1	2	3	4	5	6	7	8
1	0.75	1.89	1.05	1.31	1.36	1.52	1.44	1.43
2	0.66	1.47	0.63	0.89	0.93	1.09	1.01	1.01
3	0.60	1.63	0.63	1.25	1.30	1.46	1.02	1.37
4	0.88	1.08	0.78	0.50	0.86	0.93	0.94	0.93
5	1.48	1.88	0.45	0.90	0.66	0.70	0.43	0.79
6	1.51	1.68	1.05	0.79	0.38	0.42	0.86	0.51
7	2.06	2.29	1.19	0.93	0.52	0.56	1.01	0.65
8	2.26	2.48	1.02	1.13	0.72	0.56	0.64	0.65
9	1.50	1.58	0.99	0.49	0.56	0.66	0.98	0.63
10	1.72	2.35	1.14	1.06	0.91	0.72	0.99	0.64

---

iteratively, the greatest of the remaining similarities is found and, if this does not cross existing correlation traces, is represented by a correlation trace. This is continued until no further traces are possible. Worked example: see Box 8.15.

The advantage of this method is its simplicity; its potential disadvantage is the weight attached to the greatest similarities: if one of these is spurious, the rest of the correlations are liable to be spoilt.

Slotting

This approach is sensitive to the overall structure of the similarity matrix. The 'no crossing' constraint is incorporated by the use of the observation that any crossing correlations are represented in the matrix by a pair of elements which are disposed with a top right–bottom left offset; conversely, a top left–bottom right or a horizontal or vertical offset means no crossing. A completely correlated pair of sections, then, will be represented by a trace through the matrix with a general top left–bottom right trajectory and no reversals. The object of slotting is to find the best such route through the matrix; the criterion being that the total sum of the similarities should be

---

Box 8.15 Stratigraphic correlation from the distance matrix

By lateral tracing

We find the successively lowest coefficients (highest similarities), as indicated in order by the superscripts:

| | Borehole B | | | | | | | |
Borehole A	1	2	3	4	5	6	7	8
1	0.75	1.89	1.05	1.31	1.36	1.52	1.44	1.43
2	0.66	1.47	0.63	0.89	0.93	1.09	1.01	1.01
3	0.60^{10}	1.63	0.63	1.25	1.30	1.46	1.02	1.37
4	0.88	1.08	0.78	0.50^{6}	0.86	0.93	0.94	0.93
5	1.48	1.88	0.45^{4}	0.90	0.66	0.70	0.43^{3}	0.79
6	1.51	1.68	1.05	0.79	0.38^{1}	0.42^{2}	0.86	0.51^{7}
7	2.06	2.29	1.19	0.93	0.52^{8}	0.56^{9}	1.01	0.65
8	2.26	2.48	1.02	1.13	0.72	0.56^{9}	0.64	0.65
9	1.50	1.58	0.99	0.49^{5}	0.56^{9}	0.66	0.98	0.63
10	1.72	2.35	1.14	1.06	0.91	0.72	0.99	0.64

Correlations with superscripts 3, 5, 6, 8 and 9 on the matrix are not plotted as they cross correlations based on higher similarities.
This gives the following correlation:

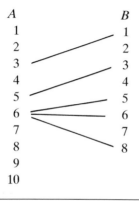

A   B
1   1
2   2
3   3
4   4
5   5
6   6
7   7
8   8
9
10

Box 8.15 *Continued*

By slotting

We find the route from top left to bottom right in the matrix that involves the lowest total distances (in **bold**):

	Borehole *B*							
Borehole *A*	1	2	3	4	5	6	7	8
1	0.75	1.89	1.05	1.31	1.36	1.52	1.44	1.43
2	0.66	1.47	0.63	0.89	0.93	1.09	1.01	1.01
3	**0.60**	1.63	0.63	1.25	1.30	1.46	1.02	1.37
4	**0.88**	**1.08**	**0.78**	0.50	0.86	0.93	0.94	0.93
5	1.48	1.88	**0.45**	**0.90**	**0.66**	0.70	0.43	0.79
6	1.51	1.68	1.05	0.79	**0.38**	**0.42**	0.86	0.51
7	2.06	2.29	1.19	0.93	0.52	**0.56**	1.01	0.65
8	2.26	2.48	1.02	1.13	0.72	**0.56**	**0.64**	**0.65**
9	1.50	1.58	0.99	0.49	0.56	0.66	0.98	**0.63**
10	1.72	2.35	1.14	1.06	0.91	0.72	0.99	0.64

This gives the following correlation:

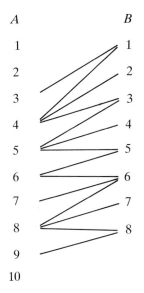

The main difference between the two results is the lateral tracing correlation between *A*6 and *B*8, which is not picked by the slotting method. The slotting solution regards the *A*6–*B*8 similarity as anomalous, in that it is not supported by the overall structure of the matrix.

maximised (or dissimilarities minimised). Starting from the top left (or bottom right) corner, each step through the matrix involves a choice of two alternatives. The number of candidate routes is impossibly large even for a modestly sized matrix; in practice, the computer (or brain!) looks for the best route for the next few steps only. A complication is the likelihood that the correct route involves entering the matrix from a side and not from the corner; this is accommodated by duplicating the edge rows and columns as extra dummy entries. Worked example: see Box 8.15.

Although more robust than lateral tracing in that it is less sensitive to anomalous values in the matrix, there are, nevertheless, problems associated with slotting.

1 Slotting involves an asymmetric approach to the data and the decision to enter the matrix from the top left or bottom right is arbitrary. Unless the data are particularly well behaved, the two approaches often yield different results. This merely reflects the fact that the pragmatic stepwise approach does not necessarily find the globally optimum solution.

2 The result of slotting is sometimes perverted by a single anomalous sample in one of the sections: this can create a wall of low similarity across the matrix which the correlation path is forced along rather than across.

8.6.4 **Cluster analysis**

8.6.4.1 **Principles of numerical classification**

Classifications of geological objects are now predominantly based on numerical criteria; the alternative of qualitative definition and assessment is subjective and potentially unscientific. Palaeontology is an exception but is a special case: species definitions are subject to special rules and focus on individual type (holotype) specimens, and higher taxa are normally classified according to evidence of phylogeny. Palaeontological classification can be enhanced by numerical analysis, but researchers face the additional difficulty of quantifying the information content of fossils.

There are two main types of numerical classification commonly used in geology.

Partitioning based on absolute values

It is common in geology to find rock type definitions which include criteria such as 'less than 10% quartz' or 'more than 30% matrix'. Conceptually, all rocks lie in an m-dimensional space, where there are m mineral types (or other components) thought to be important. This space is divided up by lines or planes into compartments, and each compartment defines a specified rock type (e.g. greywacke or syenite) and is used to classify new samples. In practice, such schemes are often expressed as triangular diagrams. This type of classification does not normally require data analysis.

Advantages

1 Robustness: the classification can be applied globally and unambiguously.
2 Partitions may be placed in positions with theoretical significance (e.g. indicating silica saturation), and/or in positions which allow easy field or laboratory diagnosis (e.g. grain- or matrix-supported).

Disadvantages

1 The positions of partitions may be proposed by experts or committees of experts, but, despite this and advantage **2** above, they are often arbitrary and unnatural.
2 It is very possible that two similar specimens will plot close to each other but on opposite sides of a partition, and so will be classified differently. Conversely, two much less similar specimens may fall within the same category. This is clearly an undesirable property of a classification scheme.
3 It becomes cumbersome to manage partitions if they are defined with respect to more than two or three variables.

Clustering based on relative sample properties

The opposite approach to the above ignores the absolute scale of the variables and focuses instead on the properties of the available sample. The individual categories within a classification are based on clusters of points (the points being objects, observations or specimens). Clusters are concentrations of points in attribute space, and two points in the same cluster tend to be more similar than two points in different clusters. This is the approach underlying the cluster analysis procedures described in Section 8.6.4.2.

Advantages

1 A coherent, unimodal group of observations will not be split up (potentially arbitrarily) among different categories.
2 The boundaries between clusters are, by definition, in regions of multivariate space where there are few points; this may be due to some aspect of the geological process, in which case the boundaries would be natural.
3 The methodology readily allows consideration of all variables.

Disadvantages

1 Instability: the addition of further observations to the analysis is likely to add new clusters and will inevitably redefine old ones. Consequently, the data used should be an unbiased sample of the population to which it is intended to be applied.

2   Allocating new specimens to clusters is not as straightforward as simply placing it on one side or other of a partition.

The balance of advantages and disadvantages listed above is in favour of partitioning for global schemes, largely because we cannot expect to have an unbiased sample from the global population, but in favour of clustering for local investigations.

Typical applications, then, are: clustering of carbonate sediments on the basis of clast types to produce a map of sediment type within a formation; clustering of wireline log responses to produce lithological logs of a suite of boreholes; classification of satellite image data to distinguish soil types; clustering of igneous trace element geochemistry to identify multiple intrusions; and classification of fossil mollusc species and genera from shell measurements.

8.6.4.2   **Clustering algorithms**

A group of objects which are classifiable together on numerical grounds will form a cluster of points in multivariate space. The variety of methods for separating clusters include divisive methods, where we find the sparse areas for positioning boundaries between groups; density methods, where clusters are found by searching multivariate space for concentrations of points; and linkage methods, where nearby points are iteratively linked together. Here we will focus on linkage methods, as these are more intuitive, computationally efficient and readily available. All of these are operations performed on the similarity matrix, and all begin with the formation of the initial cluster by linkage of the two objects with the greatest similarity (or least dissimilarity), and at each iteration the most similar pair of objects/clusters are linked. Once clusters have formed, they cannot be divided and can only amalgamate with others; consequently the result is purely hierarchical. The methods differ in the way that the similarity between such new clusters and all the other objects/clusters should be calculated.

Note that, although most are arithmetically simple, clustering algorithms cannot feasibly be done manually with more than a very small sample.

Nearest-neighbour linkage (or single linkage)

In this algorithm, the similarity between a point and a new cluster is equal to the similarity between that point and the most similar point in the cluster. For example, if we are using a distance measure and the distance coefficient between objects A and B is 3.5 and between A and C is 4.5, the distance between A and the cluster BC is regarded as being 3.5 (lowest distance meaning greatest similarity).

Worked example: see Box 8.16.

This is the minimum reasonable measure of distance between a point and a cluster (or between two clusters), and is less than the true distance for most of the contained points. Consequently, it is often described as distorting multivariate space by contraction. Under this procedure, it is easy for points to link on to the ends of straggly, elongated clusters. Points at opposite ends of such clusters will be substantially different. This effect is known as chaining, and tends to result in messy, unnatural-looking hierarchical

---

Box 8.16 Worked example: clustering algorithms

Suppose a Euclidean distance matrix (dissimilarity coefficient) for four objects is:

	A	B	C	D
A	—	2.0	4.5	3.5
B		—	2.5	3.0
C			—	4.0
D				—

The greatest similarity (least dissimilarity) is between A and B, so these are the first to be linked.

Nearest-neighbour algorithm

Having linked A and B, we need to rewrite the matrix with AB as one entity. The coefficient between the new cluster AB and C will be the greatest similarity of $A-C$ and $B-C$, which is 2.5. Similarly, the coefficient for AB and D will be the greatest similarity of $A-D$ and $B-D$, which is 3.0. So we rewrite the matrix:

	AB	C	D
AB	—	2.5	3.0
C		—	4.0
D			—

In the next iteration, the cluster AB is linked to C, and the matrix becomes:

	ABC	D
ABC	—	3.0
D		—

The result of the successive linkages can be expressed in the dendrogram in Fig. B8.16.1(a).

continued on p. 288

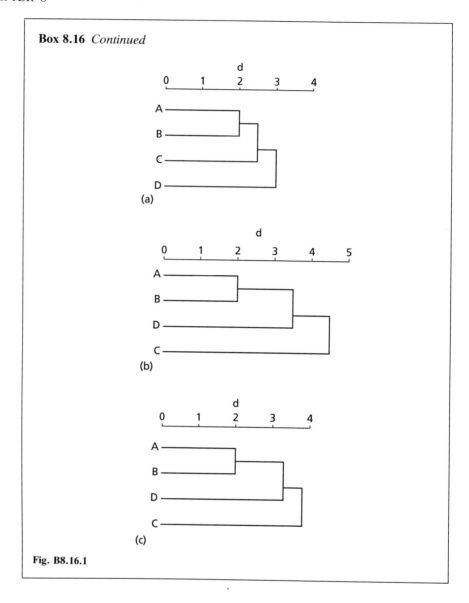

Box 8.16 *Continued*

(a)

(b)

(c)

Fig. B8.16.1

systems (Fig. 8.8a). For these reasons, the nearest-neighbour approach is among the least favoured methods. However, the criterion for similarity calculation is not unreasonable and it should not be rejected unless there is some special reason for doubting the validity of elongated clusters.

Furthest-neighbour linkage (or complete linkage)

This is the exact converse of the nearest-neighbour method: the similarity between a point and a cluster (or between two clusters) is taken to be the least of all the candidate pairwise similarities. So, with distance coefficients

Box 8.16 *Continued*

Furthest-neighbour or complete linkage algorithm

Here the coefficients associated with the first cluster *AB* are found by taking the least similarities.

First iteration:

	AB	C	D
AB	—	4.5	3.5
C		—	4.0
D			—

Second iteration:

	ABD	C
ABD	—	4.5
C		—

Dendrogram (Fig. B8.16.1b).

Average linkage algorithm

Here, the recalculated similarities are the average similarities for the cluster members.

First iteration

	AB	C	D
AB	—	3.5	3.25
C		—	4.0
D			—

Second iteration:

	ABD	C
ABD	—	3.75
C		—

Dendrogram (Fig. B8.16.1c).

Notice that the nearest-neighbour algorithm is alone in linking *C* with *AB* before *D*. *D* is generally closer to *AB*, but the short distance between *A* and *C* causes the result seen in the nearest-neighbour method.

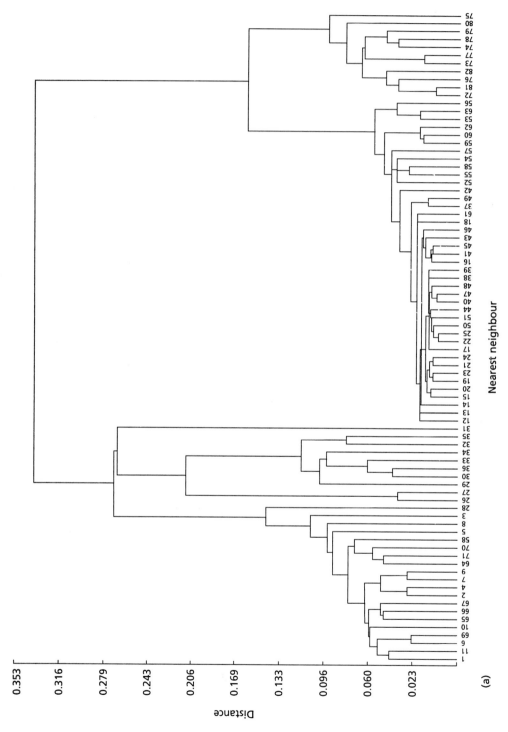

Fig. 8.8 Three linkage algorithms applied to the same data (the AFRA data set of foram measurements: Appendix 3.12), showing the characteristic features of the resulting dendrograms. The nearest-neighbour method shows straggly 'chained' clusters; the other methods produce neat, tight clusters.

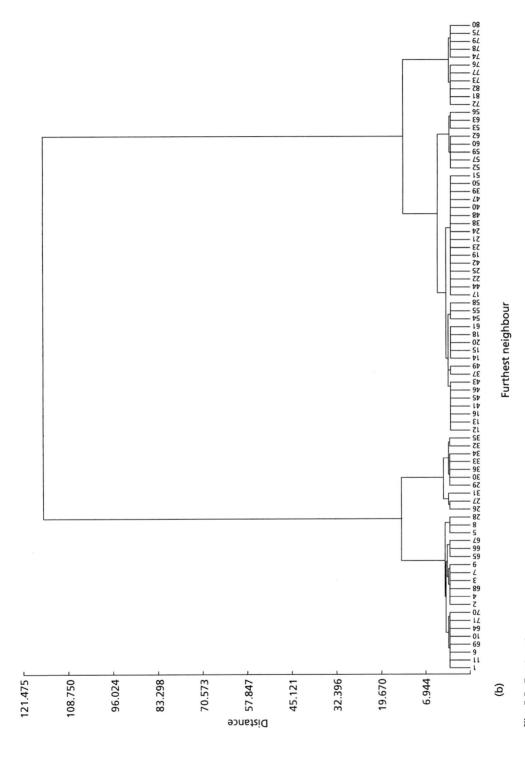

Fig. 8.8 *Continued*

(b)

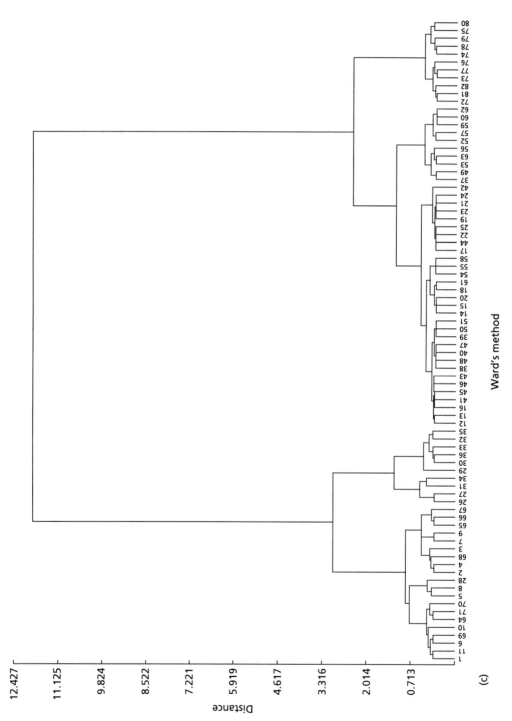

Ward's method

(c)

Fig. 8.8 *Continued*

3.5 for A to B and 4.5 for A to C, the distance for A to BC is regarded as 4.5 (greatest distance is least similarity).

Worked example: see Box 8.16.

As the apparent intercluster distances are maximised, this method dilates space and results in hyperspherical clusters. Even slightly ellipsoidal unimodal clusters of points are likely to be broken up, giving results suggesting that there are multiple modes (Fig. 8.8b).

Average linkage

Here, the criterion for similarity recalculation is intermediate between the two previous extremes. If the distance A to B is 3.5 and A to C is 4.5, it is logical for A to BC to be regarded as 4.0, and this is the result that average methods give. This result is both the mean of the two distances and the distance to the centroid of BC. However, if clusters are larger, there are more permutations possible and a range of subtly different algorithms exist. Using the above figures, suppose that B is a cluster of five points and C is a cluster of 10 points. We could either regard these as of equal status and use the answer of 4.0, or we could calculate a weighted average with cluster C twice the weight of B. The latter approach would require the calculation of similarity as:

$$((3.5 \times 5) + (4.5 \times 10))/15 = 4.167$$

There are no guidelines as to which of the average linkage methods is to be preferred, but the results tend to be similar.

Worked example: see Box 8.16.

Average linkage methods produce results which are intermediate in all respects and show a neat hierarchical structure.

Ward's method

Ward's method is somewhat different from the foregoing in that the criterion determining the choice of linkage is that there should be the least increase in the sum of squared deviations from cluster means. The error sum of squares is an almost ubiquitous measure in statistics and is a natural choice for controlling the increase in variance of clusters during linkage. Ward's method typically produces 'good-looking' results with hyperspherical clusters and a well-proportioned hierarchical structure (Fig. 8.8c). For these reasons, it has recently become a *de facto* standard or default option, but there is no reason to suppose that 'good-looking' results are any nearer the 'true' result, whatever that is!

8.6.4.3   The dendrogram and its interpretation

Dendrogram construction

The result of all of the linkage methods described above is an ordered series of linkages between specified objects/clusters, each at a specific magnitude of similarity. This information is conventionally conveyed in a dendrogram, which, as its name suggests, is a branching tree-like construction. The dendrogram is essentially one-dimensional, with a scaled horizontal axis for the similarity coefficient. A second dimension merely accommodates the array of objects and a topological display of the linkages (Fig. 8.9), with vertical bars linking the appropriate objects/clusters at the appropriate similarity value. It is important to realise that the order of the objects is to a large extent arbitrary; the only useful constraint is that the arrangement avoids branch crossovers, and there are always enormous numbers of equally valid different ways that this can be achieved.

The phenon line

The dendrogram displays the full hierarchical structure of the data, but is not in itself a useful classification. For practical purposes, perhaps for mapping, we normally need to discern a number of categories at one or perhaps two levels of the hierarchy. This can be done by drawing a vertical line – the phenon line – at a specific similarity value. The phenon line cuts dendrogram branches and thereby isolates clusters. These clusters will then have a similarity greater than the specified value within a cluster, and less than the specified value between clusters.

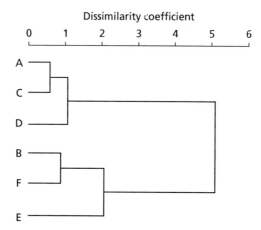

Fig. 8.9 A dendrogram showing the linkages between six objects *A* to *F*. The vertical links are drawn at positions corresponding to the dissimilarity coefficients between the linked objects or clusters. The order of *A* to *F* is unimportant: it is contrived so that lines don't cross.

A phenon line at the extreme left of a dendrogram (very high similarity) can distinguish n 'clusters' (n = number of objects); at the opposite end, it would isolate only one cluster. Both of these extremes are useless, but there are a total of n different solutions to choose from! The choice of a position for the phenon line is based on the following.

1 Pragmatic requirements: a study of the heterogeneity of the geological situation may not be well served by too many (e.g. 20) or too few (e.g. 2) clusters. The researcher may wish to select a number according to the detail required and general convenience.

2 Preconceptions: other lines of evidence may lead a worker to expect a specific number of categories. As long as the number of categories is not in itself under investigation, such a number may be as good as any.

3 'Natural' solutions: there may be levels in the hierarchy where there are large gaps between linkage events, meaning a particularly large difference between 'within cluster' and 'between cluster' similarities. In palaeontology, the genetic and other mechanisms by which species maintain their integrity provide a reason to expect this, but there is probably no mechanism in the rest of geology that is likely to systematically produce such gaps. Nevertheless, where they occur, gaps between linkages are obvious locations for phenon lines.

8.6.4.4 *q* vs. *r* mode

Cluster analysis has been described here in q mode, which means that the result clusters objects on the basis of values of variables. It is usually objects (such as rocks) that are required to be classified. In r mode, the variables are clustered on the basis of their values in objects: r mode analyses cluster together correlating variables (the correlation coefficient is usually used as the similarity index), such as ecological associations of fossils or environmental associations of carbonate clasts, but these cannot be mapped as clusters and are probably better investigated by eigenvector methods (Section 8.5).

8.6.4.5 A critique of cluster analysis

Cluster analysis, at best, is the only objective way of classifying geological objects. But, as we have seen, there are choices to be made at every stage: choice of similarity coefficient, choice of linkage method, choice of phenon line; and there are very few recommendations or guidelines to help. It is possible, then, to obtain a substantial range of different results from the same data. This undermines objectivity, because the unscrupulous researcher can pick the result that best suits his or her preconceptions, and find a justification for the methodology retrospectively. Objectivity can be restored if the choices are made arbitrarily, or on the basis of general recommendations, in advance of the analysis, and with a commitment to accept the results! Nevertheless, it is difficult to attach great significance to a result which may be only one of a number of equally valid alternatives!

Cluster analysis is not a statistical procedure, and there is no easy way of deciding whether or not data are more clustered than would be expected from a random population. There are two difficulties here: (a) ordinary tests for the significance of difference between samples are not applicable to clusters because, even with genuinely random data, the cluster algorithms ensure that clusters are inevitably discrete, non-overlapping and therefore significantly different: the null hypothesis of randomness does not equate with equality of sample means; and (b) it is not clear whether one cluster should be assessed for significance or whether the important potential non-randomness resides in the structure of the whole dendrogram. However, the pragmatic geologist who merely needs a suite of coherent categories in order to get a picture of heterogeneity does not need to know about statistical significance! Cluster analysis results are generally useful and sometimes enlightening for displaying structure in multivariate data, but profound conclusions should not be based on such uncertain foundations.

FURTHER READING

Godwin C.I. and Sinclair A.J. (1979) Application of multiple regression analysis to drill-target selection, Casino porphyry copper–molybdenum deposit, Yukon Territory, Canada. *Transactions of the Institute of Mining and Metallurgy*, **88**, B93–B106.

Hodgson C.J. and Troop D.G. (1988) A new computer-aided methodology for area selection in gold exploration: a case study from the Abitibi Greenstone Belt. *Economic Geology*, **83**, 952–977.

Zhijun Y. (1983) New method of oil prediction. *Bulletin of the American Association of Petroleum Geologists*, **67**(11), 2053–2056.

interesting case studies of multiple regression in applied geology.

Fedikow M.A.F., Parbery D. and Ferreira K.J. (1991) Geochemical target selection along the Agassiz metallotect utilizing stepwise discriminant function analysis. *Economic Geology*, **86**, 588–599.

Potter P.E., Shimp N.F. and Witters J. (1963) Trace elements in marine and freshwater argillaceous sediments. *Geochimica et Cosmochimica Acta*, **27**, 669–694.

typical examples of applied discriminant function analysis.

Saunders W.B. and Swan A.R.H. (1984) Morphology and morphologic diversity of mid-Carboniferous ammonoids in space and time. *Paleobiology*, **10**, 195–228.

Swan A.R.H. and Saunders W.B. (1987) Function and shape in late Paleozoic (mid-Carboniferous) ammonoids. *Paleobiology*, **13**, 297–311.

particularly useful examples of principal components analysis in that the properties of the data can be seen in illustrations of fossils. The 1987 paper explains the PCs.

Elueze A.A. and Olade M.A. (1985) Interpretation through factor analysis of stream sediment reconaissance data for gold exploration in Ilesha greenstone belt, S.W. Nigeria. *Transactions of the Institute of Mineralogy and Metallurgy*, **B94**, B115–B160.

Griffiths J.C. (1966) A genetic model for the interpretive petrology of detrital sediments. *Journal of Geology*, **74**, 655–671.

classic examples of factor analysis.

Woodcock N.H. (1977) Specification of fabric shapes using an eigenvalue method. *Geological Society of America Bulletin*, **88**, 1231–1236.

spherical data analysis.

Vasey G.M. and Bowes G.E. (1985) The use of cluster anlysis in the study of some non-marine bivalvia from the Westphalian D of the Sydney Coalfield, Nova Scotia, Canada. *Journal of the Geological Society of London*, **142**, 397–410.

clear example of cluster analysis.

Appendix 1: Matrix Algebra

The analysis of geological data usually requires the computation of statistics for several variables. In order to do this in an orderly manner it is convenient to arrange the data in arrays called matrices and vectors. This Appendix shows how simple calculations are performed with these structures and presents some important types of vector and matrix. No attempt is made to develop the mathematical theory.

VECTORS

A vector is a set of numbers in which the order is significant. Often, an ordered pair (x_1, x_2) represents a point in two-dimensional space, the first and second elements being distances from the origin along two perpendicular axes:

The elements of the vector $(x_1, x_2, \ldots, x_{12})$ may be percentages of oxides in a specimen of rock; it will often be helpful to think of the vector as a point in 12-dimensional space or as a line between that point and the origin.

Notation: We shall use x to represent a column of numbers (a column vector),

$$x = \begin{pmatrix} x_1 \\ x_2 \\ \vdots \\ x_p \end{pmatrix}$$

and x^T to represent the row vector $(x_1, x_2, \ldots, x_p)$. In Chapter 8 on Multivariate Analysis we make frequent use of row vectors for the sake of typographic convenience.

Such vectors may hold a set of values of several variables measured on a single individual or values of a single variable found on several individuals:

$$x = \begin{pmatrix} x_1 \\ x_2 \\ x_3 \\ x_4 \end{pmatrix} \quad \begin{array}{l} \text{Carbonate content} \\ \text{Ni content} \\ \text{Mo content} \\ \text{Organic material content} \end{array}$$

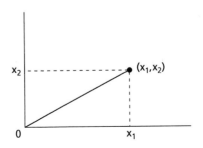

Fig. A1.1

399

or

$$x = \begin{pmatrix} x_1 \\ x_2 \\ x_3 \end{pmatrix} \quad \begin{array}{l} \text{Length of 1st belemnite} \\ \text{Length of 2nd belemnite} \\ \text{Length of 3rd belemnite} \end{array}$$

We now develop some ideas which are helpful for describing the treatment of data in these forms. In particular we need to represent operations such as addition and multiplication and to relate vectors to concepts of space such as distance and direction.

Equality of vectors

If $x^T = (x_1, x_2, \ldots, x_p)$ and $y^T = (y_1, \ldots, y_p)$ then $x^T = y^T$ if and only if $x_1 = y_1$, $x_2 = y_2$ and so on. Thus the vectors $(1,3,4)$, $(4,1,3)$ are *not* equal.

Addition

Vectors are added by summing corresponding elements:

$$x + y = \begin{pmatrix} x_1 \\ x_2 \\ \vdots \\ x_p \end{pmatrix} + \begin{pmatrix} y_1 \\ y_2 \\ \vdots \\ y_p \end{pmatrix} = \begin{pmatrix} x_1 + y_1 \\ x_2 + y_2 \\ \vdots \quad \vdots \\ x_p \quad y_p \end{pmatrix}$$

Example: x might contain percentages of Fe^{2+} ion and y might contain percentages of Fe^{3+} ion in several specimens of a rock. Then the vector $x + y$ would contain percentages of total Fe ion.

Multiplication by a scalar

From the addition rule we see that, for example,

$$x + x = \begin{pmatrix} 2x_1 \\ 2x_2 \\ \vdots \\ 2x_p \end{pmatrix}$$

and this is the same as $2x$. In general, if c is a scalar we have

$$cx = \begin{pmatrix} cx_1 \\ cx_2 \\ \vdots \\ cx_p \end{pmatrix}$$

The vectors x and cx point in the same direction but the second vector is c times the length of the first. This can be verified by, for example, drawing the vectors $(1,3)$ and $(2,6)$ using the same axes.

Scalar product of two vectors

This is the result of combining the elements of two vectors in the following way. Suppose that vectors x, y contain the elements $(x_1, x_2, \ldots, x_n)$ and $(y_1, y_2, \ldots, y_n)$ respectively; then the scalar product of the vectors is

$$x^T y = (x_1, x_2, \ldots, x_n) \begin{pmatrix} y_1 \\ y_2 \\ \vdots \\ y_p \end{pmatrix} = x_1 y_1 + x_2 y_2 + \ldots + x_n y_n$$

Note that the result is a scalar, i.e. an ordinary number.

Length of a vector

Consider the length of the line which joins the origin to the point (x_1, x_2), as in the above diagram. From Pythagoras's theorem in two dimensions, the distance of the point from the origin is given by

$$\sqrt{(x_1^2 + x_2^2)}$$

This distance, the length of vector x, is denoted by $|x|$. In p dimensions the length of a vector is, analagously,

$$|x| = \sqrt{(x_1^2 + x_2^2 + \ldots + x_p^2)}$$

Example: Consider a vector d in which the elements are differences between n sample values and their sample mean:

$$d^T = (x_1 - \bar{x}, x_2 - \bar{x}, \ldots, x_n - \bar{x}).$$

Then the length of the vector d is

$$|d| = \sqrt{\{(x_1 - \bar{x})^2 + (x_2 - \bar{x})^2 + \ldots + (x_n - \bar{x})^2\}}$$

It is the square root of the corrected sum of squares of the data and this illustrates the connection between the statistic and distance.

Frequently, we need a measure of the distance between two points, neither of them being the origin. For example, in cluster analysis we may require the distance between the points represented by vectors x, y:

The distance between points A,B is, by Pythagoras's theorem,

$$\sqrt{\{(y_1 - x_1)^2 + (y_2 - x_2)^2\}} = |y - x|$$

This is the length of the vector $(y_1 - x_1, y_2 - x_2)$.

Angle between two vectors

There is an interesting statistical interpretation of the angle between two vectors x, y in n-dimensional space: its cosine is equal to the sample correlation coefficient calculated from the data in the vectors.

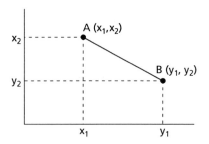

Fig. A1.2

When the correlation is low, the angle between the vectors is large, whereas the angles between vectors of values of highly correlated variables are small.

Orthogonal vectors

Orthogonal vectors are at right angles, so the consine of the angle between them is zero: if α, β are orthogonal then

$$\alpha^T\beta = 0$$

Examples:

$$(1 \quad 1)\begin{pmatrix} 1 \\ -1 \end{pmatrix} = 0$$

$$(1 \quad 3 \quad 5)\begin{pmatrix} 6 \\ -2 \\ 0 \end{pmatrix} = 0$$

Orthonormal vectors

These are orthogonal and of unit length. That is, vectors α, β are orthonormal if

$$\alpha^T\beta = 0$$

and

$$|\alpha| = 1, \qquad |\beta| = 1$$

Example: The vectors $(\sin\theta \ \cos\theta)$ and $(\cos\theta \ -\sin\theta)$ are orthonormal because

$$(\sin\theta \ \cos\theta)\begin{pmatrix} \cos\theta \\ -\sin\theta \end{pmatrix} = \sin\theta \cos\theta - \cos\theta \sin\theta = 0$$

and

$$\sin^2\theta + \cos^2\theta = 1$$

Normalisation of a vector

Normalisation is the process of finding a vector of unit length which points in the same direction as a given vector. (Recall that when a vector is multiplied by a scalar we obtain a vector pointing in the same direction as the first, but having a different length.)

For example, the vector (1 3 5) is of length $\sqrt{\{1^2 + 3^2 + 5^2\}}$, i.e. $\sqrt{35}$. We can obtain a vector of unit length by dividing all the elements of the given vector by $\sqrt{35}$: it will be

$$u^T = (1/\sqrt{35}\ 3/\sqrt{35}\ 5/\sqrt{35})$$

since $|u| = \sqrt{\{(1/\sqrt{35})^2 + (3/\sqrt{35})^2 + (5/\sqrt{35})^2\}} = 1.$

MATRICES

A m × n matrix is a rectangular array of numbers with m rows and n columns. The array is enclosed in curved brackets and is usually denoted by an upper case letter.

E.g.,

$$A = \begin{pmatrix} 2 & 5 & 3 & 4 \\ 1 & 2 & 7 & 1 \end{pmatrix}$$

is a 2 × 4 matrix.

The element in the ith row, jth column of a matrix A is labelled as a_{ij}. For instance, the elements of a 3 × 3 matrix A would be labelled as follows.

$$A = \begin{pmatrix} a_{11} & a_{12} & a_{13} \\ a_{21} & a_{22} & a_{23} \\ a_{31} & a_{32} & a_{33} \end{pmatrix}$$

For conciseness we often write $A = (a_{ij})$.
An example of a data matrix might be

	SiO_2	Al_2O_3	Fe_2O_3	...	P_2O_5
Specimen 1	43.70	18.01	2.95	...	0.42
Specimen 2	44.97	15.74	3.43	...	0.37
:	:	:	:	:	:
:	:	:	:	:	:
Specimen 30	45.33	13.32	2.77	...	0.08

Note that geochemical data are sometimes presented with data on variables in rows and data on specimens in columns. The arrangement used here (and in other statistical texts) allows for the presence of a large number of specimens (more in hope than in expectation, perhaps!).

Covariance matrices

These are denoted by Σ, often with a suffix to indicate name of the joint random variable. The covariance between X_i and X_j is the element in the ith row and jth column and also in the jth row, ith column.

$$\Sigma = \begin{pmatrix} \sigma_{11} & \sigma_{12} & \sigma_{13} & \cdots & \sigma_{1p} \\ \sigma_{21} & \sigma_{22} & \sigma_{23} & \cdots & \sigma_{2p} \\ \vdots & \vdots & \vdots & \vdots & \vdots \\ \sigma_{p1} & \sigma_{p2} & \sigma_{p3} & \cdots & \sigma_{pp} \end{pmatrix}$$

where σ_{ij} is the covariance between X_i and X_j.

This is an example of a *symmetric* matrix: both σ_{12} and σ_{21} denote the covariance between X_1 and X_2 so they are equal, and in general $\sigma_{ij} = \sigma_{ji}$. Symmetric matrices are described with the aid of a numerical example in the section below headed 'Some special types of matrix.'

Correlation matrices

$$\rho = \begin{pmatrix} \rho_{11} & \rho_{12} & \rho_{13} & \cdots & \rho_{1p} \\ \rho_{21} & \rho_{22} & \rho_{23} & \cdots & \rho_{2p} \\ \vdots & \vdots & \vdots & \vdots & \vdots \\ \rho_{p1} & \rho_{p2} & \rho_{p3} & \cdots & \rho_{pp} \end{pmatrix} = \begin{pmatrix} 1 & \rho_{12} & \rho_{13} & \cdots & \rho_{1p} \\ \rho_{21} & 1 & \rho_{23} & \cdots & \rho_{2p} \\ \vdots & \vdots & \vdots & \vdots & \vdots \\ \rho_{p1} & \rho_{p2} & \rho_{p3} & \cdots & 1 \end{pmatrix}$$

Equality of matrices

$A = B$ implies that $a_{ij} = b_{ij}$ for every i and j, i.e. all the elements in corresponding positions of A, B are equal.

Addition of matrices

$A + B = (a_{ij} + b_{ij})$

Example: If

$$A = \begin{pmatrix} 3 & 1 & 2 \\ 4 & 2 & 5 \\ 1 & 1 & 6 \end{pmatrix} \qquad B = \begin{pmatrix} 1 & -1 & 3 \\ 2 & 7 & -4 \\ 0 & 2 & 6 \end{pmatrix}$$

$$A + B = \begin{pmatrix} 4 & 0 & 5 \\ 6 & 9 & 1 \\ 1 & 3 & 12 \end{pmatrix}$$

Multiplication by a scalar

As with vectors,

$$cA = (ca_{ij})$$

Pre-multiplication by a row vector

Each column of the matrix is multiplied by the vector as if it were itself a vector, for example

$$(2 \quad 3 \quad 5) \begin{pmatrix} 4 & 5 \\ 3 & 1 \\ 2 & 8 \end{pmatrix} = (27 \quad 53)$$

The result is a row vector.

Post-multiplication by a column vector

Each row of the matrix is treated as a vector and used to multiply the column vector, for example

$$\begin{pmatrix} 2 & 3 & 7 \\ 1 & 4 & 1 \end{pmatrix} \begin{pmatrix} 1 \\ 3 \\ 2 \end{pmatrix} = \begin{pmatrix} 25 \\ 15 \end{pmatrix}$$

The result is a column vector.

Note that for pre-multiplication of a matrix by a vector, the number of rows in the matrix must be the same as the number of elements in the vector; also, for post-multiplication by a vector the number of columns in the matrix must equal the number of elements in the vector.

Multiplication of matrices by matrices

To find the product AB of matrices A, B the columns of B are multiplied by the rows of A in the manner described for scalar products of vectors.
Example: Find the product AB when

$$A = \begin{pmatrix} 1 & 4 & 2 \\ 4 & 0 & 1 \\ 7 & 3 & 2 \\ 2 & 1 & 6 \end{pmatrix} \quad B = \begin{pmatrix} 3 & 5 \\ 2 & 1 \\ 4 & 3 \end{pmatrix}$$

$$AB = \begin{pmatrix} 19 & 15 \\ 16 & 23 \\ 35 & 44 \\ 32 & 29 \end{pmatrix}$$

where, for example,

$$1 \times 3 + 4 \times 2 + 2 \times 4 = 19$$

Notice that (a) the resulting matrix has the same number of rows as A but the same number of columns as B; and (b) the product BA does not exist in this case. In general, AB is not equal to BA in matrix algebra.

Some special types of matrix

1 Diagonal matrix: Non-zero elements occur only in the leading diagaonal (the diagonal from the upper left-hand to the lower right-hand corner). Pre-multiplication of a matrix A by a diagonal matrix multiplies the elements of a row by the same number:

$$\begin{pmatrix} 3 & 0 & 0 \\ 0 & 4 & 0 \\ 0 & 0 & 2 \end{pmatrix} \begin{pmatrix} 1 & 2 \\ 3 & 4 \\ 5 & 6 \end{pmatrix} = \begin{pmatrix} 3 & 6 \\ 12 & 16 \\ 10 & 12 \end{pmatrix}$$

Post-multiplication results in the all the elements in a column of A being multiplied by the same number:

$$\begin{pmatrix} 2 & 1 & 6 \\ 3 & 4 & 1 \\ 1 & 2 & 3 \end{pmatrix} \begin{pmatrix} 3 & 0 & 0 \\ 0 & 4 & 0 \\ 0 & 0 & 2 \end{pmatrix} = \begin{pmatrix} 6 & 4 & 12 \\ 9 & 16 & 2 \\ 3 & 8 & 6 \end{pmatrix}$$

2 Identity matrix: This is a diagonal matrix whose elements in the leading diagonal are all equal to unity. For example a 3×3 identity matrix would be

$$\begin{pmatrix} 1 & 0 & 0 \\ 0 & 1 & 0 \\ 0 & 0 & 1 \end{pmatrix}$$

In multiplication it has the effect of multiplying the elements of a suitably dimensioned matrix by 1. Thus it behaves like the number 1 in ordinary arithmetic.

3 Inverse of a matrix: The inverse of matrix A is written as A^{-1} and is defined to be the matrix such that

$$AA^{-1} = A^{-1}A = I, \text{ the identity matrix}$$

Division of matrices by matrices is not defined, but we can multiply by inverse matrices.

4 Transpose of a matrix: The transpose of a matrix A, denoted by A^{T}, is obtained by writing the rows of A as columns. For example

$$A = \begin{pmatrix} 3 & 4 & 1 \\ 2 & 7 & 5 \end{pmatrix} \qquad A^{\mathrm{T}} = \begin{pmatrix} 3 & 2 \\ 4 & 7 \\ 1 & 5 \end{pmatrix}$$

5 Symmetric matrix: The first row of a symmetric matrix is the same as the first column, and similarly for the other rows and columns, for example

$$A = \begin{pmatrix} 2 & 5 & 3 \\ 5 & 1 & 4 \\ 3 & 4 & 9 \end{pmatrix}$$

When A is symmetric it is equal to its transpose.

6 Orthogonal matrices: A matrix A is said to be orthogonal if $A^{\mathrm{T}}A = I$, that is, the rows are orthogonal to its columns. For example

$$\begin{pmatrix} \cos\theta & -\sin\theta \\ \sin\theta & \cos\theta \end{pmatrix} \begin{pmatrix} \cos\theta & \sin\theta \\ -\sin\theta & \cos\theta \end{pmatrix} = \begin{pmatrix} 1 & 0 \\ 0 & 1 \end{pmatrix}$$

They are used in the rigid rotation of axes in such techniques as principal components analysis.

DETERMINANTS

A determinant is a number associated with a square matrix. It is represented as an array consisting of the elements of the matrix, with straight, instead of curved, lines down the sides. For example, the determinant of the matrix A is written $|A|$ and if A is

$$\begin{pmatrix} 2 & 3 & 7 \\ 6 & 9 & 1 \\ 4 & 6 & 3 \end{pmatrix}$$

then its determinant is

$$\begin{vmatrix} 2 & 3 & 7 \\ 6 & 9 & 1 \\ 4 & 6 & 3 \end{vmatrix}$$

We shall not give details of the computation, which is complex in all but the most trivial of cases and should be performed on a computer; but two important cases are as follows.

1 The value of a determinant is zero whenever the elements of one row or column are proportional to those in another row or column. In the above example, the values 3,6,9 in the second column are 1.5 times the values 2,4,6 in the first, so $|A| = 0$. A consequence is that the determinant of the correlation or covariance matrix of highly correlated variables will be close to zero. This leads to, among other things, difficulties with inverting the matrices and so estimating regression coefficients.

2 The determinant of an identity matrix is equal to unity. Now the correlation matrix of a set of uncorrelated variables is equal to an identity matrix, so this fact is important in testing for lack of correlation: the log of the determinant of the sample correlation matrix will be close to zero when the variables are uncorrelated.

Appendix 2: Statistical Tables

APPENDIX 2.1 BINOMIAL DISTRIBUTION

		p				
n	*x*	0.1	0.2	0.25	0.3	0.5
2	0	0.810	0.640	0.563	0.444	0.250
	1	0.180	0.320	0.375	0.444	0.500
	2	0.010	0.040	0.063	0.111	0.250
3	0	0.729	0.512	0.422	0.296	0.125
	1	0.243	0.384	0.422	0.444	0.375
	2	0.027	0.096	0.141	0.222	0.375
	3	0.001	0.008	0.016	0.037	0.125
4	0	0.656	0.410	0.316	0.198	0.063
	1	0.292	0.410	0.422	0.395	0.250
	2	0.049	0.154	0.211	0.296	0.375
	3	0.004	0.026	0.047	0.099	0.250
	4	0.000	0.002	0.004	0.012	0.063
5	0	0.590	0.328	0.237	0.132	0.031
	1	0.328	0.410	0.396	0.329	0.156
	2	0.073	0.205	0.264	0.329	0.313
	3	0.008	0.051	0.088	0.165	0.313
	4	0.000	0.006	0.015	0.041	0.156
	5	0.000	0.000	0.001	0.004	0.031
10	0	0.349	0.107	0.056	0.017	0.001
	1	0.387	0.268	0.188	0.087	0.010
	2	0.194	0.302	0.282	0.195	0.044
	3	0.057	0.201	0.250	0.260	0.117
	4	0.011	0.088	0.146	0.228	0.205
	5	0.001	0.026	0.058	0.137	0.246
	6	0.000	0.006	0.016	0.057	0.205
	7	0.000	0.001	0.003	0.016	0.117
	8	0.000	0.000	0.000	0.003	0.044
	9	0.000	0.000	0.000	0.000	0.010
	10	0.000	0.000	0.000	0.000	0.001
15	0	0.206	0.035	0.013	0.002	0.000
	1	0.343	0.132	0.067	0.017	0.000
	2	0.267	0.231	0.156	0.060	0.003
	3	0.129	0.250	0.225	0.130	0.014
	4	0.043	0.188	0.225	0.195	0.042
	5	0.010	0.103	0.165	0.214	0.092
	6	0.002	0.043	0.092	0.179	0.153
	7	0.000	0.014	0.039	0.115	0.196
	8	0.000	0.003	0.013	0.057	0.196
	9	0.000	0.001	0.003	0.022	0.153
	10	0.000	0.000	0.001	0.007	0.092
	11	0.000	0.000	0.000	0.002	0.042
	12	0.000	0.000	0.000	0.000	0.014
	13	0.000	0.000	0.000	0.000	0.003
	14	0.000	0.000	0.000	0.000	0.000
	15	0.000	0.000	0.000	0.000	0.000

APPENDIX 2.2 NORMAL DISTRIBUTION

(a) Cumulative percentage

Z	%	Z	%
−3.0	0.13	0.0	53.98
−2.9	0.19	0.1	57.93
−2.8	0.26	0.2	61.79
−2.7	0.35	0.3	65.54
−2.6	0.47	0.4	69.15
−2.5	0.62	0.5	72.57
−2.4	0.82	0.6	75.80
−2.3	1.07	0.7	78.81
−2.2	1.39	0.8	81.59
−2.1	1.79	0.9	84.13
−2.0	2.28	1.0	86.43
−1.9	2.87	1.1	88.49
−1.8	3.59	1.2	90.32
−1.7	4.46	1.3	91.92
−1.6	5.48	1.4	93.32
−1.5	6.68	1.5	94.52
−1.4	8.08	1.6	95.54
−1.3	9.68	1.7	96.41
−1.2	11.51	1.8	97.13
−1.1	13.57	1.9	97.72
−1.0	15.87	2.0	98.21
−0.9	18.41	2.1	98.61
−0.8	21.19	2.2	98.93
−0.7	24.20	2.3	99.18
−0.6	27.43	2.4	99.38
−0.5	30.85	2.5	99.53
−0.4	34.46	2.6	99.65
−0.3	38.21	2.7	99.74
−0.2	42.07	2.8	99.81
−0.1	46.02	2.9	99.87
0.0	50.00	3.0	

(b) Percentage points

%	Z	%	Z
1	−2.33	51	0.03
2	−2.05	52	0.05
3	−1.88	53	0.08
4	−1.75	54	0.10
5	−1.64	55	0.13
6	−1.55	56	0.15
7	−1.48	57	0.18
8	−1.41	58	0.20
9	−1.34	59	0.23
10	−1.28	60	0.25
11	−1.23	61	0.28
12	−1.17	62	0.31
13	−1.13	63	0.33
14	−1.08	64	0.36

continued on p. 410

Appendix 2.2 *Continued*

Z	%	Z	%
15	−1.04	65	0.39
16	−0.99	66	0.41
17	−0.95	67	0.44
18	−0.92	68	0.47
19	−0.88	69	0.50
20	−0.84	70	0.52
21	−0.81	71	0.55
22	−0.77	72	0.58
23	−0.74	73	0.61
24	−0.71	74	0.64
25	−0.67	75	0.67
26	−0.64	76	0.71
27	−0.61	77	0.74
28	−0.58	78	0.77
29	−0.55	79	0.81
30	−0.52	80	0.84
31	−0.50	81	0.88
32	−0.47	82	0.92
33	−0.44	83	0.95
34	−0.41	84	0.99
35	−0.39	85	1.04
36	−0.36	86	1.08
37	−0.33	87	1.13
38	−0.31	88	1.17
39	−0.28	89	1.23
40	−0.25	90	1.28
41	−0.23	91	1.34
42	−0.20	92	1.41
43	−0.18	93	1.48
44	−0.15	94	1.55
45	−0.13	95	1.64
46	−0.10	96	1.75
47	−0.08	97	1.88
48	−0.05	98	2.05
49	−0.03	99	2.33
50	0.00		

APPENDIX 2.3 POISSON DISTRIBUTION

	μ			
x	0.5	1.0	1.5	2.0
0	0.607	0.368	0.223	0.135
1	0.303	0.368	0.335	0.271
2	0.076	0.184	0.251	0.271
3	0.013	0.061	0.126	0.180
4	0.002	0.015	0.047	0.090
5	0.000	0.003	0.014	0.036

continued

Appendix 2.3 *Continued*

	μ			
x	0.5	1.0	1.5	2.0
6	0.000	0.001	0.004	0.012
7	0.000	0.000	0.001	0.003
8	0.000	0.000	0.000	0.001
9	0.000	0.000	0.000	0.000
10	0.000	0.000	0.000	0.000

APPENDIX 2.4 t DISTRIBUTION

	α (%)				
ν	10	5	2.5	1	0.5
1	3.078	6.314	12.706	31.821	63.656
2	1.886	2.920	4.303	6.965	9.925
3	1.638	2.353	3.182	4.541	5.841
4	1.533	2.132	2.776	3.747	4.604
5	1.476	2.015	2.571	3.365	4.032
6	1.440	1.943	2.447	3.143	3.707
7	1.415	1.895	2.365	2.998	3.499
8	1.397	1.860	2.306	2.896	3.355
9	1.383	1.833	2.262	2.821	3.250
10	1.372	1.812	2.228	2.764	3.169
11	1.363	1.796	2.201	2.718	3.106
12	1.356	1.782	2.179	2.681	3.055
13	1.350	1.771	2.160	2.650	3.012
14	1.345	1.761	2.145	2.624	2.977
15	1.341	1.753	2.131	2.602	2.947
16	1.337	1.746	2.120	2.583	2.921
17	1.333	1.740	2.110	2.567	2.898
18	1.330	1.734	2.101	2.552	2.878
19	1.328	1.729	2.093	2.539	2.861
20	1.325	1.725	2.086	2.528	2.845
22	1.321	1.717	2.074	2.508	2.819
24	1.318	1.711	2.064	2.492	2.797
26	1.315	1.706	2.056	2.479	2.779
28	1.313	1.701	2.048	2.467	2.763
30	1.310	1.697	2.042	2.457	2.750
35	1.306	1.690	2.030	2.438	2.724
40	1.303	1.684	2.021	2.423	2.704
45	1.301	1.679	2.014	2.412	2.690
50	1.299	1.676	2.009	2.403	2.678
60	1.296	1.671	2.000	2.390	2.660
70	1.294	1.667	1.994	2.381	2.648
80	1.292	1.664	1.990	2.374	2.639
90	1.291	1.662	1.987	2.368	2.632
100	1.290	1.660	1.984	2.364	2.626
200	1.286	1.653	1.972	2.345	2.601
∞	1.282	1.645	1.960	2.326	2.576

APPENDIX 2.5 F DISTRIBUTION

v_2	v_1 1	2	3	4	5	6	8	10	12	15	20	25	∞
1	161.45	199.50	215.71	224.58	230.16	233.99	238.88	241.88	243.90	245.95	248.02	249.26	.
2	18.51	19.00	19.16	19.25	19.30	19.33	19.37	19.40	19.41	19.43	19.45	19.46	.
3	10.13	9.55	9.28	9.12	9.01	8.94	8.85	8.79	8.74	8.70	8.66	8.63	.
4	7.71	6.94	6.59	6.39	6.26	6.16	6.04	5.96	5.91	5.86	5.80	5.77	.
5	6.61	5.79	5.41	5.19	5.05	4.95	4.82	4.74	4.68	4.62	4.56	4.52	.
6	5.99	5.14	4.76	4.53	4.39	4.28	4.15	4.06	4.00	3.94	3.87	3.83	.
7	5.59	4.74	4.35	4.12	3.97	3.87	3.73	3.64	3.57	3.51	3.44	3.40	.
8	5.32	4.46	4.07	3.84	3.69	3.58	3.44	3.35	3.28	3.22	3.15	3.11	.
9	5.12	4.26	3.86	3.63	3.48	3.37	3.23	3.14	3.07	3.01	2.94	2.89	.
10	4.96	4.10	3.71	3.48	3.33	3.22	3.07	2.98	2.91	2.85	2.77	2.73	.
11	4.84	3.98	3.59	3.36	3.20	3.09	2.95	2.85	2.79	2.72	2.65	2.60	.
12	4.75	3.89	3.49	3.26	3.11	3.00	2.85	2.75	2.69	2.62	2.54	2.50	.
13	4.67	3.81	3.41	3.18	3.03	2.92	2.77	2.67	2.60	2.53	2.46	2.41	.
14	4.60	3.74	3.34	3.11	2.96	2.85	2.70	2.60	2.53	2.46	2.39	2.34	.
15	4.54	3.68	3.29	3.06	2.90	2.79	2.64	2.54	2.48	2.40	2.33	2.28	.
16	4.49	3.63	3.24	3.01	2.85	2.74	2.59	2.49	2.42	2.35	2.28	2.23	.
17	4.45	3.59	3.20	2.96	2.81	2.70	2.55	2.45	2.38	2.31	2.23	2.18	.
18	4.41	3.55	3.16	2.93	2.77	2.66	2.51	2.41	2.34	2.27	2.19	2.14	.
19	4.38	3.52	3.13	2.90	2.74	2.63	2.48	2.38	2.31	2.23	2.16	2.11	.
20	4.35	3.49	3.10	2.87	2.71	2.60	2.45	2.35	2.28	2.20	2.12	2.07	.
22	4.30	3.44	3.05	2.82	2.66	2.55	2.40	2.30	2.23	2.15	2.07	2.02	.
24	4.26	3.40	3.01	2.78	2.62	2.51	2.36	2.25	2.18	2.11	2.03	1.97	.
26	4.23	3.37	2.98	2.74	2.59	2.47	2.32	2.22	2.15	2.07	1.99	1.94	.
28	4.20	3.34	2.95	2.71	2.56	2.45	2.29	2.19	2.12	2.04	1.96	1.91	.
30	4.17	3.32	2.92	2.69	2.53	2.42	2.27	2.16	2.09	2.01	1.93	1.88	.
35	4.12	3.27	2.87	2.64	2.49	2.37	2.22	2.11	2.04	1.96	1.88	1.82	.
40	4.08	3.23	2.84	2.61	2.45	2.34	2.18	2.08	2.00	1.92	1.84	1.78	.
45	4.06	3.20	2.81	2.58	2.42	2.31	2.15	2.05	1.97	1.89	1.81	1.75	.
50	4.03	3.18	2.79	2.56	2.40	2.29	2.13	2.03	1.95	1.87	1.78	1.73	.
60	4.00	3.15	2.76	2.53	2.37	2.25	2.10	1.99	1.92	1.84	1.75	1.69	.
80	3.96	3.11	2.72	2.49	2.33	2.21	2.06	1.95	1.88	1.79	1.70	1.64	.
100	3.94	3.09	2.70	2.46	2.31	2.19	2.03	1.93	1.85	1.77	1.68	1.62	.
200	3.89	3.04	2.65	2.42	2.26	2.14	1.98	1.88	1.80	1.72	1.62	1.56	.
∞	.	.	.	.	.	.	.	.	.	.	.	.	.

APPENDIX 2.6 χ^2 DISTRIBUTION

ν	α (%)				
	10	5	2.5	1	0.5
1	2.706	3.841	5.024	6.635	7.879
2	4.605	5.991	7.378	9.210	10.597
3	6.251	7.815	9.348	11.345	12.838
4	7.779	9.488	11.143	13.277	14.860
5	9.236	11.070	12.832	15.086	16.750
6	10.645	12.592	14.449	16.812	18.548
7	12.017	14.067	16.013	18.475	20.278
8	13.362	15.507	17.535	20.090	21.955
9	14.684	16.919	19.023	21.666	23.589
10	15.987	18.307	20.483	23.209	25.188
11	17.275	19.675	21.920	24.725	26.757
12	18.549	21.026	23.337	26.217	28.300
13	19.812	22.362	24.736	27.688	29.819
14	21.064	23.685	26.119	29.141	31.319
15	22.307	24.996	27.488	30.578	32.801
16	23.542	26.296	28.845	32.000	34.267
17	24.769	27.587	30.191	33.409	35.718
18	25.989	28.869	31.526	34.805	37.156
19	27.204	30.144	32.852	36.191	38.582
20	28.412	31.410	34.170	37.566	39.997
22	30.813	33.924	36.781	40.289	42.796
24	33.196	36.415	39.364	42.980	45.558
26	35.563	38.885	41.923	45.642	48.290
28	37.916	41.337	44.461	48.278	50.994
30	40.256	43.773	46.979	50.892	53.672
35	46.059	49.802	53.203	57.342	60.275
40	51.805	55.758	59.342	63.691	66.766
45	57.505	61.656	65.410	69.957	73.166
50	63.167	67.505	71.420	76.154	79.490
60	74.397	79.082	83.298	88.379	91.952
70	85.527	90.531	95.023	100.425	104.215
80	96.578	101.879	106.629	112.329	116.321
90	107.565	113.145	118.136	124.116	128.299
100	118.498	124.342	129.561	135.807	140.170
200	226.021	233.994	241.058	249.445	255.264

APPENDIX 2.7 CRITICAL VALUES OF KOLMOGOROV–SMIRNOV STATISTIC

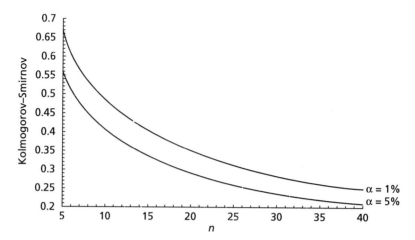

Fig. A2.1

APPENDIX 2.8 CRITICAL VALUES OF r FOR NORMAL SCORES TEST ($\alpha = 5\%$)

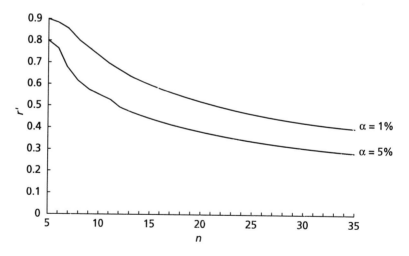

Fig. A2.2

APPENDIX 2.9 CRITICAL VALUES OF T FOR MANN–WHITNEY TEST ($\alpha = 5\%$)

m	5	6	7	8	9	10	11	12	13	14	15	16	17	18	19	20	25	30	35
n																			
3	1	1	1	2	2	3	3	4	4	5	5	6	6	7	7	8	11	13	16
	2	2	2	3	3	4	5	5	6	7	7	8	9	9	10	11	15	18	22
4	1	2	3	4	4	5	6	7	8	9	10	11	11	12	13	13	19	23	27
	2	3	4	5	6	7	8	9	10	11	12	14	15	16	17	18	24	29	34
5	2	3	5	6	7	8	9	11	12	13	14	15	17	18	19	20	27	33	39
	4	5	6	8	9	11	12	13	15	16	18	19	20	22	23	25	32	40	47
6		5	6	8	10	11	13	14	16	17	19	21	22	24	25	27	35	43	51
		7	8	10	12	14	16	17	19	21	23	25	26	28	30	32	42	51	60
7			8	10	12	14	16	18	20	22	24	26	28	30	32	34	44	54	64
			10	13	15	17	19	21	24	26	28	30	33	35	37	39	131	62	73
8				13	15	17	19	22	24	26	29	31	34	36	38	41	53	65	77
				15	18	20	23	26	28	31	33	36	39	41	44	47	63	74	87
9					17	20	23	26	28	31	34	37	39	42	45	48	62	76	90
					21	24	27	30	33	36	39	42	45	48	51	54	70	89	100
10						23	26	29	33	36	39	42	45	48	52	55	71	87	103
						27	31	34	37	41	44	48	51	55	58	62	79	97	114
11							30	33	37	40	44	47	51	55	58	62	80	98	116
							34	38	42	46	50	54	57	61	65	69	89	109	128
12								37	41	45	49	53	57	61	65	69	89	109	129
								42	47	51	55	60	64	68	72	77	99	120	142
13									45	50	54	59	63	67	72	76	98	120	143
									51	56	61	65	70	75	80	84	109	132	156
14										55	59	64	67	74	78	83	108	132	156
										61	66	71	77	82	87	92	118	144	170
15											64	70	75	80	85	90	117	143	169
											72	77	83	88	94	100	128	156	184
16												75	81	86	92	98	126	155	183
												83	89	95	101	107	138	168	198
17													87	93	99	105	136	166	197
													96	102	109	115	148	180	213
18														99	106	112	145	177	210
														109	116	123	158	192	227
19															113	119	154	189	224
															123	130	168	204	241
20																127	164	201	237
																138	178	216	256
25																	201	259	306
																	227	277	327
30																		317	376
																		338	400
35																			445

Upper and lower entries are for two-tailed and one-tailed tests respectively. As the distribution is symmetrical in (m,n) repeated values have been omitted.

APPENDIX 2.10 CRITICAL VALUES OF SPEARMAN'S RANK CORRELATION COEFFICIENT

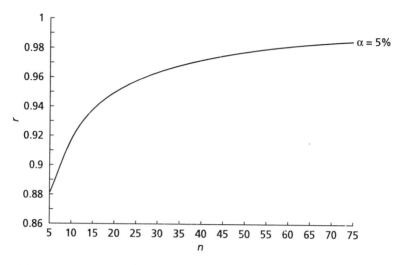

Fig. A.2.3

APPENDIX 2.11 CRITICAL VALUES OF R̄ FOR RAYLEIGH'S TEST

n	α (%)			
	10	5	2.5	1
3	0.86	0.96	1.03	1.13
4	0.75	0.84	0.91	1.03
5	0.67	0.75	0.82	0.94
6	0.62	0.69	0.76	0.87
7	0.57	0.64	0.70	0.82
8	0.53	0.60	0.66	0.77
9	0.50	0.57	0.63	0.73
10	0.48	0.54	0.59	0.70
11	0.46	0.52	0.57	0.67
12	0.44	0.49	0.54	0.64
13	0.42	0.48	0.52	0.62
14	0.40	0.46	0.51	0.60
15	0.39	0.44	0.49	0.58
16	0.38	0.43	0.47	0.56
17	0.37	0.42	0.46	0.54
18	0.36	0.41	0.45	0.53
19	0.35	0.39	0.44	0.52
20	0.34	0.38	0.42	0.50
22	0.32	0.37	0.41	0.48
24	0.31	0.35	0.39	0.46

continued

Appendix 2.11 *Continued*

	α (%)			
n	10	5	2.5	1
26	0.30	0.34	0.37	0.44
28	0.29	0.33	0.36	0.43
30	0.28	0.31	0.35	0.41
35	0.26	0.29	0.32	0.38
40	0.24	0.27	0.30	0.36
45	0.23	0.26	0.28	0.34
50	0.21	0.24	0.27	0.32
60	0.20	0.22	0.25	0.30
70	0.18	0.21	0.23	0.27
80	0.17	0.19	0.21	0.26
90	0.16	0.18	0.20	0.24
100	0.15	0.17	0.19	0.23
200	0.11	0.12	0.14	0.16

APPENDIX 2.12 VALUES OF CONCENTRATION PARAMETER κ FROM $\bar{R}$

$\bar{R}$	κ	$\bar{R}$	κ
0.00	0.00000	0.59	1.47543
0.02	0.04001	0.60	1.51574
0.04	0.08006	0.61	1.55738
0.06	0.12022	0.62	1.60044
0.08	0.16051	0.63	1.64506
0.10	0.20101	0.64	1.69134
0.12	0.24175	0.65	1.73945
0.14	0.28279	0.66	1.78953
0.16	0.32419	0.67	1.84177
0.18	0.36599	0.68	1.89637
0.20	0.40828	0.69	1.95357
0.22	0.45110	0.70	2.01363
0.24	0.49453	0.71	2.07685
0.26	0.53863	0.72	2.14359
0.28	0.58350	0.73	2.21425
0.30	0.62922	0.74	2.28930
0.31	0.65242	0.75	2.36930
0.32	0.67587	0.76	2.45490
0.33	0.69958	0.77	2.54686
0.34	0.72356	0.78	2.64613
0.35	0.74783	0.79	2.75382
0.36	0.77241	0.80	2.87129

continued on p. 418

Appendix 2.12 *Continued*

$\bar{R}$	κ	$\bar{R}$	κ
0.37	0.79730	0.81	3.00020
0.38	0.82253	0.82	3.14262
0.39	0.84812	0.83	3.30114
0.40	0.87408	0.84	3.47901
0.41	0.90043	0.85	3.68041
0.42	0.92720	0.86	3.91072
0.43	0.95440	0.87	4.17703
0.44	0.98207	0.88	4.48876
0.45	1.01022	0.89	4.85871
0.47	1.03889	0.90	5.30469
0.48	1.06810	0.91	5.85223
0.49	1.09788	0.92	6.53939
0.50	1.12828	0.93	7.42572
0.51	1.15932	0.94	8.61035
0.52	1.19105	0.95	10.27169
0.53	1.22350	0.96	12.76676
0.54	1.25672	0.97	16.92891
0.55	1.29077	0.98	25.25763
0.56	1.32570	0.99	50.25378
0.57	1.36156	1.00	∞
0.58	1.39842		
	1.43635		

Appendix 3: Data Sets

These data sets are presented here as they are the source for many of the worked examples. They may also be useful to students who wish to practice their own analyses.

APPENDIX 3.1

Copper in vein with distance along mine adit

Distance (m)	Copper (%)	Distance (m)	Copper (%)
0.0	0.67667	17.8	0.60000
0.8	0.64333	19.7	0.63667
1.0	0.70333	20.2	0.88667
1.3	0.59333	21.1	0.77667
2.7	0.64333	24.7	1.20000
4.5	0.54000	25.2	0.66000
5.1	0.43000	26.3	1.13667
6.0	0.58333	28.1	1.12333
7.2	0.41000	30.5	1.19333
8.7	0.49667	32.0	1.23333
10.2	0.16667	35.3	1.12667
11.8	0.62000	36.2	1.55667
13.1	0.51000	37.7	2.10667
15.2	0.57000	38.4	1.60000
16.3	0.90000	41.0	2.03333

APPENDIX 3.2

Age, temperature and depth of major oilfields worldwide (data from Zhijun Y. (1983) New method of oil prediction. *Bulletin of the American Association of Petroleum Geologists*, **67**(11), 2053–2056, reprinted by permission.

Age (Ma)	Temperature (°C)	Depth (m)
70	65	1200
12	115	2440
12	127	2740
180	60	1400
112	90	3300
135	72	2500
38	106	3250
105	85	2740
12	120	3050
70	80	2900
359	62	1750
32	95	3350

continued on p. 420

Appendix 3.2 *Continued*

Age (Ma)	Temperature (°C)	Depth (m)
35	93	2200
35	90	2200
110	70	1330
100	65	1230
90	63	1180
50	81	1700

APPENDIX 3.3

Depths in a borehole and sizes of two species of ostracod

Species A		Species B	
Depths (m)	Sizes (mm)	Depths (m)	Sizes (mm)
202	0.6	242	1.3
203	0.4	253	0.9
208	0.8	271	0.7
233	1.2	292	0.8
251	0.7	305	0.8
258	0.7	332	1.2
271	0.5	335	0.9
282	0.4	337	1.1
283	0.8	338	1.6
301	0.7	350	1.6
308	0.6	357	1.0
314	1.1	364	1.2
327	1.0	365	1.3
329	1.0	371	1.4
330	0.6	372	1.1
350	0.6	385	0.9
356	0.7	401	1.3
378	0.8	402	1.5
385	0.9	410	1.8
386	0.7	412	1.6
387	0.5	418	1.2
399	0.5	423	1.5
411	0.8	427	1.5
422	1.1	429	1.7
428	1.1	432	1.9
446	0.9	446	1.5
		451	1.6
		454	1.2
		460	1.6
		470	1.7
		474	1.8
		481	1.8
		497	1.3

APPENDIX 3.4

Soil geochemistry from Masson Hill, Derbyshire, England, a, traverse number; b, reference number

a	b	East	North	Zn	Pb	Cd	Mg	Ca	Cu	Ag	Sr	Ba
1	1	5.4	99.9	564.2	1026.0	20.14	10180	28770	57.22	1.45	12.24	610.8
1	3	6.8	89.5	697.3	1389.0	18.34	5689	29130	53.99	1.79	10.03	690.3
1	4	7.5	83.6	470:6	1219.0	14.01	20430	3968	20.49	*	61.43	736.0
1	5	8.4	77.2	3317.0	4870.0	76.28	22980	51630	45.41	1.66	15.7	1996.0
1	7	9.7	62.4	522.4	1750.0	23.85	8993	27400	48.11	1.46	7.39	522.5
1	8	9.5	56.0	1136.0	1384.0	5.83	14250	5552	17.22	*	48.98	246.8
1	9	10.3	47.8	920.1	1224.0	27.59	1800	11180	44.88	1.49	1.28	281.1
1	10	9.2	41.2	947.7	4233.0	22.07	8640	77020	40.28	1.28	8.57	498.3
1	11	9.0	33.0	394.9	924.4	21.55	13840	20080	51.91	1.31	8.33	319.2
1	12	7.4	24.3	303.0	339.5	21.31	15020	17170	71.42	1.47	8.45	355.3
1	13	3.7	14.3	904.4	793.2	16.98	26530	5527	30.78	*	60.09	147.9
1	14	5.0	5.0	202.3	637.0	12.1	1456	7996	57.75	1.63	0.5	280.1
2	2	8.8	93.8	813.0	2082.0	21.29	7450	45340	64.56	2.09	13.69	892.0
2	4	10.0	83.3	2318.0	709.0	44.98	9561	85950	71.77	2.57	34.87	592.0
2	6	11.8	70.4	2688.0	7168.0	42.66	8749	58790	57.64	2.40	17.42	1177.0
2	7	13.2	64.4	789.2	4307.0	2.72	17110	13530	18.09	*	65.59	1059.0
2	8	13.2	55.4	1189.0	4317.0	24.56	8767	56280	59.76	2.39	14.03	699.8
2	9	14.3	49.0	1306.0	6634.0	10.33	26970	31660	23.27	0.0	80.37	723.6
2	10	15.1	41.4	4840.0	25260.	86.17	1626	119500	380.47	18.82	30.46	2470.0
2	11	12.0	27.8	886.0	3509.0	6.83	16680	16050	19.66	*	62.65	405.4
2	12	10.0	14.6	291.3	522.2	19.64	16220	20150	145.00	1.10	12.01	295.9
3	1	11.1	100.8	1018.0	1715.0	23.87	10920	43850	64.34	2.03	14.9	1074.0
3	3	13.7	85.2	382.2	610.2	21.51	13370	20380	65.89	1.58	8.95	670.1
3	4	15.1	76.9	1253.0	4107.0	27.89	18420	56680	66.79	1.94	21.84	1011.0
3	5	16.8	68.7	913.0	3767.0	36.60	22590	15060	20.87	*	67.88	1331.0
3	6	17.8	62.0	2416.6	3975.2	44.04	14260	21220	52.51	1.64	7.13	1544.3
3	7	18.5	55.5	357.3	1274.0	14.4	14780	4303	17.54	*	57.34	551.3
3	9	19.5	43.9	527.2	1663.0	24.12	1830	36200	68.60	1.45	10.46	470.0
3	10	19.1	38.4	289.9	2346.0	2.94	2070	82800	118.82	8.02	30.85	289.6
3	11	18.5	32.0	524.8	1203.0	16.00	9373	13830	52.61	1.71	1.70	329.3
3	12	18.2	23.5	624.0	636.6	24.3	18410	21130	70.38	1.68	7.75	309.0
3	14	17.0	11.5	3355.3	525.9	61.59	15240	30500	69.15	1.7	16.22	744.3
4	2	14.5	95.3	3515.9	863.3	64.19	9090	23050	65.95	1.53	7.98	626.0
4	3	16.6	85.1	657.7	1574.0	5.62	17280	13880	19.78	*	64.74	804.5
4	5	19.5	75.5	870.3	3704.0	13.35	24640	15820	20.93	*	75.75	1304.0
4	6	19.8	67.1	721.7	2253.0	24.49	11840	37910	50.80	1.61	13.39	757.8
4	7	21.0	59.5	10450.	18370.	83.76	12500	95100	29.9	*	123.9	5330.0
4	8	21.4	51.2	781.3	2621.0	23.67	10940	28860	47.68	1.71	7.39	656.7
4	9	21.6	42.8	392.6	928.3	24.61	23200	34120	58.02	1.72	13.45	304.2
4	10	22.7	34.4	1145.0	14240.	46.1	33180	42500	21.61	*	114.9	593.0
4	11	22.5	25.6	3298.8	266.8	63.34	12660	13170	42.79	1.38	3.90	269.1
4	12	22.6	17.5	916.1	3150.0	30.84	14480	36550	55.08	2.21	18.53	345.4
4	13	22.7	8.7	3948.0	3522.0	49.41	32800	6431	18.84	*	84.98	1631.7
5	2	19.6	94.4	372.2	450.6	21.33	13520	18360	66.26	1.82	8.30	385.4
5	4	22.3	82.0	740.7	1277.0	24.41	14430	32240	52.32	1.95	14.65	805.6
5	6	24.4	69.7	453.6	1040.0	21.81	15430	27110	55.87	1.51	12.53	478.1
5	7	25.5	63.7	3192.0	30510.	38.33	20740	68850	21.94	*	145.5	2257.7
5	8	26.1	56.4	2629.3	4762.0	44.21	17880	95140	38.92	1.87	32.39	1446.0
5	9	26.6	50.0	143.0	1024.0	35.1	32590	61820	19.3	*	59.87	205.7
5	10	27.4	42.7	2060.0	7417.0	55.16	13040	74930	56.36	2.87	22.96	528.2
5	11	28.0	34.8	2859.5	739.0	17.81	35220	21770	42.22	1.86	21.48	242.1
5	12	29.2	28.0	463.3	1053.0	26.26	13750	21720	54.19	1.99	11.65	323.5
5	13	30.0	20.0	362.5	422.0	25.81	11710	17200	57.52	2.11	9.67	300.8
5	14	30.5	13.0	59.5	481.4	17.13	3180	5684	19.55	*	88.56	59.4
6	1	20.9	103.0	205.6	248.4	9.42	2594	4800	9.75	1.80	7.73	425.3
6	3	24.6	88.5	375.3	446.5	21.91	15240	15860	50.06	1.98	4.35	419.6
6	4	26.3	80.2	233.5	919.1	16.1	25790	6259	20.28	*	57.11	537.4
6	5	27.5	74.9	943.6	2209.0	24.35	11500	35930	54.11	1.76	10.71	746.7

Continued on p. 422

Appendix 3.4 *Continued*

a	b	East	North	Zn	Pb	Cd	Mg	Ca	Cu	Ag	Sr	Ba
6	7	30.2	61.2	1657.0	6650.0	83.29	24490	119500	32.66	*	162.2	7426.0
6	8	31.4	48.4	3215.7	846.7	5.3	25670	51990	49.18	2.01	18.91	360.6
6	10	33.5	37.2	635.2	528.0	9.35	30450	15510	19.98	*	64.08	377.4
6	11	35.6	30.0	541.4	2155.0	25.29	18490	40320	52.36	2.13	17.97	398.4
6	13	36.5	15.6	946.1	3368.0	9.69	20210	12780	18.68	*	61.23	309.2
7	2	25.7	94.6	324.7	334.6	20.7	9249	12010	44.44	1.79	1.74	347.3
7	3	28.2	85.3	404.2	556.4	18.43	18222	14320	22.24	*	63.42	450.3
7	4	28.9	80.6	474.7	704.4	23.76	14560	21830	48.84	1.68	10.47	587.5
7	5	31.0	73.0	449.3	1557.0	17.07	26710	6326	19.68	*	62.64	640.1
7	6	32.3	66.5	556.7	929.1	23.20	11430	24800	68.34	1.61	9.76	547.3
7	7	34.2	60.0	126.2	741.1	12.04	239	4728	15.57	*	58.57	353.2
7	8	35.8	51.8	2334.0	3281.0	70.40	10610	26110	46.82	1.82	2.44	1435.0
7	9	36.2	44.8	335.9	1534.0	17.88	25990	5911	17.57	*	55.28	386.7
7	10	38.4	37.5	359.7	764.4	20.80	18180	21330	71.92	1.77	17.02	324.5
7	12	41.0	25.0	392.8	803.1	28.59	14350	24270	51.11	1.99	11.92	364.9
7	14	42.8	15.0	864.2	2844.0	10.44	22340	11644	19.63	*	52.24	280.6

APPENDIX 3.5

Current directions from ripple forests from two sandstone formations in degrees from North

Formation A

216	118	223	305	242	198	172	222	155	233	269	238	189	219	111
217	141	201	201	260	276	182	212	245	221	177	248	192	210	222
251	214	228	217	262	280	234	244	218	208	191				

Formation B

123	217	171	251	288	135	163	35	106	349	333	15	118	205	271
72	341	148	190	313	255	96	93	320	215	237	105	179	180	223
216	222	242	196	300	227									
255	257	278	290	275										

APPENDIX 3.6

Current directions from trough cross bedding in Middle Jurassic Bearraraig Sandstone, Isle of Skye, Scotland, in degrees from North. (Data from Raglan P. (1989), unpublished BSc thesis, Kingston University)

121	113	97	113	100	118	354	256	220	192	283	128	145	335	333	6	342	45	54
169	172	160	146	177	179	169	338	321	335	22	338	128	44	59	25	4	28	30
24	58	199	208	175	197	199	208	215	176	85	295	299	1	16	334	328	339	33
14																		

APPENDIX 3.7

Organic matter (%) in successive intervals from the Kimmeridge Shale (data from Dunn 1974, see Chapter 6 references)

14.4	14.8	8.9	6.4	9.2	7.0	9.0	15.1	17.9	16.2	9.9	30.0	6.6	6.9	10.9
8.9	6.1	27.3	9.6	10.1	17.2	32.1	17.7	18.0	6.0	7.8	7.2	9.1	61.6	39.4
48.7	19.6	15.2	9.0	7.0	5.3	10.1	10.5	7.1	9.2	6.6	19.1	31.8	19.9	14.1
16.3	31.3	43.2	18.4	14.8	8.6	6.2	33.8	21.4	11.2	10.9	11.6	18.8	17.8	13.4
28.2	31.0	21.6	11.0											

APPENDIX 3.8

Widths of successive growth bands on a Silurian Nautiloid (residuals from linear regression)

4.89	−4.11	−4.12	0.86	1.86	−2.14	−2.15	−0.16	−4.16	−3.17
1.81	−5.19	−2.19	−7.20	2.78	2.77	2.76	−0.23	−0.24	−1.20
−3.26	−3.26	1.72	1.71	0.70	−0.29	−0.30	−1.31	3.67	3.67
2.66	−0.34	−1.35	−0.36	−1.36	−7.37	−0.38	−1.39	−1.39	3.59
−1.41	6.57	0.57	−1.43	−0.44	−1.45	−2.45	4.53	1.52	3.51
2.50	3.50	2.49	1.48	−0.52	−5.52	−3.53	0.45	3.44	3.44
2.43	2.42	1.41	1.41	−1.59	−0.60	5.38	1.37	2.37	1.36
0.35	3.34	3.34	−0.66	−1.67	−4.68	−0.68	2.30	4.29	5.28
1.28	3.27	0.26	−2.74	−3.75	−3.75	−4.76	−2.77	−5.78	4.21
4.20	4.19	−3.81	−3.81	−2.82	−3.83	2.15	4.15	1.14	2.13
4.12	2.11	7.11	−0.89	−0.90	−1.91	−2.91	−5.92	−2.93	3.05
3.05	3.04	0.03	0.02	1.02	1.01	1.00	−1.00	−0.01	−7.01
−5.02	−5.03	3.95	−1.04	−0.05	−2.06	−6.07	0.92		

APPENDIX 3.9

Thickness and grain size of an alluvial sandstone at various positions with specified co-ordinates

East	North	Thickness (m)	Grain size (ϕ)	East	North	Thickness (m)	Grain size (ϕ)
4	89	5	−2	100	102	10	−2
10	41	3	−1.75	101	77	11	−1
21	121	0	*	122	151	0	*
21	71	13	−1.75	112	141	3	−2.5
22	10	7	−2	121	121	5	−2
41	141	0	*	128	97	12	−1.5
47	106	2	−1.5	117	40	7	−0.5
49	70	17	−1.5	112	10	0	*
44	53	8	−1.5	144	101	4	−1.0
36	28	12	−1.75	134	77	21	−1.0

continued on p. 424

Appendix 3.9 *Continued*

East	North	Thickness (m)	Grain size (ϕ)	East	North	Thickness (m)	Grain size (ϕ)
56	25	8	−1.5	150	60	23	−0.5
60	136	2	−3.0	132	24	3	0
79	155	8	−3.5	183	91	0	*
71	51	17	−1.25	172	67	7	0
75	15	3	−1	169	26	8	0.5
91	137	13	−3	162	5	0	*
89	61	17	−1	198	22	20	1.0
99	122	16	−2.5	165	135	0	*

APPENDIX 3.10

Petrophysical parameters of the Triassic Sherwood Sandstone (from Olurenfemi M.O. (1985) Statistical relationships among some formation parameters for Sherwood Sandstone, England. *Mathematical Geology*, **17**, 8, 845–852). a, permeability (mm/sec); b, porosity (ϕ); c, matrix conductivity; d, true formation factor; e, induced polarisation.

a	b	c	d	e	a	b	c	d	e
0.032	0.204	0.020	12.9	4.61	1.864	0.231	0.012	13.0	7.28
0.262	0.234	0.013	10.1	6.67	0.022	0.173	0.018	23.1	3.69
1.400	0.226	0.011	10.8	6.32	0.121	0.163	0.007	13.5	6.91
0.072	0.205	0.019	13.7	4.31	0.078	0.161	0.013	14.6	6.45
2.168	0.242	0.008	9.0	7.80	0.052	0.219	0.017	13.3	7.07
0.485	0.222	0.013	11.0	6.84	6.692	0.269	0.010	5.8	8.54
2.600	0.238	0.009	9.1	8.80	12.240	0.273	0.009	4.9	8.46
2.538	0.245	0.009	9.6	8.41	8.950	0.296	0.016	5.6	7.17
0.048	0.224	0.027	10.6	3.45	0.226	0.227	0.024	6.2	4.25
0.326	0.254	0.016	9.1	7.07	2.950	0.258	0.012	8.1	7.19
1.889	0.212	0.007	8.5	10.79	0.564	0.213	0.021	8.3	4.27
0.031	0.166	0.020	10.1	4.00	0.900	0.235	0.013	7.1	5.43
2.190	0.284	0.014	6.6	7.75	0.119	0.213	0.020	8.2	4.40
0.130	0.233	0.026	8.2	3.96	0.059	0.204	0.027	8.8	2.78
0.760	0.247	0.019	7.6	6.18	16.110	0.255	0.008	7.5	7.58
0.314	0.208	0.020	8.8	5.62	6.750	0.251	0.011	7.9	6.65
0.088	0.183	0.014	10.0	6.68	1.460	0.241	0.011	8.4	6.33
0.328	0.225	0.013	12.3	5.78	0.520	0.217	0.017	9.0	4.55
0.694	0.196	0.012	10.4	7.76	2.160	0.241	0.011	8.3	6.36
0.104	0.177	0.014	10.9	6.99	5.670	0.253	0.011	7.6	6.16

APPENDIX 3.11

Measurements of fossil oreodont mammal skulls. a, width of braincase; b, length of cheek tooth; c, length of bulla; d, depth of bulla. Data from Miller R.L. and Kahn J.S. (1962) *Statistical Analysis in the Geological Sciences*. Wiley, New York.

a	b	c	d	a	b	c	d
58	129	26	16	46	84	16	6
52	126	27	18	51	87	21	8
50	122	28	22	46	80	17	7
52	123	29	18	50	90	18	8
60	138	33	17	46	85	16	7
61	122	28	17	48	85	15	7
54	132	30	17	47	85	17	8
65	131	32	18	49	83	18	8
55	130	32	17	43	79	15	7
64	125	26	16	47	87	19	8
56	124	28	16	46	87	18	8
45	91	16	8	37	88	17	4
46	93	17	7	43	79	14	4
48	92	19	5	43	84	19	4
46	91	19	6	42	80	17	5
45	86	15	7	39	83	12	5
51	93	19	8	39	87	15	5
47	92	16	5	40	86	18	5
48	89	18	7	34	77	16	5
47	91	18	6	35	82	15	5
50	91	17	7	45	88	17	5
48	91	19	8	33	80	15	4
49	93	18	7	42	85	13	4
49	87	17	7	60	114	27	20
49	91	19	8	60	118	31	19
78	165	35	18	60	111	31	21
77	165	37	19	58	102	30	20
65	148	30	20	55	116	28	20
74	163	31	15	59	117	29	17
65	169	31	16	59	114	24	17
70	176	34	23	60	121	25	19
69	161	28	13	47	99	26	15
67	178	31	14	42	93	26	16
65	174	34	18	40	90	22	13
64	168	28	13	46	100	22	11
68	166	32	15	46	96	24	16
42	81	15	8	42	88	26	15
48	83	18	9	43	89	23	14
45	87	18	9	44	78	23	13
48	83	17	8	44	90	25	11
				47	99	27	15
				47	92	27	13

APPENDIX 3.12

Measurements of the foram *Afrabolivina* at various depths in two formations (see Reyment R.A. (1978). Graphical display of growth-free variation in the Cretaceous benthonic foraminifer *Afrobolivina afra. Palaeogeography, Palaeoclimatology, Palaeoecology,* **25**, 267–276). a, sample number; b, stratigraphic reference number; c, length; d, breadth; e, breadth of final chamber; f, height of final chamber; g, height of 2nd last chamber; h, diameter of proloculus; i, width; j, width of foramen; k, breadth of non-overlapped part of final chamber.

a	b	c	d	e	f	g	h	i	j	k
1.00	1.00	73.91	37.26	26.01	15.40	14.18	9.78	19.74	4.56	12.48
2.00	1.00	75.30	32.46	23.32	14.41	13.63	9.11	19.04	4.03	9.86
3.00	1.00	48.21	24.67	18.66	12.96	11.89	8.37	17.24	3.58	8.06
4.00	1.00	87.44	33.55	26.62	15.90	15.23	9.45	20.71	4.00	10.43
5.00	1.00	65.11	31.11	24.51	15.93	14.28	9.17	20.02	3.75	10.23
6.00	1.00	76.62	35.25	27.88	18.67	17.64	12.08	22.90	4.21	11.01
7.00	1.00	61.64	31.34	24.47	15.49	15.36	10.56	20.24	4.11	10.43
8.00	1.00	77.46	30.71	24.49	15.84	14.14	10.66	20.32	3.94	9.96
9.00	1.00	73.54	31.72	25.26	16.00	15.42	11.38	21.62	2.46	9.88
10.00	1.00	75.25	35.36	29.62	19.91	18.85	11.10	27.63	5.90	10.80
11.00	1.00	81.68	35.09	28.66	20.41	19.71	11.67	26.30	5.13	10.27
12.00	1.00	64.17	32.57	25.30	17.21	16.20	10.17	24.42	4.33	10.80
13.00	1.00	62.39	29.06	23.08	18.14	14.72	9.72	20.19	3.62	8.56
14.00	1.00	70.52	28.95	23.81	16.62	15.76	9.62	22.69	3.90	8.42
15.00	1.00	69.75	27.94	23.25	16.62	15.11	9.34	20.59	3.79	7.91
16.00	1.00	68.61	27.72	23.51	16.55	15.82	9.33	21.54	3.88	9.55
17.00	1.00	81.01	34.72	27.97	20.39	17.02	11.86	23.65	4.52	10.27
18.00	1.00	113.76	39.96	33.63	22.80	21.27	14.07	30.85	4.33	11.34
19.00	1.00	73.01	34.98	28.75	20.71	19.91	13.04	25.08	4.04	10.54
20.00	1.00	72.81	31.57	26.01	20.28	18.77	12.38	22.62	3.75	9.54
21.00	1.00	89.11	39.70	31.57	22.35	20.96	14.15	29.58	3.79	11.96
22.00	1.00	73.78	32.07	27.45	20.05	18.24	11.76	24.74	3.88	10.24
23.00	1.00	85.54	33.70	28.31	20.26	18.54	10.67	25.88	4.18	10.23
24.00	1.00	70.33	32.70	26.42	20.47	18.52	11.31	23.29	3.89	9.31
25.00	1.00	68.19	34.13	28.04	22.01	19.39	12.39	23.32	3.85	9.72
26.00	1.00	73.45	34.92	27.96	20.60	19.70	11.00	25.40	3.49	10.28
27.00	1.00	71.05	31.11	26.16	19.23	18.46	10.67	25.17	3.42	9.42
28.00	1.00	81.06	32.12	28.17	21.21	19.56	11.08	25.23	3.47	9.35
29.00	1.00	72.09	30.71	26.50	20.79	17.97	11.06	24.65	3.54	9.34
30.00	1.00	73.99	30.67	27.00	20.86	19.03	10.53	23.95	3.64	9.45
31.00	1.00	60.56	28.51	24.21	18.69	17.30	11.00	22.20	3.54	9.30
32.00	1.00	92.26	36.72	30.40	22.23	20.83	13.34	28.84	3.73	11.20
33.00	1.00	64.85	29.60	24.77	17.83	17.44	10.22	22.64	3.72	8.75
34.00	1.00	81.08	33.50	28.21	20.84	20.69	11.82	28.18	4.24	11.12
35.00	1.00	70.47	32.42	25.04	18.17	18.08	19.54	23.92	3.88	10.08
36.00	1.00	65.14	29.89	25.48	19.09	16.82	10.93	23.00	4.05	9.32
37.00	1.00	79.04	32.70	28.34	20.57	19.29	9.63	25.39	3.84	9.68
38.00	1.00	78.59	30.57	26.02	20.54	18.95	10.04	23.46	3.32	8.63
39.00	1.00	50.43	28.02	22.57	17.31	15.75	9.91	19.74	3.61	8.27
40.00	1.00	58.86	28.94	22.10	18.31	16.13	9.04	20.50	3.37	8.58
41.00	1.00	64.79	29.59	24.20	19.30	17.60	9.75	22.75	3.75	9.53
42.00	1.00	67.67	30.61	24.90	20.49	17.76	10.01	24.21	3.69	10.68
43.00	1.00	111.26	40.06	34.19	25.71	20.86	14.78	31.07	3.70	13.14
44.00	1.00	114.70	40.60	33.78	26.06	23.75	11.45	33.20	4.04	13.06
45.00	1.00	96.02	37.27	30.50	22.55	21.10	12.68	29.55	3.98	11.68
46.00	1.00	74.76	32.21	26.63	20.72	19.07	12.50	26.91	3.97	10.07
1.00	2.00	90.15	39.98	32.44	20.93	19.60	16.53	30.07	3.83	11.40
2.00	2.00	83.66	37.41	29.41	20.41	19.74	16.56	27.99	3.76	10.13
3.00	2.00	76.24	36.70	28.81	20.07	18.81	17.44	27.60	3.61	9.68
4.00	2.00	54.60	28.82	23.72	17.75	16.73	12.60	21.03	3.42	8.35

continued

Appendix 3.12 *Continued*

a	b	c	d	e	f	g	h	i	j	k
5.00	2.00	80.32	37.04	29.67	19.41	18.31	14.96	29.08	3.39	9.33
6.00	2.00	70.54	34.17	27.58	18.54	18.58	13.50	25.10	3.03	9.42
7.00	2.00	77.89	36.65	28.18	20.00	20.00	13.43	27.05	3.16	9.48
8.00	2.00	87.75	39.35	30.81	22.06	19.80	13.30	30.20	3.40	9.20
9.00	2.00	64.20	38.07	29.23	19.87	19.33	15.20	29.00	3.07	10.33
10.00	2.00	73.93	36.02	29.02	20.54	20.17	13.83	28.42	3.37	9.58
11.00	2.00	99.32	42.29	34.39	23.69	22.19	15.66	31.06	3.81	11.19
12.00	2.00	82.30	42.48	32.00	22.00	21.00	15.09	31.18	3.50	11.45
13.00	2.00	79.42	36.16	29.89	21.27	20.51	14.56	27.89	3.06	8.89
14.00	2.00	62.53	33.46	27.17	20.62	19.42	14.33	26.75	3.42	8.54
15.00	2.00	99.16	41.63	34.99	25.55	24.92	14.98	31.65	3.60	11.62
16.00	2.00	78.38	38.38	31.20	22.83	20.96	14.58	31.67	3.42	10.58
17.00	2.00	78.60	42.15	34.19	23.93	22.07	15.53	33.00	3.49	9.80
18.00	2.00	84.94	39.07	32.41	22.64	21.82	14.55	31.00	3.36	9.09
19.00	2.00	69.78	33.06	26.71	18.00	17.62	14.29	28.58	3.08	8.96
20.00	2.00	86.87	42.49	33.19	22.93	23.14	15.14	32.71	3.46	11.14
21.00	2.00	73.33	38.28	29.60	22.00	20.42	14.75	31.50	3.32	9.42
22.00	2.00	65.69	37.22	30.28	21.07	19.25	14.50	30.29	3.44	10.29
23.00	2.00	84.86	46.14	37.93	24.01	23.46	16.14	35.07	3.43	10.93
24.00	2.00	106.61	46.18	36.22	26.56	24.00	16.33	37.78	3.61	10.56
25.00	2.00	92.36	34.86	28.57	19.61	19.29	13.14	24.57	2.91	8.71
26.00	2.00	85.54	36.38	30.25	19.42	18.67	14.54	23.92	2.78	9.33
27.00	2.00	85.04	42.46	33.04	23.69	22.94	14.59	33.06	3.37	10.44
28.00	2.00	89.19	44.02	35.77	25.78	23.92	14.17	33.69	3.28	10.38
29.00	2.00	89.08	42.47	35.25	24.48	23.25	14.38	31.66	3.13	9.28
30.00	2.00	90.62	42.32	33.12	23.16	21.59	14.22	32.97	3.20	11.45
31.00	2.00	79.47	39.66	31.02	23.59	22.22	14.84	30.19	3.21	11.41
32.00	2.00	107.28	46.40	37.45	26.81	25.37	15.69	33.73	3.35	10.60
33.00	2.00	75.22	37.05	30.19	23.57	21.71	13.27	30.67	3.16	10.53
34.00	2.00	97.50	42.16	35.37	26.42	24.67	14.25	33.79	3.07	12.33
35.00	2.00	79.29	39.48	31.43	22.43	20.73	15.23	32.96	3.39	10.92
36.00	2.00	86.60	37.32	29.90	20.56	19.25	12.56	28.66	2.99	9.75
37.00	2.00	88.93	37.84	30.79	21.12	19.81	12.97	28.56	3.02	10.16
38.00	2.00	84.95	37.21	28.63	20.83	19.76	12.45	28.09	2.93	9.95
39.00	2.00	85.79	36.07	29.59	20.59	19.68	12.66	28.97	3.00	10.00
40.00	2.00	93.94	38.03	32.12	21.14	19.67	13.28	29.08	3.10	9.12
41.00	2.00	102.91	37.39	31.01	21.62	19.60	14.73	30.27	3.11	10.90
42.00	2.00	97.43	36.15	29.26	20.12	19.75	12.85	28.17	2.98	9.27
43.00	2.00	88.52	35.58	29.37	20.37	19.54	13.18	28.48	3.31	10.34
44.00	2.00	106.10	41.51	34.52	23.21	21.48	12.01	32.50	3.34	10.49
45.00	2.00	120.77	45.40	36.77	24.61	22.10	13.15	34.15	3.60	11.34

APPENDIX 3.13

Counts of specimens of 5 species of fossil (a to e) from 10 samples in section A and 8 from section B

Section A						Section B					
Sample	Species					Sample	Species				
	a	b	c	d	e		a	b	c	d	e
1	8	2	1	0	0	1	5	3	3	1	0
2	3	2	1	0	0	2	3	8	4	8	1
3	5	3	0	3	0	3	1	1	0	6	0

continued on p. 428

Appendix 3.13 *Continued*

	Section A						Section B				
Sample	Species					Sample	Species				
	a	b	c	d	e		a	b	c	d	e
4	1	4	2	10	0	4	1	1	2	9	2
5	0	2	0	15	0	5	0	1	1	11	3
6	0	3	1	17	3	6	0	0	1	13	3
7	0	1	1	23	5	7	0	0	0	10	1
8	0	0	0	20	5	8	0	0	1	10	3
9	0	1	2	7	6						
10	0	0	1	0	7						

APPENDIX 3.14

Trace element geochemistry of lake sediments in Carswell area, Saskatchewan, Canada. This area is of interest as a site of a meteorite impact structure and Uranium mines. Data from Dunn C.E. (1980) Lake sediment and water geochemistry of the Carswell structure, Northwest Saskatchewan. *Saskatchewan Energy and Mines Report*, **224**.

Site	East	North	U	Pb	Ni	Co	Zn	Mn	Cu	V	Mo	Cr	Th	Ag
1	595.80	466.85	5.1	6.0	11.0	6.0	51.0	137.0	18.0	56.0	3.0	28.0	1.0	20.0
2	598.45	470.40	5.4	11.0	10.0	14.0	38.0	1360.0	13.0	91.0	2.0	21.0	1.0	26.0
3	600.95	476.33	3.7	3.0	4.0	2.0	16.0	34.0	3.0	20.0	1.0	8.0	1.0	1.0
4	603.62	480.63	1.8	4.0	9.0	4.0	53.0	28.0	5.0	29.0	1.0	13.0	1.0	3.0
5	600.65	480.33	2.8	7.0	13.0	10.0	107.0	119.0	17.0	30.0	2.0	15.0	3.0	5.0
6	604.60	478.12	2.9	7.0	12.0	7.0	35.0	105.0	7.0	66.0	1.0	32.0	4.0	3.0
7	593.97	470.17	1.1	4.0	7.0	6.0	55.0	108.0	6.0	26.0	1.0	11.0	3.0	4.0
8	595.35	473.23	0.7	12.0	10.0	12.0	72.0	234.0	9.0	18.0	2.0	12.0	1.0	8.0
9	595.15	475.20	1.8	7.0	13.0	6.0	60.0	77.0	16.0	13.0	4.0	7.0	1.0	1.0
10	594.03	477.40	1.5	10.0	12.0	8.0	78.0	173.0	13.0	33.0	5.0	16.0	3.0	2.0
11	582.80	470.15	1.7	5.0	15.0	10.0	62.0	131.0	8.0	21.0	1.0	17.0	1.0	4.0
12	583.20	472.45	3.5	5.0	18.0	17.0	102.0	364.0	11.0	25.0	2.0	18.0	3.0	1.0
13	585.17	473.77	0.6	6.0	31.0	17.0	105.0	464.0	11.0	12.0	1.0	15.0	1.0	1.0
14	587.17	475.00	2.6	13.0	20.0	26.0	82.0	3020.0	24.0	53.0	7.0	20.0	1.0	3.0
15	579.15	462.55	27.7	11.0	23.0	19.0	62.0	314.0	31.0	30.0	2.0	22.0	1.0	5.0
16	580.78	463.73	6.6	12.0	21.0	16.0	44.0	387.0	28.0	37.0	2.0	23.0	1.0	1.0
17	581.50	464.52	5.4	14.0	24.0	31.0	36.0	1140.0	45.0	78.0	5.0	27.0	1.0	6.0
18	581.78	465.38	8.6	6.0	42.0	28.0	54.0	286.0	50.0	24.0	6.0	15.0	6.0	4.0
19	582.87	465.10	5.1	12.0	19.0	13.0	111.0	242.0	20.0	51.0	2.0	39.0	7.0	3.0
20	583.15	466.80	1.1	3.0	5.0	4.0	12.0	1020.0	3.0	5.0	1.0	6.0	11.0	1.0
21	585.80	470.00	174.0	36.0	74.0	35.0	127.0	219.0	74.0	63.0	3.0	71.0	21.0	12.0
22	585.03	472.65	6.4	6.0	29.0	17.0	90.0	865.0	29.0	27.0	3.0	23.0	4.0	1.0
23	591.15	472.40	2.3	5.0	28.0	16.0	78.0	103.0	28.0	22.0	3.0	17.0	32.0	3.0
24	591.40	469.77	0.4	7.0	14.0	18.0	144.0	245.0	14.0	18.0	2.0	15.0	1.0	1.0
25	589.72	469.23	0.8	10.0	12.0	13.0	86.0	317.0	12.0	55.0	1.0	26.0	1.0	6.0
26	609.08	484.20	2.8	12.0	21.0	10.0	89.0	40.0	21.0	33.0	2.0	35.0	8.0	3.0
27	602.85	493.10	1.0	10.0	10.0	17.0	136.0	245.0	10.0	57.0	3.0	33.0	1.0	1.0
28	599.38	491.27	2.5	20.0	9.0	15.0	57.0	477.0	9.0	31.0	3.0	17.0	1.0	2.0
29	597.28	488.08	1.5	4.0	11.0	17.0	34.0	1340.0	11.0	22.0	3.0	10.0	1.0	10.0
30	602.45	489.77	1.5	16.0	9.0	7.0	63.0	52.0	9.0	22.0	3.0	12.0	1.0	3.0
31	599.40	485.42	2.4	9.0	16.0	7.0	95.0	90.0	16.0	34.0	4.0	10.0	1.0	3.0
32	596.00	483.58	2.2	7.0	12.0	7.0	44.0	42.0	12.0	13.0	4.0	12.0	1.0	3.0
33	592.40	481.25	4.2	7.0	22.0	7.0	67.0	46.0	22.0	26.0	6.0	18.0	1.0	3.0

continued

Appendix 3.14 *Continued*

Site	East	North	U	Pb	Ni	Co	Zn	Mn	Cu	V	Mo	Cr	Th	Ag
34	593.03	486.35	2.9	8.0	11.0	8.0	121.0	52.0	11.0	20.0	4.0	13.0	1.0	2.0
35	590.60	478.85	1.2	6.0	15.0	9.0	77.0	222.0	15.0	30.0	5.0	16.0	1.0	2.0
36	586.60	486.40	2.1	5.0	7.0	12.0	67.0	125.0	10.0	12.0	8.0	7.0	1.0	5.0
37	584.47	486.95	1.0	8.0	12.0	16.0	213.0	336.0	7.0	29.0	2.0	16.0	1.0	2.0
38	584.05	483.40	7.3	9.0	11.0	10.0	130.0	15.0	23.0	38.0	4.0	19.0	1.0	2.0
39	585.97	481.83	1.5	5.0	9.0	6.0	99.0	107.0	6.0	17.0	3.0	9.0	1.0	1.0
40	583.97	480.92	4.9	8.0	26.0	8.0	78.0	65.0	38.0	21.0	2.0	17.0	4.0	2.0
41	581.03	480.52	15.8	4.0	24.0	6.0	58.0	26.0	12.0	12.0	9.0	11.0	4.0	3.0
42	579.95	478.35	5.5	9.0	11.0	17.0	117.0	689.0	6.0	44.0	3.0	19.0	1.0	2.0
43	576.85	474.83	7.3	12.0	22.0	23.0	72.0	453.0	14.0	69.0	2.0	29.0	1.0	20.0
44	574.22	475.10	0.5	5.0	7.0	7.0	78.0	207.0	6.0	14.0	2.0	8.0	1.0	18.0
45	574.38	472.02	1.5	12.0	22.0	18.0	109.0	749.0	29.0	64.0	2.0	44.0	1.0	2.0
46	578.92	468.42	2.1	7.0	6.0	8.0	81.0	72.0	6.0	24.0	2.0	18.0	1.0	1.0
47	610.90	473.17	0.9	5.0	3.0	5.0	47.0	48.0	3.0	36.0	1.0	17.0	1.0	1.0
48	611.58	476.02	1.3	4.0	8.0	7.0	66.0	141.0	8.0	30.0	1.0	16.0	1.0	3.0
49	593.13	492.15	3.4	8.0	7.0	4.0	40.0	78.0	6.0	12.0	2.0	6.0	1.0	4.0
50	588.63	488.63	2.7	7.0	9.0	10.0	66.0	287.0	14.0	19.0	4.0	10.0	1.0	1.0
51	585.38	497.15	3.5	14.0	10.0	13.0	58.0	280.0	7.0	56.0	3.0	28.0	1.0	1.0
52	586.33	492.42	1.4	5.0	7.0	4.0	97.0	33.0	7.0	14.0	3.0	5.0	1.0	4.0
53	582.03	491.85	0.4	9.0	11.0	15.0	106.0	370.0	8.0	20.0	1.0	12.0	1.0	5.0
54	575.92	489.37	1.6	12.0	10.0	21.0	63.0	5860.0	8.0	31.0	2.0	16.0	1.0	22.0
55	573.30	486.00	15.3	8.0	13.0	11.0	47.0	278.0	12.0	51.0	5.0	26.0	1.0	16.0
56	577.38	487.05	10.9	7.0	11.0	5.0	80.0	23.0	6.0	24.0	2.0	15.0	1.0	3.0
57	573.85	483.77	1.7	4.0	10.0	6.0	58.0	36.0	9.0	14.0	2.0	8.0	2.0	6.0
58	572.13	480.90	3.2	11.0	13.0	18.0	59.0	539.0	15.0	29.0	3.0	17.0	1.0	8.0
59	581.30	486.60	2.1	6.0	14.0	8.0	119.0	32.0	11.0	19.0	3.0	10.0	1.0	6.0
60	578.45	482.55	9.5	8.0	14.0	8.0	59.0	67.0	15.0	27.0	4.0	18.0	1.0	2.0
61	576.52	480.23	2.0	6.0	11.0	7.0	54.0	94.0	20.0	20.0	3.0	10.0	1.0	2.0
62	574.92	477.83	1.2	5.0	8.0	6.0	87.0	123.0	6.0	20.0	3.0	8.0	2.0	1.0
63	582.72	475.98	10.5	18.0	29.0	29.0	94.0	4960.0	32.0	99.0	4.0	34.0	1.0	14.0
64	577.05	470.20	4.3	8.0	17.0	16.0	79.0	255.0	21.0	98.0	3.0	53.0	1.0	6.0
65	572.25	469.75	2.3	6.0	12.0	7.0	147.0	50.0	9.0	15.0	6.0	8.0	1.0	3.0
66	578.33	466.30	6.9	4.0	8.0	5.0	50.0	36.0	7.0	18.0	7.0'	6.0	1.0	6.0
67	576.55	464.77	1.4	9.0	12.0	9.0	54.0	141.0	11.0	31.0	6.0	12.0	1.0	2.0
68	576.75	467.15	2.3	10.0	12.0	8.0	84.0	77.0	11.0	16.0	5.0	8.0	1.0	4.0
69	574.15	467.23	4.2	5.0	10.0	6.0	72.0	215.0	18.0	14.0	3.0	8.0	1.0	6.0
70	577.58	461.17	14.1	13.0	21.0	13.0	75.0	231.0	28.0	36.(	9.0	17.0	1.0	5.0
71	585.90	469.90	69.0	30.0	33.0	16.0	83.0	199.0	37.0	28.0	3.0	26.0	2.0	54.0

Index

Note: page numbers in *italic* indicate figures; those in **bold** indicate tables